I0054756

Bistatic Radar Polarimetry

Theory and Principles

Zbigniew H. Czyz
Author

Michael Nyden
Editor

Wexford
2008

Disclaimer: This book is intended only for academic informational purposes and not to be taken as a manual of operation, a repair guide, or as direction for any procedure that may be described herein.

CONTENTS

FOREWORD

Polarimetry enriches vision of electromagnetic waves by introducing new dimensions to their vectorial representations and, consequently, by applying polarization matrices instead of scalar coefficients describing propagation or scattering properties of media on their path (see, e.g., Kong [98], Ulaby et al.[137], Wanielik[138], Giuli[78]).

Radar polarimetry differs from its optical predecessor by dealing with waves propagating in opposite directions (Boerner et al.: [13], [24]). However, it is very important and most convenient to use exactly the same vectors representing polarization independently of direction of propagation. For instance, we used to speak that optimum transmission between two antennas is being achieved when they are *identically* (!) polarized, though they have to look towards each other, thus being oriented in *opposite* directions. There are also other even more essential reasons for introducing such a unique representation of polarization which appears necessary when analysing 'geometrical' representations of transmission by applying the Poincare sphere transformations. Such representations appear physically fully justified in virtue of the fundamental property of electromagnetic plane waves which, being solutions of Maxwell equations, demonstrate their invariance under time reversal called also the time reversal symmetry or T-symmetry (Brosseau [28]).

To follow the above requirement, vectors independent of direction of propagation will be introduced which, however, determine not only polarizations themselves but also phases. Moreover, what is essential, those phases will be not always temporal but rather spatial ones, expressing the wave's 'locations' along the propagation path in a 'frozen' time, independently of direction of propagation. That appears necessary for two reasons, both equally important. Namely, the temporal phase of the voltage received during one- or two-way transmission depends not only on physical distance between the receiving and transmitting antenna or scattering object but also on the difference between the two corresponding 'spatial phases', of the receiving antenna and the incoming wave (see, e. g., Section 4.5). That dependence is of special importance for problems of polarimetric interferometry. Another reason is connected with an orthogonal, generally elliptical, polarization basis. Its vectors cannot be determined without assuming their (spatial) phases. Even in the case of a linear polarization basis the null phases of its vectors are being tacitly assumed. Basis vectors can be presented as tangential polarization (TP) phasors on the Poincare sphere. Their orientations, determined by double spatial phases (Section 4.3), enable one to precisely express TP phasors of waves and to analyze their scattering transformations.

That kind of analysis is possible when considering the Poincare sphere model of the scattering matrix. Such a model explains transformation of any illumination, its amplitude polarization and phase, by inversion of the incident waves' phasors through a determined 'inversion point' situated inside the sphere and by subsequent rotation of the sphere of such 'inverted phasors' about a determined axis by a determined angle. The cornerstones of the model are: its inversion point, and the characteristic coordinate system (CCS) in which location of that point is determined in some allowed regions of the sphere interior. Coordinates of that point in the CCS determine (sometimes two solutions are possible) all elements of the scattering Sinclair and Kennaugh matrices, axis and angle of rotation after inversion and, consequently, location of all special polarization points for those matrices. The scattered wave's intensity is proportional to the distance between the incident polarization and inversion points. In such a CCS, equivalent to the characteristic for the radar target polarization basis, the form of matrices becomes most simple ('canonical'). Transformation of any orthogonal polarization basis to that characteristic basis is unique, except of special cases for which an ambiguity is immaterial.

This monograph is especially devoted to the bistatic radar polarimetry which includes monostatic and forward scattering representing its special cases. Among them the monostatic polarimetry exhibits a particular feature. Its scattering matrices are symmetrical. On the other hand forward scatterings, employing non-symmetrical matrices, can be considered as belonging to strictly bistatic polarimetry, without any distinguishing attributes. That is one of the reasons for which the bistatic polarimetry is of practical significance and deserves thorough studies. For example, the transmittance matrices of the polarimetric two-ports could not be analyzed without theoretical backgrounds of the bistatic polarimetry. The difference between the Poincare sphere models of the mono- and bistatic scattering matrices is essential. In the case of the monostatic scattering matrix the allowed region for the inversion point contains only the negative part of the first axis ('Q-axis') of its CCS inside the model's sphere. In cases of bistatic scattering the inversion point leaves that semiaxis but remains inside the so-called (upper) 'small hemisphere' of diameter coinciding with semiaxis for the monostatic scattering case and additionally covers a bounded region above that hemisphere (in that additional region two solutions for determination of the scattering matrix are possible).

Such transformations are possible owing to the fact that not only the incident waves but also those scattered, as well as receiving antennas, can be represented by polarization and spatial phase (PP) vectors or their tangential polarization phasors on one Poincare sphere. That is evident because they all do not depend neither on direction of propagation nor on antenna orientation, contrary to waves' complex amplitudes (CA) or antenna effective heights with their conjugate values for opposite directions of propagation and with temporal phases.

Active transformations of inversion and rotation in the three Stokes parameter space, and passive rotation known as the change of basis, are not the only Poincare sphere transformations being used in applications. However, all other transformations can be obtained as their special cases or their superpositions with those special cases. For instance, the active transformation of orthogonality (Section 6.1) may serve as an example of a special case of inversion; it is (with minus sign) an inversion through the center of the polarization sphere. Very important passive transformation is reversal of spatial coordinate system by its rotation (Section 6.2). It can be obtained as superposition of orthogonality transformation and special rotation (about a chosen linear polarization axis) in the three Stokes parameter space (Section 7.2). Formally, it is equivalent to the active transformation reversing an antenna versus its spatial coordinate system. As passive, that transformation allows to obtain the propagation Jones or Mueller matrices from their Sinclair and Kennaugh scattering counterparts (Section 7.11). The procedure is called the 'change of alignment'.

Passive transformation changing the order of vectors - members of the polarization basis - is of a different kind (Section 7.7). That transformation reverses orientation of all the tangential phasors and handedness of their polarizations. It can be used to compare representations of the PP or CA vectors and scattering/propagation matrices assumed by different authors applying opposite orders of basis vectors.

It is hoped that the new here presented approach to the traditional vector based methods of analysis indicates one of possible ways of filling the hitherto existing gap caused by the insufficient set of basic concepts thus removing all appearing inconsistencies and ensures firm ground to solution of fundamental theoretical and computational problems of bistatic radar polarimetry.

1. Introduction

Recognition of targets becomes more effective when using radars which are fully polarimetric. It means that when coherent scattering they should measure all elements of the 2 x 2 complex Sinclair matrix of a target, altogether 7 real parameters of the nonsymmetrical matrix when neglecting the absolute phase depending on the distance. However, the elements themselves have no direct physical meaning. They depend on the polarization basis in which the matrix has been expressed. That is why parameters independent of the polarization basis should be specified.

When using properly selected the so-called 'characteristic' polarization and phase basis, the scattering matrix takes simple 'canonical' form. Then, the number of its real parameters, excluding absolute phase and target's magnitude, is three only for the coherent scattering case and nonsymmetrical matrices (this is in the bistatic scattering case, often being met in practice). These three parameters determine all inherent polarimetric properties of a target. They can be used to build up a geometrical model of the target's scattering matrix. It will take the form of the Poincare sphere of incident polarizations with an 'inversion point' inside. Three rectangular coordinates of such a point in the 'characteristic' coordinate system (CCS), corresponding to the characteristic basis, can represent these three parameters. The model shows how the incident polarizations and phases are being transformed during scattering. At first, the sphere is inverted through the inversion point, and then rotated. Both axis and angle of rotation are functions of the inversion point coordinates in the CCS.

Some incident polarization points are of special properties. There are points which do not change their locus after inversion and rotation of the sphere (they are called the 'eigenpolarizations'). Some correspond to a maximum and minimum scattered power. Two points become antipodal (orthogonal) against themselves after rotation and inversion, thus resulting in no received voltage when the antenna of the same polarization has been used for transmission and reception. There is one point corresponding to a transmit-receive antenna polarization for which the voltage received from the scattered wave is of maximum value. There are also two the so called 'mutual polarizations' (No.1 and No.2): if the polarization No.1 is radiated and No.2 scattered, then No.2, if radiated, produces No.1 scattered. Some points correspond to a maximum or minimum received voltage when the receiving antenna is orthogonally polarized.

All these special polarization points of the model are functions of the three parameters mentioned. Some of them can be determined by simple geometrical constructions based on the inversion point's locus in the CCS.

There is an allowed region inside the Poincare sphere scattering matrix model in which the inversion point can be located. That location can serve to classify targets for their polarimetric properties.

The whole model can be variously oriented when rotating together with its characteristic polarization basis. The three additional real parameters, of the basis rotation, can be considered another kind of target's polarimetric parameters. They are also suitable for target classification purposes.

What should be stressed here, it is the importance of an exact definition of the polarization basis in which the scattering matrix has to be determined. In the technical literature the orthogonal polarization basis is usually being defined by two antipodal points on the Poincare sphere. However, for polarimetric purposes, it is insufficient. The orthogonal null-phase (ONP) polarization basis should be introduced instead. It consists of two phasors which are tangent to the polarization sphere at the antipodal points. They are collinear, i.e. oriented in the same direction along a great circle of the sphere, and their order is essential. Phasors representing waves, not antennas, rotate in time. They can be considered rotating with a 'double speed' 2ω: in the clockwise direction for waves propagating along +z axis, or in the counter-clockwise direction for waves traveling along $-z$ axis. In a fixed time, they change their orientation after multiplication their polarization and phase (PP) vectors by an exponential *spatial phase* factor. Their rotation is counter-clockwise for waves delayed in space (shifted towards -z coordinates, independently of the direction of propagation). Basis phasors become no more collinear when both are multiplied by the same spatial phase factor. That is why the basis introduced has been called 'null-phase'.

The orthogonality transformation should be applied *four* times in order to bring the first basis phasor to its original value (after two transformations it changes its sign for opposite one). That is why the polarization sphere of tangential phasors should be considered to be a kind of the two-folded Riemann surface.

The use of that newly developed polarization phasor approach with the appropriate notation makes the whole theory, based on matrix calculus, very simple and provides powerful, indispensable tool for solving practical problems of radar polarimetry.

2. Remarks on the Existing Approaches to the Theory of Radar Polarimetry

Roots of radar polarimetry are in much earlier formulated optical polarimetry. Main difference between the two theories has been caused by the necessity of consideration, in radar applications, also waves traveling in the backward direction along an established propagation axis.

Following postulates of optical polarimetry for the monochromatic plane waves, standard polarimetric radar theory is based on the concept of the *Jones vector* [92], or *complex amplitude* (CA) column vector, as an entity defining the wave's amplitude, polarization and phase in a chosen orthonormal polarization basis. Complex amplitude corresponds to the so-called *polarization ellipse* usually determined in the *xy* (z=0) plane, perpendicular to the propagation z axis of the spatial Cartesian *xyz* coordinate system. The ellipse is traced by a tip of the electric vector rotating in time. According to the IEEE definition [90] the corresponding polarization is called right-handed or left-handed if the rotation is in the clockwise or counter-clockwise direction, correspondingly, when looking at the ellipse along the positive direction of propagation. Having determined its two angles, of the tilt and ellipticity, the polarization can be presented by a point on the Poincare sphere, in its upper or lower part - depending on the polarization handedness.

The problem arises, however, with definition of the same polarization for a wave traveling backwards. Watching the polarization ellipse of the returning wave - when looking along the same +z direction - one can see the same tilt angle but the opposite handedness. It suggests to place the point of the same polarization on the opposite part of the Poincare sphere versus its equator of linear polarizations. As a consequence, misunderstandings appear in the literature regarding representation of the oppositely propagating waves on the Poincare sphere, causing sometimes an improper formulation of transformation rules. Solution to that difficulty has been found by introducing: two kinds of mutually conjugate *'directional' Jones vectors* [112]. These directional Jones vectors stand for complex amplitudes (and represent polarization ellipses of both senses of rotation) for waves of the same polarization when taking into account two opposite directions of propagation. However, *dual formulae* are required for transformation of directional Jones vectors:

- under change of polarization basis, what follows from the relations (3.22) and (3.37).

$$\left(e_A^P\right)^+ = C_H^B\left(e_B^P\right)^+ \quad \text{and} \quad \left(e_A^P\right)^- = \widetilde{C}_B^H\left(e_B^P\right)^-$$

  because of the unique change of basis transformation rule (5.9) for the polarization and phase (PP) unit column vectors, u, and their dependence on the CA unit column vectors, e,

$$u = e^+, \qquad u^* = e^-,$$

- when reversing the propagation z axis by 180^0 rotation of the spatial coordinate system,

$$\left(e_B^{Po}\right)^- = C_B^{o}\left(e_B^P\right)^+ \quad \text{and} \quad \left(e_B^{Po}\right)^+ = C_B^{o}{}^*\left(e_B^P\right)^-$$

  what directly follows from the uniqe rule (6.11) for the PP vectors, and

- for obtaining the corresponding Stokes vectors determining coordinates of polarization points on the Poincare sphere independently of waves direction of propagation,

$$\begin{aligned} \mathsf{P}_B^P &= \widetilde{\mathsf{U}}^*[(e_B^P)^+ \otimes (e_B^P)^{+*}] = \widetilde{\mathsf{U}}[(e_B^P)^- \otimes (e_B^P)^{-*}] \\ &= \widetilde{\mathsf{U}}^*(u_B^P \otimes e_B^{P*}), \end{aligned}$$

  according to (7.8).

Another serious problem often being met in the literature refers to the polarization bases of the Jones vectors. These bases are usually being determined with an insufficient precision. They may differ:

- by the time-convention $\exp\{\pm j\omega t\}$, because the column PP unit vector (see (5.1) and (5.3)) of circular polarization is

$$u_H^C = \frac{1}{\sqrt{2}}\left(1_x + j1_y\right) = \begin{bmatrix} 1_x & 1_y \end{bmatrix}\begin{bmatrix} \cos 45^0 \\ \sin 45^0 e^{j90^0} \end{bmatrix} \Rightarrow \begin{cases} \left.\begin{matrix} 2\gamma_H^C = 90^0 \\ 2\delta_H^C = 90^0 \end{matrix}\right\} \Rightarrow \begin{cases} \text{left - handed circular} \\ \text{polarization for } e^{+j\omega t} \text{ conv.} \end{cases} \\ \left.\begin{matrix} 2\gamma_H^C = 90^0 \\ 2\delta_H^C = -90^0 \end{matrix}\right\} \Rightarrow \begin{cases} \text{right - handed circular} \\ \text{polarization for } e^{-j\omega t} \text{ conv.} \end{cases} \end{cases}$$

- by their phase, because (see (6.5) with (6.2) and Section 7.5) for, e.g.,

$$u_H^L = \frac{1}{\sqrt{2}}\begin{bmatrix}1\\j\end{bmatrix}_{(x,y)} \text{ and } u_H^R = u_H^{L\times} = \begin{bmatrix}0 & -1\\1 & 0\end{bmatrix} u_H^{L}{}^* = \frac{1}{\sqrt{2}}\begin{bmatrix}j\\1\end{bmatrix}_{(x,y)}$$

$$\Rightarrow \begin{bmatrix}\boldsymbol{u}^L & \boldsymbol{u}^{L\times}\end{bmatrix} - \begin{cases}\text{the ONP PP basis or}\\ \text{'collinear phasor' basis}\end{cases}$$

$$u_H^{L'} = u_H^L \times e^{\pm j45^0} \text{ and } u_H^{R'} = u_H^{L'\times} \times e^{\pm j45^0}$$

$$\Rightarrow \begin{bmatrix}\boldsymbol{u}^{L'} & \boldsymbol{u}^{L'\times}\end{bmatrix} - \begin{cases}\text{the orthogonal } \pm 45^0 \text{ PP basis}\\ \text{or 'parallel phasor' basis}\end{cases}$$

- by rotation in the 2-dim. complex space, what is evident because, e.g., the two bases, $\begin{bmatrix}\boldsymbol{u}^B & \boldsymbol{u}^{B\times}\end{bmatrix}$ and $\begin{bmatrix}\boldsymbol{u}^K & \boldsymbol{u}^{K\times}\end{bmatrix}$, with mutual dependence (see (5.23) and (5.22))

$$u_H^K = C_{B,H}^K u_H^B = C_H^K C_B^H u_H^B$$

are different, and

- by the order of components, because (see (3.22)) for

$$u_H^B = \begin{bmatrix}\cos\gamma e^{-j(\delta+\varepsilon)}\\ \sin\gamma e^{j(\delta-\varepsilon)}\end{bmatrix}^B_{(x,y) \text{ or } (y,x)} \quad \text{there is} \quad \begin{cases}2\gamma_{(x,y)}^B - 90^0 - 2\gamma_{(y,x)}^B\\ 2\delta_{(x,y)}^B = -2\delta_{(y,x)}^B\\ 2\varepsilon_{(x,y)}^B = 2\varepsilon_{(y,x)}^B\end{cases}$$

Moreover, in most applications, change of basis should be combined with the change of phase of basis vectors, what fact used to be overlooked in practice.

Conjugate directional Jones vectors are determined in conjugate bases. That is an additional problem referring to presentation of Jones vectors of opposite directivity by the same point on one Poincare sphere. Therefore the next improvement has been introduced here, which makes a very small formal step but of essential significance. Instead of using the directional Jones vectors, application of the polarization and phase (PP) vectors has been proposed.

Amplitude transformation matrices operate on complex amplitudes which form their domains and ranges. These CA vectors are being expressed in the literature by the ordinary Jones vectors, they may be expressed by the directional Jones vectors, and it is proposed to express them by the PP vectors which are identical with the Jones vectors (also directional) for waves propagating in the +z direction, and take conjugate values for waves propagating in the opposite direction.

Differences between these approaches can be shown on a simple example of an *evolution* of the radar scattering equation with the Sinclair matrix, *S*. The following relations will be presented between incident and scattered electric vectors by successively applying:

- the ordinary Jones CA vectors, $\boldsymbol{E}$ (formerly expressed by the unit vector $\boldsymbol{e}$; $\boldsymbol{E} = E_0\boldsymbol{e},\ E_0^2 = \boldsymbol{E}^*\cdot\boldsymbol{E}$),
- the directional Jones vectors, $\boldsymbol{E}^{\pm}$ ($\boldsymbol{E}^{\pm} = E_0\boldsymbol{e}^{\pm}$), and
- the PP vectors, $\boldsymbol{E}_0$ (further being expressed by the unit vector $\boldsymbol{u}$; $\boldsymbol{E}_0 = E_0\boldsymbol{u},\ E_0^2 = \boldsymbol{E}_0{}^*\cdot\boldsymbol{E}_0$),

and obtaining:

$$\boldsymbol{E}^s = S\boldsymbol{E}^i \;\rightarrow\; \boldsymbol{E}^{s-} = S\boldsymbol{E}^{i+} \;\rightarrow\; \boldsymbol{E}_0^s{}^* = S\boldsymbol{E}_0^i \tag{2.1}$$

Waves represented by those vectors, propagating in the $\pm z$ directions, appropriately are:

$$\boldsymbol{E}^{\pm}(t,z) = \boldsymbol{E}\, e^{j(\omega t \mp kz)} \;\rightarrow\; \begin{cases}\boldsymbol{E}^+(t,z) = \boldsymbol{E}^+\, e^{j(\omega t - kz)}\\ \boldsymbol{E}^-(t,z) = \boldsymbol{E}^-\, e^{j(\omega t + kz)}\end{cases} \;\rightarrow\; \begin{cases}\boldsymbol{E}^+(t,z) = \boldsymbol{E}_0\, e^{j(\omega t - kz)}\\ \boldsymbol{E}^-(t,z) = \boldsymbol{E}_0{}^*\, e^{j(\omega t + kz)}\end{cases} \tag{2.2}$$

Here $\omega = 2\pi/T$ means the *angular frequency* with the time period T, and $k = 2\pi/\lambda$ is the *wave number* with the *wavelength* λ. The asterisk denotes complex conjugation.

The corresponding ('full') Stokes four-vectors expressed in the form of Kronecker products with the auxiliary unitary U matrix,

$$\mathrm{U} = \frac{1}{\sqrt{2}} \begin{bmatrix} 1 & 1 & 0 & 0 \\ 0 & 0 & 1 & -j \\ 0 & 0 & 1 & j \\ 1 & -1 & 0 & 0 \end{bmatrix} \tag{2.3}$$

successively are:

$$\mathbf{I}_{0E}^{\pm} = \sqrt{2}\,\tilde{\mathrm{U}}^*(\boldsymbol{E}^{\pm} \otimes \boldsymbol{E}^{\pm *}) \;\rightarrow\; \mathbf{I}_0^{\pm} = \sqrt{2} \begin{cases} \tilde{\mathrm{U}}^*(\boldsymbol{E}^{+} \otimes \boldsymbol{E}^{+*}) \\ \tilde{\mathrm{U}}\,(\boldsymbol{E}^{-} \otimes \boldsymbol{E}^{-*}) \end{cases} \;\rightarrow\; \mathbf{I}_0^{\pm} = \sqrt{2}\,\tilde{\mathrm{U}}^*(\boldsymbol{E}_0 \otimes \boldsymbol{E}_0^*) \tag{2.4}$$

Of course, the two $\mathbf{I}_{0E}^{\pm}$ Stokes four-vectors are different for the same polarization of the wave propagating in the opposite directions. The next, $\mathbf{I}_0^{\pm}$ vectors, present correctly that polarization for two directions of propagation, but expressed in terms of directional Jones vectors require application of dual formulae and are related to mutually conjugate polarization bases. Therefore their representation on one Poincare sphere can be disputed. The third representation, in terms of the PP vectors, is free of those two inconveniences and will be solely used in the approach here applied .

Limited amount of simple formulae and the Poincare sphere geometrical models of scattering matrices can be obtained when applying to radar polarimetry the 'polarization and phase (PP) vector approach' introduced here, based on the time-symmetry of Maxwell equations and followed by a 'polarization phasor notation' which uses those phasors as upper and lower indices for vectors and matrices.

See also: Appendix O, 'Comments about relations to the existing works', and Appendix R, 'Maxwell equations in radar polarimetry'.

3. The Definition of Polarization for Monochromatic Plane Waves

3. 1. Mutual Relation Between the CA and PP Vectors

The PP unit vector $\boldsymbol{u}$ should be distinguished from the complex amplitude (CA) vector $\boldsymbol{e}$, the last being always expressed in terms of $\boldsymbol{u}$ in the PP vector approach. Their mutual dependence can be best explained on the example of two completely polarized monochromatic plane waves corresponding to the same vector $\boldsymbol{u}$ but propagating in opposite directions.

Analyzing the time reversal symmetry (see [28]) of those waves under $t \to -t$ transformation we observe that if

$$\mathcal{E}(t,z) = \mathrm{Re}\left\{E_0 \boldsymbol{u}\, e^{j(\omega t - kz)}\right\} \tag{3.1}$$

is a solution of Maxwell equations, then

$$\mathcal{E}(-t,z) = \mathrm{Re}\left\{E_0 \boldsymbol{u}\, e^{j(-\omega t - kz)}\right\} = \mathrm{Re}\left\{E_0 \boldsymbol{u}^*\ e^{j(\omega t + kz)}\right\} \tag{3.2}$$

is also their solution. It presents a wave propagating in opposite direction but of the same polarization and of the same *spatial phase* for time t=0.

The two above presented real vectorial functions can also be interpreted as helices moving, without rolling, in two opposite directions with a velocity of light. Such *polarization helices* represent both polarization and spatial phase of a plane monochromatic wave more properly than polarization ellipses do, the sense of which depends on direction of propagation and which can indicate the temporal phase only.(Temporal phase is equal to spatial phase for waves propagating in positive z direction, and changes its sign for opposite direction of propagation).

Indicating by 'plus' or 'minus' *directional indices* along the propagation z-axis of a spatial *xyz* coordinate system we arrive at the following expressions for waves in a complex form:

$$\begin{aligned} \boldsymbol{E}^+(t,z) &= E_0 \boldsymbol{e}^+\ e^{j(\omega t - kz)} = E_0 \boldsymbol{u}\, e^{j(\omega t - kz)} \\ \boldsymbol{E}^-(t,z) &= E_0 \boldsymbol{e}^-\ e^{j(\omega t + kz)} = E_0 \boldsymbol{u}^* e^{j(\omega t + kz)} \end{aligned} \tag{3.3}$$

Here, $\boldsymbol{e}^+$ and $\boldsymbol{e}^-$ are directional unit complex amplitudes equal to $\boldsymbol{u}$ and $\boldsymbol{u}^*$, accordingly, where $\boldsymbol{u}$ is the unit complex PP vector. Column vectors of those waves can be presented as in (3.4) by the Cayley-Klein complex parameters a and b:

$$\left.\begin{aligned} E^+(t,z) &= E_0 u\, e^{j(\omega t - kz)} = E_0 \underbrace{\overbrace{u^+}^{PP(+)}}_{CA(+)}\ e^{j(\omega t - kz)} \\ E^-(t,z) &= E_0 u^*\ e^{j(\omega t + kz)} = E_0 \underbrace{\overbrace{u^-}^{PP(-)} {}^*}_{CA(-)} e^{j(\omega t + kz)} \end{aligned}\right\};\quad u = \begin{bmatrix} a \\ b \end{bmatrix},\quad aa^* + bb^* = 1 \tag{3.4}$$

where E_0 means the real magnitude of the CA. The value of the unit PP vector u *does not depend on direction of wave's propagation.* It is equal to the $CA(+)$, the value of the unit CA vector for 'positive' direction of propagation of the wave along the z-axis, and to the conjugate $CA(-)$ value of the unit CA vector for 'negative' direction of wave's propagation:

$$\underbrace{\overbrace{u^+}^{PP(+)}}_{CA(+)} = \underbrace{\overbrace{u^-}^{PP(-)}}_{[CA(-)]^*} = \underbrace{\overbrace{u}^{PP(\pm)}}_{CA(+)\&[CA(-)]^*} \tag{3.5}$$

3. 2. Time-dependent PP vectors and polarization phasors

Tangential polarization phasors (see Chapter 4.) correspond to u^+ and u^- PP vectors of antennas or waves and they do not depend on time. The time-dependent PP vectors of waves are:

$$u^+ e^{j\omega t}, \quad \text{and} \quad u^- e^{-j\omega t}, \quad \text{with} \quad u^+ = u^- . \tag{3.6}$$

They appear in expressions:

$$E^+(t,z) = E_0(u^+ e^{j\omega t})e^{-jkz}, \quad \text{and} \quad E^-(t,z) = E_0(u^- e^{-j\omega t}) * e^{+jkz}. \tag{3.7}$$

As seen from the above, when considered as phasors, they rotate in time with 2ω angular velocity in opposite directions, and are equal to each other for $2\omega t = 0$ and $2\omega t = 4\pi$. For $2\omega t = 2\pi$, though coinciding in orientation, these phasors (and the corresponding PP vectors) have values of opposite sign. Thus, the polarization sphere of tangential phasors demonstrates properties of a kind of the two-folded 'Riemann surface' (see [130]).

3. 3. A complex received voltage

When neglecting coefficients independent of polarization, the transmission equation can be presented as a complex received voltage expressed by Hermitian product of two PP column vectors u and u' (with a tilde denoting transposition):

$$V_r = \tilde{u}' u^* = \underbrace{\overbrace{\tilde{u}'^+}^{PP(+)}}_{CA(+)} \underbrace{\overbrace{u^-}^{PP(-)} *}_{CA(-)} \tag{3.8}$$

One of those PP vectors, no matter which one, corresponds to the incoming wave and the other to the receiving antenna. Though only the wave's phasor rotates in time, the opposite directivity of antenna's phasor is also essential. In that Hermitian product, conjugate is always the PP vector with 'minus' index. Thus, equation (3.8) presents also an ordinary product of two unit CA vectors with opposite directional indices. One of them represents complex amplitude of the incoming wave, and the other stands for the complex height of the receiving antenna. However, there is an important condition which the two PP column vectors have to fulfill for correctness of the above equation. They should be expressed in the same PP basis corresponding to common spatial coordinate system.

3. 4. The polarization ellipse and polarization helix

The monochromatic plane wave can be geometrically modeled as a polarization helix (or screw) moving, without rolling, in a positive or negative direction along the z axis of an xyz rectangular coordinate system. Such a helix, passing the $z = 0$ plane, traces on it a polarization ellipse (Fig.3.1). The left-handed helix corresponds to the right-handed elliptical polarization, and vice versa. The helix determines the two polarization parameters and the spatial phase of the wave. That spatial phase is determined by the position of the moving helix along the z axis in time $t = 0$, independently of direction of the wave propagation.

There is one-to-one correspondence between the CA vector and the polarization ellipse, as well as between the PP vector and the polarization helix. Only *one* helix moving along the z axis in both directions sufficiently represents the *two* waves of the same polarization and spatial phase traveling in opposite directions. This is unlike the polarization ellipses which differ by their *sense* depending on the direction of the wave's propagation.

The *polarization ellipses,* in the z = 0 plane and normalized to $E_0 = 1$, can be expressed for the two opposite directions of propagation by the following vectorial functions of t:

$$r_e^+(t) = \mathrm{Re}\, E^+(t,0)/E_o = \mathrm{Re}\{u e^{j\omega t}\} \tag{3.9a}$$

and

$$r_e^-(t) = \mathrm{Re}\, E^-(t,0)/E_o = \mathrm{Re}\{u * e^{j\omega t}\} = r_e^+(-t) \tag{3.9b}$$

For the *polarization helix*, at $t = 0$ and also normalized to $E_0 = 1$, the only model representing the two waves is of the form

$$r_h^{\pm}(z) = kz\mathbf{1}_z + \operatorname{Re} E^{\pm}(0,z)/E_0 = kz\mathbf{1}_z + \operatorname{Re}\{\mathbf{u}e^{-jkz}\} = kz\mathbf{1}_z + \operatorname{Re}\{\mathbf{u}^* e^{+jkz}\} \qquad (3.10)$$

It is seen from the above equalities that there are the PP vectors $\mathbf{u}$, or polarization helices only, and not the CA vectors, or polarization ellipses, that define unambiguously the polarization and spatial phase of a wave *independently of its direction of propagation or antenna orientation.*

That is a very important property of the PP vectors that will be used when considering, e.g., the direct transmission between two antennas. The *complex* received voltage for such a transmission will be found depending on the mutual positions (locations and orientations) of two *phasors*, standing for the PP vectors, representing the two antennas looking at each other (*oriented in the opposite directions*) and, of course, on the phase factor accounting for the distance between the antennas. The phasors will be located in planes tangent to one, *common* polarization and spatial phase sphere (the PP sphere) at points corresponding to their polarizations. Orientations of the phasors will reflect the spatial phases of waves radiated by the two antennas in a transmit regime, reduced to the 'null distance' from each antenna.

That is why the PP vectors, and not the CAs, ought to be represented by phasors on such a sphere, and that is one of several reasons for which the PP sphere of tangential phasors is an indispensable tool for the analysis of the polarimetric transmission equations.

3. 5. Three groups of polarization and phase parameters

In radar polarimetry authors are using different polarization and phase (PP) bases, and for complex representation of waves they are applying different time-dependence conventions, $\exp(\pm j\omega t)$. To explain mutual relations between the existing standard approaches and the new one here presented, some notions, explained more fully in the next part of this monograph, will be in advance introduced now, at the very beginning.

Parameters describing completely polarized (elliptically in general) monochromatic plane wave, propagating in an established direction and presented by its electric vector, will be divided into three groups:

1° wave-dependent parameters
2° component-dependent parameters, and
3° PP-basis-dependent parameters.

If complex representation of waves is being applied, all those parameters are independent of the time convention.

The component-dependent parameters correspond to basis vectors but do not depend on their order in that basis. They can be determined by basis-dependent parameters but differently for bases of mutually reversed order of their vectors.

In the new notation proposed here, parameters belonging to the second or third group will be labelled with a lower index specifying the component or basis, accordingly, and may be labelled with an upper index identifying the polarization and phase of a wave (or of an antenna emitting such a wave).

3. 6. Wave dependent parameters

The following parameters belong to that group:

- angular (radian) frequency, ω, corresponding to the wave number $k = \omega c$, with c equal to the speed of light,
- absolute value (magnitude) of the wave's electric vector amplitude, E_0, and
- the PP parameter of an absolute phase and 'complete' polarization, not yet specified, and represented by, e.g., two-dimensional complex unit vector $\mathbf{u}$ or, equivalently, by the 'polarization phasor'.

The polarization phasor can be interpreted as a 'vector' tangent to the *Poincare polarization sphere* at the wave's polarization point, and of orientation:

- representing wave's absolute phase for time $t = 0$, uniquely determined versus an established phasor which is known as being abslutely of null phase,
- changing within the range of 4π (after 2π rotation the 'vector's' value becomes of opposite sign, though its orientation looks the same), and

- same for both mutually opposite directions of wave's propagation in time $t = 0$ (with time, phasors of waves rotate in opposite directions with the angular velocity of $\pm 2\omega$, and sense of their rotation depends on the direction of propagation along the z axis of a right-handed Cartesian xyz coordinate system).

(Remark: orientation of all the polarization phasors, though independent of rotation of the polarization bases, becomes reversed with the reversal of the order of basis vectors. In such a sense the 1:1 correspondence between the PP vectors and polarization phasors is not entirely true if the the PP bases of two orders are being considered. One PP vector corresponds then to two oppositely oriented polarization phasors belonging to two mutually reversed polarization bases composed of the same PP basis vectors.)

Parameters belonging to the first group can be used to expess the electric vector of a wave propagating in the $+z$ or $-z$ directions. Applying the $\exp\{j\omega t\}$ convention we will write:

$$E^{+}(t,z) = E_0 \boldsymbol{u}\, e^{j(\omega t - kz)} \tag{3.11a}$$

$$E^{-}(t,z) = E_0 \boldsymbol{u}^{*}\, e^{j(\omega t + kz)} \tag{3.11b}$$

Similar expressions in the $\exp\{-j\omega t\}$ convention would be of the form

$$\begin{aligned} E^{+}(t,z) &= E_0 \boldsymbol{u}^{*}\, e^{-j(\omega t - kz)} \\ E^{-}(t,z) &= E_0 \boldsymbol{u}\, e^{-j(\omega t + kz)} \end{aligned} \tag{3.12}$$

However, only the $\exp(+j\omega t)$ convention, approved by the IEEE Standard [90], will be used further. Values of all expressions in the other convention should just be taken conjugate. Formally, it can be performed by exchanging everywhere the imaginary unit, j, by another imaginary unit, $i = -j$.

In the above formulae the upper indices, '+' or '-', mean direction of propagation versus the z axis orientation, and the asterisk denotes complex conjugation. All these waves are of the same polarization and spatial phase expressed by the PP $\boldsymbol{u}$ vector. (The term 'spatial phase' will be explained in the next section. Also the use of mutually conjugate amplitudes for waves propagating in opposite directions will be later justified by the time-symmetry of Maxwell equations.)

In the new notation applied here, the absolute amplitude, E_0, and the complex unit PP vector, $\boldsymbol{u}$, may be labelled with the upper index identifying the polarization and phase of the wave or antenna. Only the column vector version of the PP vector can also be labelled with the lower index specifying its PP basis.

3. 7. Component-dependent analytical parameters. Spatial and temporal phase delays

Real representation of waves will be chosen at first to show that the PP parameters are independent of the time convention.

To determine the component-dependent parameters, the xyz *right*-handed Cartesian spatial coordinate system (SCS) will be chosen. Waves will be assumed to propagate along its z-axis in both directions $(\pm z)$. The two non-vanishing x- and y- real components of their electric vectors can be presented as follows:

$$\begin{aligned} \mathcal{E}_x^{\pm}(t,z) &= a_x \cos\{\omega t \mp (kz + v_x)\} \\ \mathcal{E}_y^{\pm}(t,z) &= a_y \cos\{\omega t \mp (kz + v_y)\}. \end{aligned} \tag{3.13}$$

Here: $k = \omega c = 2\pi / \lambda$ (c is the velocity of light, λ – the wave's length), and a_x, a_y, v_x, v_y will be called the component-dependent analytical parameters. They describe:

- amplitudes of the two components, a_x and a_y,
- spatial phases (delays) of those components, v_x and v_y.

The term '*spatial* phase delay' can be explained when considering $z = z_{0x}$ and $z = z_{0y}$ polarization planes in which each component obtains maximum value at the time $t = 0$. E.g.:

$$\mathcal{E}_{x\,\max}^{\pm} = a_x = \mathcal{E}_x^{\pm}(t = 0, z_{0x}). \tag{3.14}$$

That maximum appears for the cosine argument

$$kz_{0x} + v_x = 0, \tag{3.15}$$

or for

$$kz_{0x} = -\nu_x . \tag{3.16}$$

So, ν_x / k represents spatial shift of the $\mathcal{E}_x^{\pm}(t=0,z)$ component in the -z direction, and is called the 'spatial delay' of that component versus its position for $\nu_x = 0$, exhibiting maximum value at $z = 0$. Therefore ν_x, in radians, can be called the spatial phase delay.
Similar considerations apply to the $\mathcal{E}_y^{\pm}(t=0,z)$ component.

Evidently, ν_x and ν_y parameters mean also *temporal* phase delays, but only for 'forward propagating' waves, in the +z direction. For backward propagating waves they mean temporal phase advance. Indeed, for $z = 0$,

$$\mathcal{E}_{x\max}^{\pm} = a_x = \mathcal{E}_x^{\pm}(t_{0x}, z=0) \tag{3.17}$$

for

$$\omega t_{0x} \mp \nu_x = 0 , \tag{3.18}$$

or for

$$\omega t_{0x} = \pm \nu_x . \tag{3.19}$$

Here ν_x / ω represents temporal delay of the $\mathcal{E}_x^{+}(t,z=0)$ component, or temporal advance of $\mathcal{E}_x^{-}(t,z=0)$, when considering their positions in time corresponding to maximum values at $t = 0$ for $\nu_x = 0$. Therefore ν_x, in radians, can be called the temporal phase delay or advance, correspondingly.
And again, similar considerations apply also to the $\mathcal{E}_y^{\pm}(t,z=0)$ component.

The component-dependent parameters, defined by the electric vector components (3.13), represent the same polarization and spatial phase of the two waves propagating in the opposite directions. That fact is of especial importance. It can be considered an immediate result of the time-symmetry of Maxwell equations for plane waves (see, for example, Brosseau [28]).

The two of those parameters can serve to determine the wave's absolute amplitude,

$$E_0 = \sqrt{a_x^2 + a_y^2} = a_0 , \tag{3.20}$$

which is represented geometrically by the Monge radius of the polarization ellipse described by the tip of the electric vector rotating with time in the plane $z = 0$.

3. 8. Basis-dependent analytical parameters

It is convenient to define the basis-dependent parameters by using the electric vector in its matrix form, in the linear basis (x,y) or (y,x), with the order of components corresponding to the order of the basis vectors, $\mathbf{1}_x$ and $\mathbf{1}_y$:

$$\mathcal{E}_{(x,y)}^{\pm}(t,z) = \begin{bmatrix} \mathcal{E}_x^{\pm}(t,z) \\ \mathcal{E}_y^{\pm}(t,z) \end{bmatrix} = \begin{bmatrix} 0 & 1 \\ 1 & 0 \end{bmatrix} \mathcal{E}_{(y,x)}^{\pm}(t,z) \tag{3.21}$$

Introducing the PP basis-dependent analytical parameters $\gamma, \delta, \varepsilon$ defined as follows,

$$\begin{aligned} \tan\gamma_{(x,y)} &= a_y / a_x = \cot\gamma_{(y,x)} \\ \delta_{(x,y)} &= (\nu_x - \nu_y)/2 = -\delta_{(y,x)} \\ \varepsilon_{(x,y)} &= (\nu_x + \nu_y)/2 = \varepsilon_{(y,x)} \end{aligned} \tag{3.22}$$

and exchanging, for basis-dependent, the component-dependent parameters of expressions (1.13), representing components in (3.21),

$$\begin{aligned} a_x &= E_0 \cos\gamma_{(x,y)} = E_0 \sin\gamma_{(y,x)} \\ a_y &= E_0 \cos\gamma_{(y,x)} = E_0 \sin\gamma_{(x,y)} \\ v_x &= \delta_{(x,y)} + \varepsilon_{(x,y)} = -\delta_{(y,x)} + \varepsilon_{(y,x)} \\ v_y &= \delta_{(y,x)} + \varepsilon_{(y,x)} = -\delta_{(x,y)} + \varepsilon_{(x,y)}, \end{aligned} \tag{3.23}$$

we arrive at the following forms of the real column electric vectors for waves propagating in two opposite directions, expressed in the two bases of mutually reversed orders:

$$\mathcal{E}^{\pm}_{(x,y)}(t,z) = E_0 \begin{bmatrix} \cos\gamma_{(x,y)} \cos\{\omega t \mp (kz + \delta_{(x,y)} + \varepsilon_{(x,y)})\} \\ \sin\gamma_{(x,y)} \cos\{\omega t \mp (kz - \delta_{(x,y)} + \varepsilon_{(x,y)})\} \end{bmatrix} \tag{3.24a}$$

and

$$\mathcal{E}^{\pm}_{(y,x)}(t,z) = E_0 \begin{bmatrix} \cos\gamma_{(y,x)} \cos\{\omega t \mp (kz + \delta_{(y,x)} + \varepsilon_{(y,x)})\} \\ \sin\gamma_{(y,x)} \cos\{\omega t \mp (kz - \delta_{(y,x)} + \varepsilon_{(y,x)})\} \end{bmatrix} \tag{3.24b}$$

Several important conclusions result from the above equalities, for example:

- the 2δ parameter presents *spatial phase delay of the first component versus the second one*,
- the right-handed elliptical polarizations correspond to the negative 2δ parameters expressed in the (x,y) basis, and to their positive values in the (y,x) basis (after the IEEE Standard [28], according to which the polarization is right-handed if the electric vector is seen rotating clockwise when looking in direction of wave propagation),
- the above relation between polarization handedness and the sign of the 2δ parameter in the linear polarization bases has been established for real electric vectors, and therefore does not depend on the choice of the time-dependence convention, $\exp(\pm j\omega t)$,
- circular polarizations correspond to $2\gamma = 90^0$ and linear polarizations to $2\delta = 0^0$ or 180^0 in all linear polarization bases, independently of the order of the basis components,
- average spatial phase delay of the two wave's components, ε, also does not depend on the order of the polarization basis (though it depends on the polarization basis vectors and, therefore, should be used with the lower index denoting the basis).

So, only three real parameters sufficiently define the wave's polarization and spatial phase. However, they depend on the polarization basis, including the order of its vectors.

3. 9. Basis-dependent geometrical parameters

According to commonly accepted convention, the tilt angle, β, of the major axis of the polarization ellipse is measured from the first axis of the linear polarization basis in direction to its second axis. So: the $\beta_{(x,y)}$ angle is measured from the x axis in direction to the y axis, and the $\beta_{(y,x)}$ angle from the y axis in direction to the x axis. That results in the relation between the tilt angles similar to that for the γ angles:

$$\beta_{(x,y)} = 90^0 - \beta_{(y,x)}. \tag{3.25a}$$

The ellipticity angles are of course of opposite signs, similarly as δ angles:

$$\alpha_{(x,y)} = -\alpha_{(y,x)}. \tag{3.25b}$$

Only the phase angles of both components along the axes of the polarization ellipse remain independent of the order of components, similarly as ε angles :

$$\chi_{(x,y)} = \chi_{(y,x)}. \tag{3.25c}$$

Rotating the original linear basis in two directions by the angles $\beta_{(x,y)}$ and $\beta_{(y,x)}$ to cover with the ellipse axes we arrive at two new right-handed coordinate systems, $\xi\eta z$, with the following relation between their axes:

$$\xi(x,y) = \eta(y,x)$$

and
$$\eta(x,y) = -\xi(y,x)\,. \tag{3.26}$$

Components of the electric vector in one of those coordinate systems are

$$\begin{aligned}\mathcal{E}^{\pm}_{\xi(x,y)}(t,z) &= a_0 \cos\alpha_{(x,y)} \cos\{\omega t \mp (kz + v_{\xi(x,y)})\}\\ \mathcal{E}^{\pm}_{\eta(x,y)}(t,z) &= a_0 \sin\alpha_{(x,y)} \cos\{\omega t \mp (kz + v_{\eta(x,y)})\},\end{aligned} \tag{3.27}$$

with

$$\begin{aligned}2v_{\xi(x,y)} &= 2\delta_{\xi(x,y)} + 2\varepsilon_{\xi(x,y)}\\ &= \tfrac{\pi}{2} + 2\chi_{(x,y)} - \tfrac{\pi}{2} = 2\chi_{(x,y)}\end{aligned}$$

and

$$\begin{aligned}2v_{\eta(x,y)} &= 2\delta_{\eta(x,y)} + 2\varepsilon_{\eta(x,y)}\\ &= -\tfrac{\pi}{2} + 2\chi_{(x,y)} - \tfrac{\pi}{2} = 2\chi_{(x,y)} - \pi\end{aligned} \tag{3.28}$$

That yields

$$\begin{aligned}\mathcal{E}^{\pm}_{\xi(x,y)}(t,z) &= a_0 \cos\alpha_{(x,y)} \cos\{\omega t \mp (kz + \chi_{(x,y)})\}\\ \mathcal{E}^{\pm}_{\eta(x,y)}(t,z) &= \mp a_0 \sin\alpha_{(x,y)} \sin\{\omega t \mp (kz + \chi_{(x,y)})\}\end{aligned} \tag{3.29a}$$

Similarly, in the other coordinate system,

$$\begin{aligned}\mathcal{E}^{\pm}_{\eta(y,x)}(t,z) &= a_0 \cos\alpha_{(y,x)} \cos\{\omega t \mp (kz + \chi_{(y,x)})\}\\ \mathcal{E}^{\pm}_{\xi(y,x)}(t,z) &= \mp a_0 \sin\alpha_{(y,x)} \sin\{\omega t \mp (kz + \chi_{(y,x)})\}\end{aligned} \tag{3.29b}$$

Here $\alpha_{(x,y)} < 0$ and $\alpha_{(y,x)} > 0$ correspond to the right-handed elliptical polarizations.

3. 10. Mutual relations between analytical and geometrical polarization parameters

The same $\mathcal{E}^{\pm}_x$ and $\mathcal{E}^{\pm}_y$ components depend on two sets of $\mathcal{E}^{\pm}_\xi$ and $\mathcal{E}^{\pm}_\eta$ components as follows:

$$\begin{bmatrix}\mathcal{E}^{\pm}_x(t,z)\\ \mathcal{E}^{\pm}_y(t,z)\end{bmatrix} = \begin{bmatrix}\cos\beta & -\sin\beta\\ \sin\beta & \cos\beta\end{bmatrix}_{(x,y)} \begin{bmatrix}\mathcal{E}^{\pm}_\xi(t,z)\\ \mathcal{E}^{\pm}_\eta(t,z)\end{bmatrix}_{(x,y)} \tag{3.30a}$$

or

$$\begin{bmatrix}\mathcal{E}^{\pm}_y(t,z)\\ \mathcal{E}^{\pm}_x(t,z)\end{bmatrix} = \begin{bmatrix}\cos\beta & -\sin\beta\\ \sin\beta & \cos\beta\end{bmatrix}_{(y,x)} \begin{bmatrix}\mathcal{E}^{\pm}_\eta(t,z)\\ \mathcal{E}^{\pm}_\xi(t,z)\end{bmatrix}_{(y,x)} \tag{3.30b}$$

Comparing them with the components (24a) and (24b) we arrive (see Appendix A1) at the Stokes four-vectors expressed in terms of analytical and geometrical parameters. They are identical for both opposite directions of wave's propagation (or antenna orientation) but differ in sign of their second and fourth component for bases of the reversed order:

$$\mathbf{I}^{\pm}_0 = \begin{bmatrix}\mathrm{I}\\ \mathrm{Q}\\ \mathrm{U}\\ \mathrm{V}\end{bmatrix}_{(x,y)} = a_0^2 \begin{bmatrix}1\\ \cos 2\gamma\\ \sin 2\gamma \cos 2\delta\\ \sin 2\gamma \sin 2\delta\end{bmatrix}_{(x,y)} = a_0^2 \begin{bmatrix}1\\ \cos 2\alpha \cos 2\beta\\ \cos 2\alpha \sin 2\delta\\ \sin 2\alpha\end{bmatrix}_{(x,y)} = \begin{bmatrix}\mathrm{I}\\ \text{-Q}\\ \mathrm{U}\\ \text{-V}\end{bmatrix}_{(y,x)} \tag{3.31}$$

Other relations between analytical and geometrical parameters, also including phases, can be found from the same formulae or from equations of spherical trigonometry. They are valid in both bases, (x,y) and (y,x):

$$\begin{gathered}\tan[\delta-(\chi-\varepsilon)]=\tan\alpha\tan\beta\\ \tan[\delta+(\chi-\varepsilon)]=\tan\alpha\cot\beta\\ \tan(2\chi-2\varepsilon)=\cos 2\gamma\tan 2\delta=\sin 2\alpha\cot 2\beta\end{gathered}\tag{3.32}$$

Relation between analytical and geometrical parameters can be seen also when comparing Fig.3.1 and Fig.3.2. Tangential polarization (TP) phasors introduced in Fig. 3.2 enable one to present on the Poincare sphere not only polarization but also spatial phase, ε. That spatial phase may change in the range between 0 and 2π. That is why its double angle will change in the range of 4π, and the TP phasor rotated by the 2π angle takes the negative of its initial value, though identically oriented. Therefore in the figure, to avoid an ambiguity, the 2ε angle should always indicate the value from that doubled range. The reason for the necessity of using the double phase angle will be explained further, in Section 4. 3.

See Appendix M, and especially Fig. M.1, for other angular parameters of the TP phasors.

3. 11. Complex form of the PP vectors

In order to obtain the complex matrix representation of electric vector waves, complexification of their real form is needed. Using earlier obtained formulae, in terms of analytical parameters it can be done as follows:

$$\begin{gathered}\left[\mathcal{E}^{\pm}(t,z)\right]_{(x,y)\text{or}(y,x)}=E_0\begin{bmatrix}\cos\gamma\cos\{\omega t\mp(kz+\delta+\varepsilon)\}\\ \sin\gamma\cos\{\omega t\mp(kz-\delta+\varepsilon)\}\end{bmatrix}_{(x,y)\text{or}(y,x)}\Rightarrow\\ \Rightarrow\left[E^{\pm}(t,z)\right]_{(x,y)\text{or}(y,x)}=E_0\begin{bmatrix}\cos\gamma e^{\mp j(\delta+\varepsilon)}\\ \sin\gamma e^{\pm j(\delta-\varepsilon)}\end{bmatrix}_{(x,y)\text{or}(y,x)}\cdot e^{j(\omega t\mp kz)}\end{gathered}\tag{3.33}$$

Similarly, in terms of geometrical parameters we obtain

$$\begin{aligned}\left[\mathcal{E}^{\pm}(t,z)\right]_{(x,y)\text{or}(y,x)}&=E_0\left\{\begin{bmatrix}\cos\beta & -\sin\beta\\ \sin\beta & \cos\beta\end{bmatrix}\begin{bmatrix}\cos\alpha\cos\{\omega t\mp(kz+v_\xi)\}\\ \sin\alpha\cos\{\omega t\mp(kz+v_\eta)\}\end{bmatrix}\right\}_{(x,y)\text{or}(y,x)}\Rightarrow\\ &\Rightarrow E_0\begin{bmatrix}\cos\alpha\cos\beta e^{\mp jv_\xi}-\sin\alpha\sin\beta e^{\mp jv_\eta}\\ \cos\alpha\sin\beta e^{\mp jv_\xi}+\sin\alpha\cos\beta e^{\mp jv_\eta}\end{bmatrix}_{(x,y)\text{or}(y,x)}e^{j(\omega t\mp kz)}\\ &=E_0\left\{\begin{bmatrix}\cos\alpha\cos\beta\mp j\sin\alpha\sin\beta\\ \cos\alpha\sin\beta\pm j\sin\alpha\cos\beta\end{bmatrix}e^{\mp j\chi}\right\}_{(x,y)\text{or}(y,x)}e^{j(\omega t\mp kz)}\end{aligned}\tag{3.34}$$

It means that for electric vectors

$$\begin{gathered}\boldsymbol{E}^{+}(t,z)=E_0\boldsymbol{u}\,e^{j(\omega t-kz)}\\ \boldsymbol{E}^{-}(t,z)=E_0\boldsymbol{u}^{*}\,e^{j(\omega t+kz)}\end{gathered}\tag{3.35}$$

and for $\boldsymbol{u}$ vector expressed by its column form u:

$$\boldsymbol{u}=\begin{bmatrix}\boldsymbol{1}_x & \boldsymbol{1}_y\end{bmatrix}\begin{bmatrix}a\\ b\end{bmatrix}_{(x,y)}=\begin{bmatrix}\boldsymbol{1}_y & \boldsymbol{1}_x\end{bmatrix}\begin{bmatrix}a\\ b\end{bmatrix}_{(y,x)};\quad aa^{*}+bb^{*}=1,\quad\begin{bmatrix}a\\ b\end{bmatrix}\equiv u\tag{3.36}$$

the following PP column unit vectors have been obtained (compare [3]):

$$u_{(x,y)\text{or}(y,x)}=\begin{bmatrix}\cos\gamma e^{-j(\delta+\varepsilon)}\\ \sin\gamma e^{j(\delta-\varepsilon)}\end{bmatrix}_{(x,y)\text{or}(y,x)}=\left\{\begin{bmatrix}\cos\alpha\cos\beta-j\sin\alpha\sin\beta\\ \cos\alpha\sin\beta+j\sin\alpha\cos\beta\end{bmatrix}e^{-j\chi}\right\}_{(x,y)\text{or}(y,x)}\tag{3.37}$$

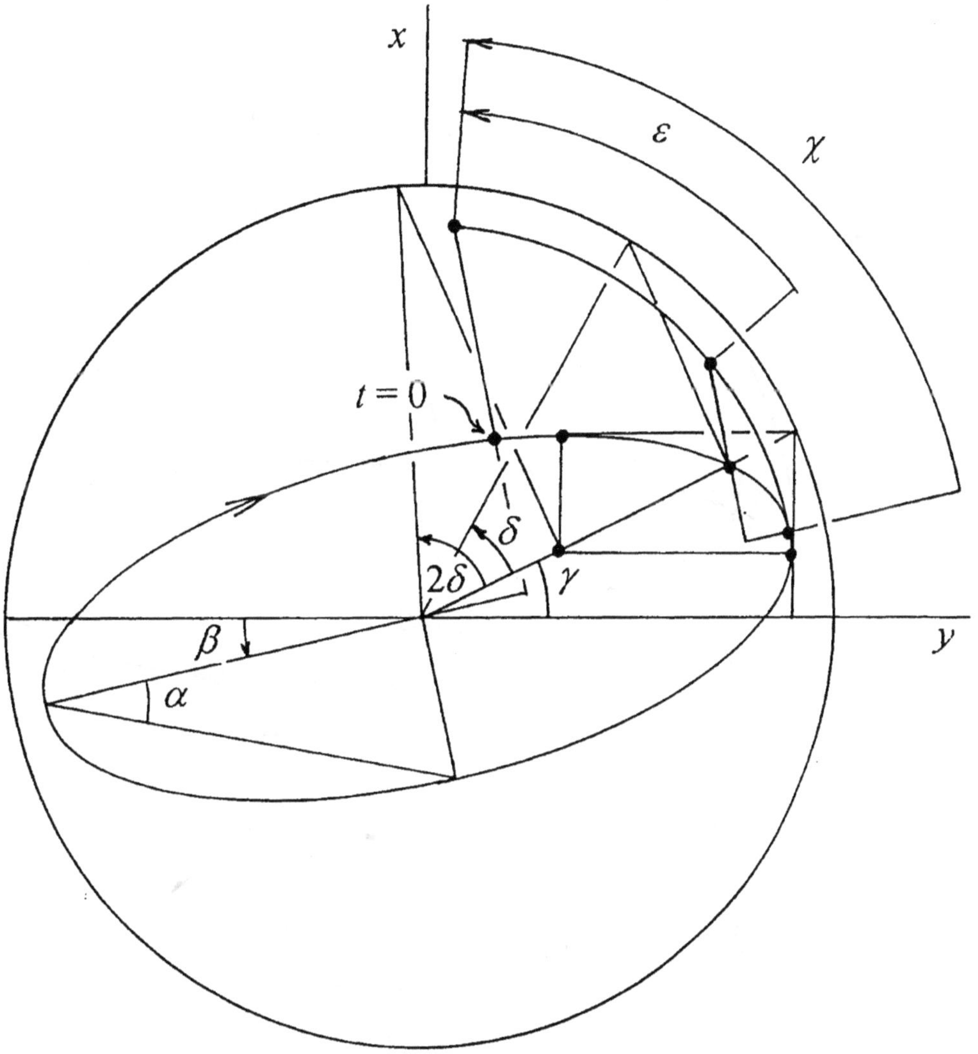

Fig. 3.1. Geometrical, α, β, χ, and analytical, γ, δ, ε, angular parameters of an oriented (right-handed) polarization ellipse in the (y,x) basis of the reversed order.

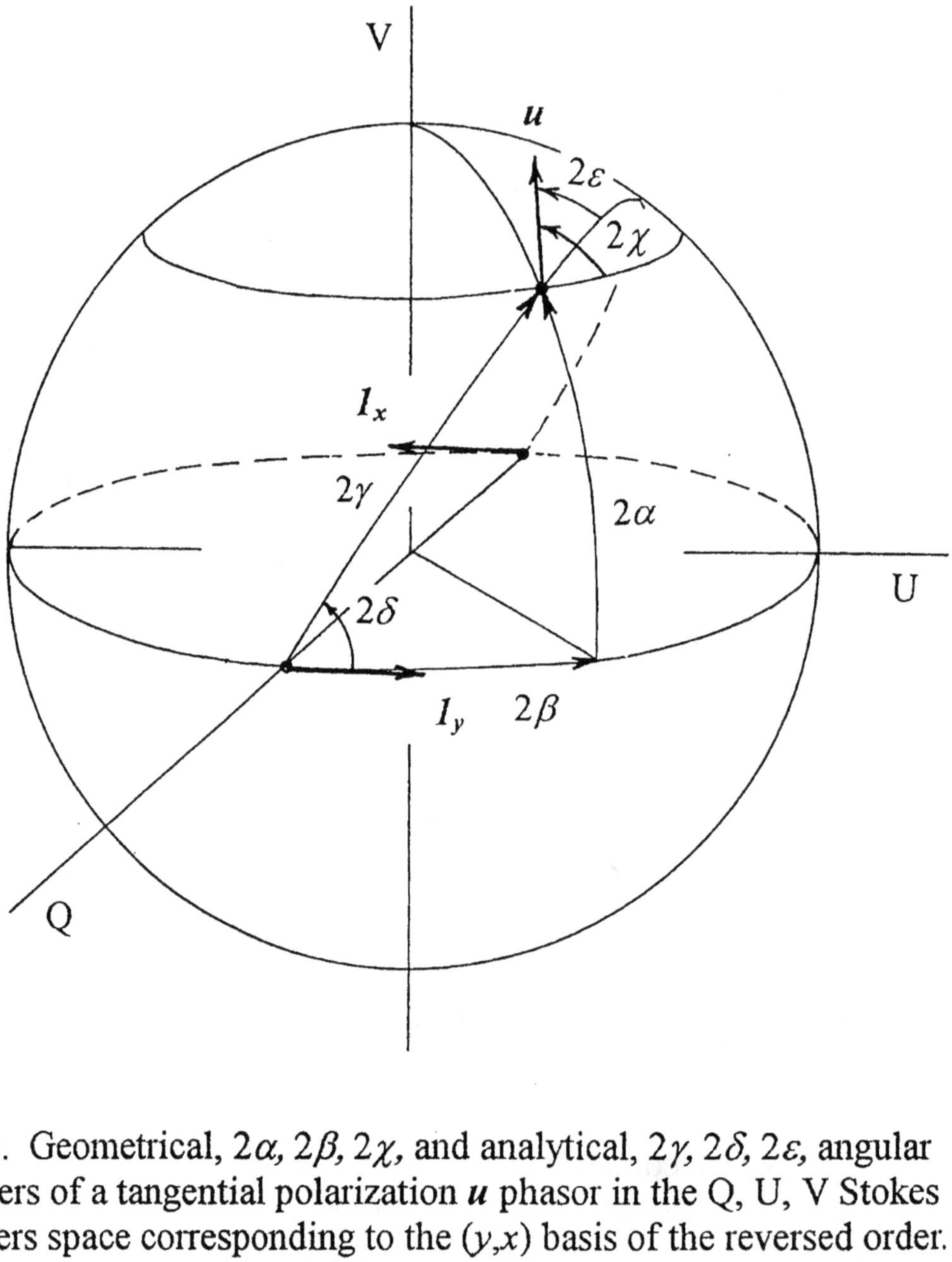

Fig. 3.2. Geometrical, 2α, 2β, 2χ, and analytical, 2γ, 2δ, 2ε, angular parameters of a tangential polarization $\boldsymbol{u}$ phasor in the Q, U, V Stokes parameters space corresponding to the (y,x) basis of the reversed order.

4. Polarization Sphere of Tangential phasors

4. 1. The Poincare sphere

Together with the above 'amplitude' representation of the PP unit vectors, also their 'power' or 'intensity' representation is possible leading to another comparison of analytical and geometrical parameters of polarized waves though without the phase information. The so called *normalized Stokes parameters* can be expressed as three components, without the first one, of the previously obtained Stokes four-vector after its normalization:

$$\begin{aligned} \mathrm{q} &= aa^* - bb^* &&= \cos 2\gamma &&= \cos 2\alpha \cos 2\beta \\ \mathrm{u} &= ab^* + ba^* &&= \sin 2\gamma \cos 2\delta &&= \cos 2\alpha \sin 2\beta \\ \mathrm{v} &= j(ab^* - ba^*) &&= \sin 2\gamma \sin 2\delta &&= \sin 2\alpha \end{aligned} \tag{4.1}$$

They satisfy the normalization equality

$$\mathrm{q}^2 + \mathrm{u}^2 + \mathrm{v}^2 = 1. \tag{4.2}$$

These parameters can be considered as rectangular coordinates of points on the Poincare *polarization sphere of unit radius* with the equator (great circle for v = 0) denotig linear polarizations, and with the poles of circular polarizations (for v = ± 1).

According to the convention applied here the *northern* (*upper*) pole, for v = +1, represents *right*-handed circular polarization and the *southern* (*lower*) pole, for v = −1, the *left*-handed circular polarization. That corresponds to the (y,x) basis of the reversed order and to the assumption commonly being applied that the second, here x-component, leads the first one by the spatial phase angle $2\delta = +\pi/2$, independently of direction of wave propagation.

The concept of the *Poincare sphere of polarization points* will then be extended to the *polarization and (spatial) phase sphere* (*the PP sphere*) by inclusion of phase information when introducing the *tangential polarization phasors.*

4. 2. Tangential polarization phasors. An introduction

Tangential phasors represent polarization helices of plane, monochromatic, completely polarized EM waves. Such helices, considered as models of waves, can be shifted with the velocity of light in two opposite directions along a propagation z-axis. Spatial phase of the wave can be defined by the position of the moving helix, in time t = 0, versus the z-axis coordinates, independently of direction of wave's propagation. The double value of that spatial phase has been represented by the angle of phasor's orientation. It means that phasors rotate in time with angular velocity of 2ω in two opposite directions depending on direction of wave's propagation.

The PP sphere will be considered as a kind of the two-folded Riemann surface. The concept of the ONP PP basis will be introduced and the orientation of their phasors explained, together with the rules of phasors' multiplication and addition. Advantages of the proposed notation will be shown on examples of various transformations. In the Appendix some useful formulae of spherical trigonometry of special value for polarimetry are attached.

Contemporary theories of electromagnetic polarimetry usually consider polarization and phase of waves separately [3], [13], [121]. That is entirely impractical way in cases when obtaining special canonical forms for transformation matrices, e.g. for Sinclair or Kennaugh matrices, is desired [69]. Then, orthogonal polarization bases of those matrices require determination of specially adjusted phases for their both vectors.

The definition of phase for elliptically polarized waves is a nontrivial problem. Polarization and phase (PP) vector approach to the theory of polarimetry proposed by this author introduces one unambiguous space of of the PP vectors for both opposite directions of wave propagation (or antenna orientation) as regular 2-dim. complex space, and establishes one-to-one correspondence relating all its vectors versus phasors tangent to the polarization sphere considered as a two-folded Riemann surface in the 3-dim. real Stokes parameters' space. It means that the same direction of phasors of the same polarization, but tangent to two different branches of the surface, corresponds to the PP vectors of opposite phase. *In such a sense, the phasor's direction defines an absolute phase* of the elliptically polarized wave. Its numerical value depends on the polarization parameters chosen and the polarization itself. So, two waves of the same absolute phase, but of different polarizations, may exhibit different phase factors. Also absolute phases of three phasors, which are not tangent to the sphere along the same great circle arc, cannot be all equal. Those difficulties, in analytical expressing the absolute phase, led to definition of the PP vectors bearing the information about the phase which appears unseparable from description of the polarization state.

In the past, different authors (e.g., see [138]) used tangential phasors to present the phase of waves but it was not the absolute phase in the above explained sense, and they did not interpret the polarization sphere as the two-folded Riemann surface. Their phasors used to rotate in time with an angular velocity ω, instead of 2ω as in the here proposed PP vector approach. Such doubled angular velocity makes the phase difference between two PP vectors equal to one half of their phasors' direction angle difference. It enables one to present two phasors of different elliptical polarizations, and their sum, by properly rotated, with the polarization sphere, equally spaced in polarization and phase two phasors, e.g., of linear polarization, and their sum. It is of essential significance for establishing geometrical rules of phasors' addition.

4. 3. Polarization and phase sphere as the two-folded Riemann surface

The fundamental question is how phasors should be oriented. The aim of introduction of the tangential phasors on the polarization sphere is to represent the phase by the direction of the phasor. That is why the phasor's direction has not to be changed when changing its polarization only by shifting the phasor along a great circle arc of the sphere without altering its phase. When moving the phasor along the equator of linear polarizations or, equivalently, when rotating the sphere about its polar axis together with the phasor under consideration, with the coordinate system being at rest, the *double* rotation is needed to arrive at the same polarization *and phase*. This is because after the first full 2π rotation one arrives at an 'opposite polarization' (see Fig.4.1) if the phase is being considered constant during the rotation. The need of the second rotation suggests that it is essential to indicate the *direction* of rotation which brings the phasor to its orthogonal *co-phased* position (e.g., from $\mathit{1}_y$ to $\mathit{1}_x$, and not to $-\mathit{1}_x$, what would be reached when moving in the opposite direction). For that reason the two phasors, $\mathit{1}_y$ and $\mathit{1}_x$, *both* should be directed *along the equator* and *in the same direction* as shown in Fig. 3.2 or 4.1. Also the phasor being a sum of those two phasors, located in the half-way between them, should be similarly oriented.

It will be natural to have the rule of addition for phasors independent of rotation of the PP sphere. So, a phasor of the right circular polarization, which can be obtained by *addition of* $\mathit{1}_y$ *and* $\mathit{1}_x$ *phasors after their rotation* with the sphere by the $\pi/2$ angle about the OQ axis, should be also equal to the similarly *rotated sum* of $\mathit{1}_y$ and $\mathit{1}_x$ phasors. Indeed, defining the unit column Jones vectors, 1_y and 1_x, by the matrix equalities

$$\mathit{1}_y = [\mathit{1}_y, \mathit{1}_x]\, 1_y \qquad \text{with} \quad 1_y = \begin{bmatrix} 1 \\ 0 \end{bmatrix}$$
$$\mathit{1}_x = [\mathit{1}_y, \mathit{1}_x]\, 1_x \qquad \text{with} \quad 1_x = \begin{bmatrix} 0 \\ 1 \end{bmatrix} \tag{4.3}$$

the linearly polarized rotated phasors by

$$\mathit{u}^Y = \mathit{1}_y e^{-j\pi/4} = [\mathit{1}_y, \mathit{1}_x]\, u^Y \qquad \text{with} \quad u^Y = \begin{bmatrix} e^{-j\pi/4} \\ 0 \end{bmatrix}$$
$$\mathit{u}^X = \mathit{1}_x e^{+j\pi/4} = [\mathit{1}_y, \mathit{1}_x]\, u^X \qquad \text{with} \quad u^X = \begin{bmatrix} 0 \\ e^{+j\pi/4} \end{bmatrix} \tag{4.4}$$

sums of phasors by

$$\sqrt{2}\,1_{45^\circ} = 1_y + 1_x = \sqrt{2}[1_y, 1_x]\, 1_{45^\circ} \qquad \text{with} \qquad 1_{45^\circ} = \frac{1}{\sqrt{2}}\begin{bmatrix}1\\1\end{bmatrix}$$

$$\sqrt{2}\, u^R = u^Y + u^X = \sqrt{2}[1_y, 1_x]\, u^R \qquad \text{with} \qquad u^R = \frac{1}{\sqrt{2}}\begin{bmatrix}e^{-j\pi/4}\\ e^{+j\pi/4}\end{bmatrix} \tag{4.5}$$

and the rotation matrix by

$$C = \frac{1}{\sqrt{2}}\begin{bmatrix}1-j & 0\\ 0 & 1+j\end{bmatrix} = \begin{bmatrix}e^{-j\pi/4} & 0\\ 0 & e^{+j\pi/4}\end{bmatrix} \tag{4.7d}$$

one arrives at the expected result for the rotation as a linear operation:

$$\sqrt{2}\, u^R = u^Y + u^X = C1_y + C1_x = \begin{bmatrix}e^{-j\pi/4}\\ e^{+j\pi/4}\end{bmatrix} = C(1_y + 1_x) = C(\sqrt{2}\; 1_{45^\circ}) \tag{4.6}$$

What follows is (see Fig.4.2) that the orientation angles of the rotated 1_y and 1_x phasors, equal to $+\pi/2$ and $-\pi/2$, are of *doubled values* of their spatial phase delay angles $+\pi/4$ and $-\pi/4$ respectively. That is also the general rule governing relations between angles of phasor direction and its spatial phase because it remains valid for any sum of phasors and any rotation of the PP sphere.

The just obtained phasor of the right circular polarization is not in phase with the 1_y vector before its rotation. To arrive at the in-phase phasor of circular polarization two ways can be chosen. You may shift 1_y along the great circle arc to the north pole of the sphere, or you may rotate the previously obtained phasor of circular polarization by $-\pi/2$ angle to the desired position. The last operation makes the rotated phasor to be advanced in phase by $\pi/4$. Generally, when rotating the sphere about its polar OV axis, the phasor of circular polarization changes its phase only and obtains its original value after two full rotations. Its total phase change is then 2π. Phasors of elliptical polarization change both their polarization and phase when rotating about that axis. Then after the next 2π rotation only they arrive at their original values.

The above presented rule of obtaining the original value of phasors can be generalized to rotations about any other axis of the sphere. Thus, the PP sphere can be considered as the two-folded Riemann surface with its branch point on any rotation axis. For the q, u, and v rectangular coordinates of the branch point, defined by the equalities (4.1), the rotation of the sphere by the 2ϕ angle can be obtained when using the rotation matrix of the form:

$$C = C(\mathrm{q},\mathrm{u},\mathrm{v};\; 2\phi) = \begin{bmatrix}\cos\phi - j\mathrm{q}\sin\phi & (-\mathrm{v} - j\mathrm{u})\sin\phi\\ (\mathrm{v} - j\mathrm{u})\sin\phi & \cos\phi + j\mathrm{q}\sin\phi\end{bmatrix} \tag{4.7}$$

That matrix is unitary and *unimodular*. Therefore it depends on three only real parameters (see also (4.14)). So, any u_1 column Jones vector representing the corresponding phasor can be found to be after the rotation:

$$u_2 = C\, u_1 \tag{4.8}$$

Of course, $u_2 = u_1$ for $2\phi = 0$ or $\pm 4\pi$, and $u_2 = -u_1$ for $2\phi = \pm 2\pi$.

The often used special cases of the rotation matrix (4.7) are:

$$C(0,0,1;2\phi) = \begin{bmatrix}\cos\phi & -\sin\phi\\ \sin\phi & \cos\phi\end{bmatrix} \tag{4.7a}$$

$$C(0,-1,0;2\phi) = \begin{bmatrix} \cos\phi & j\sin\phi \\ j\sin\phi & \cos\phi \end{bmatrix} \tag{4.7b}$$

and

$$C(1,0,0;2\phi) = \begin{bmatrix} \cos\phi - j\sin\phi & 0 \\ 0 & \cos\phi + j\sin\phi \end{bmatrix} \tag{4.7c}$$

The last one for $2\phi = \pi/2$ *is* given in (4.7d).

The following example shows two rotations of the PP sphere. They allow to obtain the right circular PP vector, which is in phase with the horizontal linear null phase vector. The first rotation changes the vector's polarization, while the second its phase only (see Fig.4):

$$\begin{aligned} u^{R'} &= C(0,-1,0;\pi/2)1_y = \frac{1}{\sqrt{2}}\begin{bmatrix} 1 & j \\ j & 1 \end{bmatrix}\begin{bmatrix} 1 \\ 0 \end{bmatrix} \\ &= C(0,0,1;-\pi/2)u^R = \frac{1}{\sqrt{2}}\begin{bmatrix} 1 & 1 \\ -1 & 1 \end{bmatrix}\frac{1}{\sqrt{2}}\begin{bmatrix} e^{-j\pi/4} \\ e^{+j\pi/4} \end{bmatrix} = \frac{1}{\sqrt{2}}\begin{bmatrix} 1 \\ j \end{bmatrix} \end{aligned} \tag{4.10}$$

4. 4. The orthogonal null-phase (ONP) polarization basis

By rotating the original linear polarization basis with the Poincare sphere, one can obtain every other orthogonal basis represented by two colinear phasors tangent to the antipodal points of the sphere. Such a basis will be called *null-phase* because, after multiplication of its both vectors by the same phase factor, their phasors usually become not collinear.

Worth noticing is that to specify any ONP polarization basis it is sufficient to take its first vector only, because its second vector is automatically defined by the same rotation rule as applied to the first one.

To present the rotation, the previously proposed matrix (4.7) can be used yielding the new basis vectors of the form:

$$u^B = C^B 1_y \qquad u^{Bx} = C^B 1_x \tag{4.11}$$

When using simpler expression for the matrix:

$$C^B = \begin{bmatrix} a & -b^* \\ b & a^* \end{bmatrix}^B \tag{4.12}$$

and the old basis vectors as in (4.3), the new basis will take the form

$$u^B = \begin{bmatrix} a \\ b \end{bmatrix}^B \qquad u^{Bx} = \begin{bmatrix} -b^* \\ a^* \end{bmatrix}^B \tag{4.13}$$

with

$$\begin{aligned} a^B &= \cos\phi - jq\sin\phi &&= \cos\gamma^B e^{-j(\delta^B+\varepsilon^B)} \\ b^B &= (\mathrm{v} - j\mathrm{u})\sin\phi &&= \sin\gamma^B e^{+j(\delta^B-\varepsilon^B)} \end{aligned} \tag{4.14}$$

For any desired analytical parameters of the first new basis vector u^B, the above equations determine both the angle of rotation 2ϕ, and the coordinates q,u,v of the sphere's branch point on the axis of rotation.

4. 5. The Hermitian product of phasors

The desirable property of phasors is the possibility of their addition and multiplication. The addition is important when considering interference of waves, whereas the multiplication allows one to find the received voltage for given phasors of the receiving antenna and incoming wave. The multiplication is necessary also to decompose a phasor into a sum of two other phasors. The rules of these operations appear to be independent of any rotation of the polarization sphere.

There is a very simple method enabling one to find the rule of multiplication for phasors. At first, the fact should be used of the one-to-one correspondence between phasors and PP vectors in the sense that the result of any operation for the PP vectors should agree with that for phasors. Then, one can immediately see that very simple rule of multiplication applicable to any two phasors on the equator of linear polarizations remains valid also for any other pair of phasors of the same mutual positions, i.e. for phasors obtained from the original pair by any rotation of the polarization sphere.

Distinguishing the two PP vectors of linear polarizations by upper indices, their Hermitian product can be presented in the form

$$\begin{aligned} \boldsymbol{u}^A \cdot \boldsymbol{u}^B * &= \tilde{u}^A u^B * \\ &= [\cos\gamma^A, \sin\gamma^A]\exp(-j\varepsilon^A)\begin{bmatrix}\cos\gamma^B \\ \sin\gamma^B\end{bmatrix}\exp(+j\varepsilon^B) \\ &= \cos(\gamma^A - \gamma^B)\exp\{-j(\varepsilon^A - \varepsilon^B)\} \\ &= \cos\frac{\mathrm{AB}}{2}\, e^{-j2\Delta_B^A} \end{aligned} \tag{4.15}$$

Here AB the means angular distance between the polarization points A and B of the two PP vectors and $2\,\Delta_B^A$ is one half of the angular difference in orientations of their phasors on the equator of linear polarizations and means the spatial phase delay of the $\boldsymbol{u}^A$ versus $\boldsymbol{u}^B$ vector.

A rotation of the sphere can now be considered. Using the rotation matrix *C*, the new PP vectors will be obtained:

$$u^{A'} = Cu^A \quad \text{and} \quad u^{B'} = Cu^B \tag{4.16}$$

Their Hermitian product is as before the rotation:

$$\tilde{u}^{A'} u^{B'} * = \tilde{u}^A \tilde{C} C * u^B * = \tilde{u}^A u^B * = \cos\frac{\mathrm{AB}}{2}\, e^{-j2\Delta_B^A} \tag{4.17}$$

because *C* is unitary, satisfying the equality

$$\tilde{C}C* = \begin{bmatrix}1 & 0 \\ 0 & 1\end{bmatrix} \tag{4.18}$$

The only difference is that the detailed result will be given by a more complex expressions. They will be found when considering the general forms of the two PP vectors in their Hermitian product. When using the *analytical* parameters of the vectors one obtains:

$$\boldsymbol{u}^A \cdot \boldsymbol{u}^{B*} = \tilde{u}^A u^{B*}$$

$$= [\cos\gamma^A e^{-j(\delta^A+\varepsilon^A)}, \sin\gamma^A e^{+j(\delta^A-\varepsilon^A)}] \begin{bmatrix} \cos\gamma^B e^{+j(\delta^B+\varepsilon^B)} \\ \sin\gamma^B e^{-j(\delta^B-\varepsilon^B)} \end{bmatrix}$$

$$= \{\cos(\gamma^A - \gamma^B)\cos(\delta^A - \delta^B) - j\cos(\gamma^A + \gamma^B)\sin(\delta^A - \delta^B)\}\, e^{-j(\varepsilon^A - \varepsilon^B)} \tag{4.19}$$

$$= \cos\frac{\mathrm{AB}}{2} e^{-j2\Delta^A_B}$$

The phase delay of the $\boldsymbol{u}^A$ versus $\boldsymbol{u}^B$ vector can be presented successively, using formulae of spherical trigonometry (see Appendix A for comparison), as:

$$2\Delta^A_B = \varepsilon^A - \varepsilon^B + \arctan\left[\frac{\cos(\gamma^A + \gamma^B)}{\cos(\gamma^A - \gamma^B)}\tan(\delta^A - \delta^B)\right] \tag{4.20}$$

$$= \varepsilon^A - \varepsilon^B + \delta^A - \delta^B - \tfrac{1}{2} E_{\mathrm{YBA}} \tag{4.21a}$$

$$= \varepsilon^A - \varepsilon^B + \tfrac{1}{2}[\pi - (\hat{A} + \hat{B})] \tag{4.21b}$$

$$= \varepsilon^A - \varepsilon^B + \tfrac{1}{4}(E'_{\mathrm{YBA}} - E_{\mathrm{YBA}}) \tag{4.21c}$$

It should be observed that for phasors located outside the equator their phase difference depends not only on their ε parameters but also is enlarged, *e.g.*, by a quarter of the difference of the excesses of two oriented colunar spherical triangles XAB and YBA (see Fig. 4.3).

The magnitude of the product (4.19) can be found by applying simple trigonometric formulae

$$\left.\begin{matrix} \cos^2\alpha \\ \sin^2\alpha \end{matrix}\right\} = \tfrac{1}{2}(1 \pm \cos 2\alpha)$$

and

$$\cos(\alpha \pm \beta) = \cos\alpha\cos\beta \mp \sin\alpha\sin\beta$$

thus obtaining

$$\cos\frac{\mathrm{AB}}{2} = \sqrt{\cos^2(\gamma^A - \gamma^B)\cos^2(\delta^A - \delta^B) + \cos^2(\gamma^A + \gamma^B)\sin^2(\delta^A - \delta^B)} \tag{4.22a}$$

$$= \sqrt{\tfrac{1}{2}[1 + \cos 2\gamma^A \cos 2\gamma^B + \sin 2\gamma^A \sin 2\gamma^B \cos(2\delta^A - 2\delta^B)]} \tag{4.22b}$$

$$= \sqrt{\tfrac{1}{2}[1 + \mathrm{q^A q^B + u^A u^B + v^A v^B}]} \tag{4.22c}$$

$$= \sqrt{\tfrac{1}{2}[1 + \cos(\mathrm{AB})]} \tag{4.22d}$$

The rectangular coordinates of the two polarization points A and B appearing in (4.22c) are given by the formulae (4.1).

The just obtained expression (4.22b) in comparison with (4.22d) yields the well-known formula of spherical trigonometry for the YBA triangle:

$$\cos(\mathrm{AB}) = \cos(\mathrm{YA})\cos(\mathrm{YB}) + \sin(\mathrm{YA})\sin(\mathrm{YB})\cos\hat{Y} \tag{4.23}$$

The Hermitian multiplication of the PP vectors in terms of the *geometrical* parameters leads to similar results:

$$\begin{aligned}
\boldsymbol{u}^A \cdot \boldsymbol{u}^B * &= \tilde{u}^A u^B * \\
&= [\cos\alpha^A \cos\beta^A - j\sin\alpha^A \sin\beta^A,\ \cos\alpha^A \sin\beta^B + j\sin\alpha^A \cos\beta^B]e^{-j\chi^A} \times \\
&\qquad \times \begin{bmatrix} \cos\alpha^B \cos\beta^B + j\sin\alpha^B \sin\beta^B \\ \cos\alpha^B \sin\beta^B - j\sin\alpha^B \cos\beta^B \end{bmatrix} e^{+j\chi^B} \\
&= \{\cos(\alpha^A - \alpha^B)\cos(\beta^A - \beta^B) - j\sin(\alpha^A + \alpha^B)\sin(\beta^A - \beta^B)\}\, e^{-j(\chi^A - \chi^B)} \\
&= \cos\frac{\mathrm{AB}}{2} e^{-j2\Delta_B^A}
\end{aligned} \tag{4.24}$$

with (see Appendix A):

$$2\Delta_B^A = \chi^A - \chi^B + \arctan\left[\frac{\sin(\alpha^A + \alpha^B)}{\cos(\alpha^A - \alpha^B)}\tan(\beta^A - \beta^B)\right] \tag{4.25}$$

$$= \chi^A - \chi^B + \beta^A - \beta^B + \tfrac{1}{2}E_{\mathrm{RAB}} \tag{4.26a}$$

$$= \chi^A - \chi^B - \tfrac{1}{2}[\pi - (\hat{A}' + \hat{B}')] \tag{4.26b}$$

$$= \chi^A - \chi^B - \tfrac{1}{4}\left(E'_{\mathrm{RAB}} - E_{\mathrm{RAB}}\right) \tag{4.26c}$$

and with

$$\cos\frac{\mathrm{AB}}{2} = \sqrt{\cos^2(\alpha^A - \alpha^B)\cos^2(\beta^A - \beta^B) + \sin^2(\alpha^A + \alpha^B)\sin^2(\beta^A - \beta^B)} \tag{4.27a}$$

$$= \sqrt{\tfrac{1}{2}[1 + \sin 2\alpha^A \sin 2\alpha^B + \cos 2\alpha^A \cos 2\alpha^B \cos(2\beta^A - 2\beta^B)]} \tag{4.27b}$$

$$= \sqrt{\tfrac{1}{2}[1 + \mathrm{q}^A\mathrm{q}^B + \mathrm{u}^A\mathrm{u}^B + \mathrm{v}^A\mathrm{v}^B]} \tag{4.22c}$$

$$= \sqrt{\tfrac{1}{2}[1 + \cos(\mathrm{AB})]} \tag{4.22d}$$

Additional explanation of formulae employing analytical and geometrical parameters can be found in Fig 4.3.

4. 6. Decomposition of the PP vector into two orthogonal components

The just described Hermitian multiplication can be used to decompose the PP vector into two parts corresponding to any desired ONP polarization basis defined by its first $\boldsymbol{u}^B$ vector as follows

$$
\begin{aligned}
\boldsymbol{u}^{A} &= (\tilde{u}^{A} u^{B} *)\boldsymbol{u}^{B} + (\tilde{u}^{A} u^{B\times} *)\boldsymbol{u}^{B\times} \\
&= \cos\frac{\mathrm{AB}}{2} e^{-j2\Delta^{A}_{B}} \boldsymbol{u}^{B} + \sin\frac{\mathrm{AB}}{2} e^{-j2\Delta^{A}_{B\times}} \boldsymbol{u}^{B\times} \\
&= a^{A}_{B} \boldsymbol{u}^{B} + b^{A}_{B} \boldsymbol{u}^{B\times}
\end{aligned} \tag{4.28}
$$

All column u vectors without the lower index are, as before, in the original real basis of linear polarizations while the lower and upper indices of the a and b components in the last equality denote phasors of the ONP polarization basis and of the vector being decomposed, respectively. The same rule when applied to the column vectors can be expressed as follows:

$$
\begin{aligned}
\boldsymbol{u}^{A} &= [\boldsymbol{u}^{B}, \boldsymbol{u}^{B\times}]\, u^{A}_{B} \\
\boldsymbol{u}^{Ax} &= [\boldsymbol{u}^{B}, \boldsymbol{u}^{B\times}]\, u^{Ax}_{B}
\end{aligned} \tag{4.29}
$$

with

$$
\begin{aligned}
u^{A}_{B} &= \begin{bmatrix} a \\ b \end{bmatrix}^{A}_{B} \equiv \begin{bmatrix} a^{A}_{B} \\ b^{A}_{B} \end{bmatrix}; \qquad aa^{*} + bb^{*} = 1 \\
u^{Ax}_{B} &= \begin{bmatrix} -b^{*} \\ a^{*} \end{bmatrix}^{A}_{B} \equiv \begin{bmatrix} -b^{A}_{B}{}^{*} \\ a^{A}_{B}{}^{*} \end{bmatrix}
\end{aligned} \tag{4.30}
$$

Now, the Cayley-Klein rotation parameters a and b of the PP vector $\boldsymbol{u}^{A}$ in the $\boldsymbol{u}^{B}$ basis can be expressed in the form:

$$
\begin{aligned}
a^{A}_{B} &= \cos\gamma^{A}_{B} e^{-j(\delta^{A}_{B} + \varepsilon^{A}_{B})} \\
b^{A}_{B} &= \sin\gamma^{A}_{B} e^{+j(\delta^{A}_{B} - \varepsilon^{A}_{B})}
\end{aligned} \tag{4.31}
$$

The correctness of the decomposition can be checked easily by multiplying both parts of the sum by the $\boldsymbol{u}^{B}*$ or $\boldsymbol{u}^{B\times}*$ term and observing both normalization and orthogonality conditions:

$$
\begin{aligned}
\boldsymbol{u}^{B} \cdot \boldsymbol{u}^{B}* &= \boldsymbol{u}^{B\times} \cdot \boldsymbol{u}^{B\times}* = 1 \\
\boldsymbol{u}^{B} \cdot \boldsymbol{u}^{B\times}* &= \boldsymbol{u}^{B\times} \cdot \boldsymbol{u}^{B}* = 0
\end{aligned} \tag{4.32}
$$

4. 7. The addition of phasors on the PP sphere

The sum of the two PP vectors of different polarizations and spatial phases will be considered in the form:

$$
E^{C}_{o} \boldsymbol{u}^{C} = E^{A}_{o} \boldsymbol{u}^{A} + E^{B}_{o} \boldsymbol{u}^{B} \tag{4.33}
$$

Here the upper indices denote the phasors of the three PP vectors, and E_{o} are the magnitudes of those phasors or the corresponding PP vectors. The result of addition can be found by decomposing the first vector into orthogonal constituents as in (4.28), one of them being of the same polarization and spatial phase as the second vector of the sum:

$$
\begin{aligned}
E^{C}_{o} \boldsymbol{u}^{C} &= E^{A}_{o} (a^{A}_{B} \boldsymbol{u}^{B} + b^{A}_{B} \boldsymbol{u}^{B\times}) + E^{B}_{o} \boldsymbol{u}^{B} \\
&= (E^{A}_{o} a^{A}_{B} + E^{B}_{o}) \boldsymbol{u}^{B} + E^{A}_{o} b^{A}_{B} \boldsymbol{u}^{B\times}
\end{aligned} \tag{4.34}
$$

Now the resultant intensity of the wave corresponding to the above PP vector can be found as

$$
\begin{aligned}
I^C &= (E_o^C \boldsymbol{u}^C)\cdot(E_o^C \boldsymbol{u}^C)^* = (E_o^C)^2 \\
&= (E_o^A a_B^A + E_o^B)(E_o^A a_B^A + E_o^B)^* + (E_o^A b_B^A)(E_o^A b_B^A)^* \\
&= I^A + I^B + 2\sqrt{I^A I^B}\cos\gamma_B^A \cos(\delta_B^A + \varepsilon_B^A)
\end{aligned} \tag{4.35}
$$

The last formula can be rewritten in the previous notation as

$$
I^C = I^A + I^B + 2\sqrt{I^A I^B}\cos\frac{\mathrm{AB}}{2}\cos 2\Delta_B^A \tag{4.36}
$$

with

$$
\frac{\mathrm{AB}}{2} = \gamma_B^A \tag{4.37}
$$

and

$$
2\Delta_B^A = \delta_B^A + \varepsilon_B^A \tag{4.38}
$$

with the evident equalities:

$$
\gamma_B^A = \gamma_A^B \tag{4.39}
$$

but

$$
2\Delta_B^A = -2\Delta_A^B \tag{4.40}
$$

Now, the location and orientation of the resultant $\boldsymbol{u}^C$ phasor should be determined. This can be done by inspection of the *ABC* triangle on the PP sphere inside the small circle through the points A, B, and C (Fig.4.4). The ABC circle (with the C point on it, of unknown location yet) will be chosen that way as to have the phasors, corresponding to the $\boldsymbol{u}^A$ and $\boldsymbol{u}^B$ vectors, tangent to the circle. This can always be done by simultaneous rotation of the two phasors by the same double phase angle 2Δ.

The 2Δ angle can be found from simple formula:

$$
2\Delta = \delta_{B'}^{A'} - \varepsilon_{B'}^{A'} \tag{4.41}
$$

when designating by $\boldsymbol{u}^{A'}$ and $\boldsymbol{u}^{B'}$ the two vectors before the rotation. The problem is how to express the $\boldsymbol{u}^{A'}$ phasor in the $\boldsymbol{u}^{B'}$ basis to obtain the angles on the right side of (4.41). This will be explained later, after introducing the change of basis techniques based on the 'polarization phasor notation' discussed in more detail in the next sections.

Starting from the expression (4.33) one obtains

$$
\begin{aligned}
E_o^C \boldsymbol{u}^C &= (E_o^A \boldsymbol{u}^A \cdot \boldsymbol{u}^{C*} + E_o^B \boldsymbol{u}^B \cdot \boldsymbol{u}^{C*})\,\boldsymbol{u}^C \\
&= \left(E_o^A \cos\frac{\mathrm{AC}}{2} e^{-j2\Delta_C^A} + E_o^B \cos\frac{\mathrm{BC}}{2} e^{-j2\Delta_C^B} \right) \boldsymbol{u}^C
\end{aligned} \tag{4.42}
$$

Having

$$
2\Delta_B^A = 2\Delta_{B'}^{A'} = 2\varepsilon_{B'}^{A'} + 2\Delta = \delta_{B'}^{A'} + \varepsilon_{B'}^{A'} \tag{4.43a}
$$

one can find the two remaining phase difference angles, as shown in Fig.7:

$$
\begin{aligned}
2\Delta_C^A &= 2\Delta_B^A - \hat{A} \\
2\Delta_B^C &= 2\Delta_B^A - \hat{B} = -2\Delta_C^B
\end{aligned} \tag{4.44}
$$

and very useful geometrical equality

$$2\Delta_B^A = \pi - \hat{C} + \tfrac{1}{2}E_{\text{ABC}} = \pi - \tfrac{1}{2}E'_{\text{ABC}} \tag{4.43b}$$

resulting from (4.44), because

$$2\Delta_B^C + 2\Delta_C^A = 4\Delta_B^A - (\hat{A} + \hat{B}) = \pi - \hat{C}$$

and

$$\hat{A} + \hat{B} = \pi - \hat{C} + E_{\text{ABC}}$$

All three angles of the spatial phase delay should be of the same sign according to the arrows on the great circle arcs joining points A, B, and C, and indicating directions of spatial phase delay. These angles determine common orientation of the three phasors all tangent to the same small circle of the sphere.

It is worth while to observe that, according to the rule accepted here, all angles of rotation *to the left* are *positive*, and so are the three phase difference angles as well as the two spherical excesses shown in Fig.4.4.

Another important rule stems from Fig. 4.4 or 4.5. The polarization point C of the sum of the two phasors is *on the right side* of an oriented BA great circle arc. The orientation is shown by the arrow directed towards the phase delay.

However the caution is recommended: the rule is true then only, when the right-elliptical polarizations correspond to the upper part of the polarization sphere, what has been proposed here. As an example confirming the rule, the locus can serve of the *right circular* polarization point. It is on the right side (at the north pole) versus the oriented semi-circle of linear polarizations beginning at the *vertical* polarization point and directed through the *45° linear* to the *horizontal* polarization point. This is the path of the (spatial) *phase lag* of 90°. The opposite direction, also from vertical to horizontal linear polarization, means the phase advance on the PP sphere considered as the two-folded Riemann surface.

The exact location of the C point can be found when demanding mutual cancellation of the orthogonally polarized Cx components of both interfering waves. The following equation should be analyzed:

$$\begin{aligned} 0 &= (E_o^C \boldsymbol{u}^C \cdot \boldsymbol{u}^{Cx}*)\boldsymbol{u}^{Cx} = (E_o^A \boldsymbol{u}^A \cdot \boldsymbol{u}^{Cx}* + E_o^B \boldsymbol{u}^B \cdot \boldsymbol{u}^{Cx}*)\boldsymbol{u}^{Cx} \\ &= (E_o^A \sin\frac{\text{AC}}{2} e^{-j2\Delta_{Cx}^A} + E_o^B \sin\frac{\text{BC}}{2} e^{-j2\Delta_{Cx}^B})\boldsymbol{u}^{Cx} \\ &= (E_o^A \sin\frac{\text{AC}}{2} - E_o^B \sin\frac{\text{BC}}{2})\boldsymbol{u}^{Cx} \end{aligned} \tag{4.45}$$

The solution is

$$4\Delta_{Cx}^A - 4\Delta_{Cx}^B = \pm 2\pi \quad \text{with} \quad 4\Delta_{Cx}^A = 0 \quad \text{and} \quad 4\Delta_{Cx}^B = \mp 2\pi \tag{4.46}$$

and

$$E_o^A \sin\frac{\text{AC}}{2} = E_o^B \sin\frac{\text{BC}}{2} \tag{4.47}$$

There are two orthogonal families of circles on which the C point can be located. The circles of the one of those families, determined by the phase difference $2\Delta_B^A$ as in Fig.6, are passing through the A and B points. If that phase difference is equal to zero or π then the C point is located on the AB arc or on the remaining part of the great circle through A and B outside that arc, respectively. The circle of another family, determined according to (4.47) by the ratio

$$\frac{\sin\frac{\text{AC}}{2}}{\sin\frac{\text{BC}}{2}} = \frac{E_o^B}{E_o^A} \tag{4.48}$$

when crossing the $2\,\Delta_B^A$ = const circle, finally indicates the C point location.

4. 8. The addition of *n* phasors

The formulae (4.34-36) can be generalized to the addition of *n* PP vectors (phasors) $E_o^{A_i} \boldsymbol{u}^{A_i}$; $i = 1,\dots,n$. The resulting intensity takes the form

$$I = \sum_i I^{A_i} + \sum_{i<j} 2\sqrt{I^{A_i} I^{A_j}} \cos\frac{\mathrm{A}_i\mathrm{A}_j}{2} \cos 2\Delta_{A_i}^{A_j}; \quad i,j = 1,\dots,n \tag{4.49}$$

with

$$I^{A_i} = \left(E_o^{A_i}\right)^2 \tag{4.50}$$

$$\frac{\mathrm{A}_i\mathrm{A}_j}{2} = \gamma_{A_i}^{A_j} \tag{4.51}$$

and

$$2\,\Delta_{A_i}^{A_j} = \delta_{A_i}^{A_j} + \varepsilon_{A_i}^{A_j}\,. \tag{4.52}$$

The derivation of formulae (4.36) and (4.49) can be found in Pancharatnam's paper [121], but without analytical parameters of the PP vectors involved.

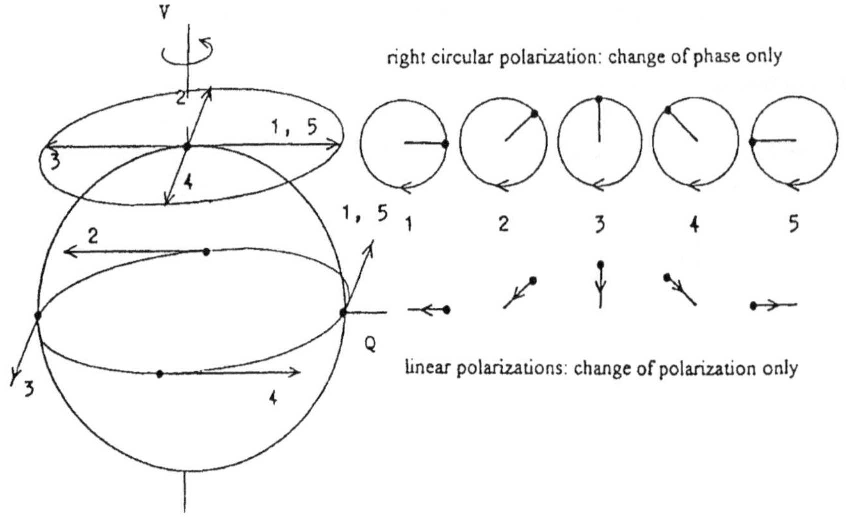

Fig. 4.1. Change of phasor with the PP sphere rotation about its polar axis. Phasors at the poles are changing their phase only; phasors at the equator are changing their polarization only; phasors between the pole and equator are changing both polarization and phase; the total change is π after the 2π rotation.

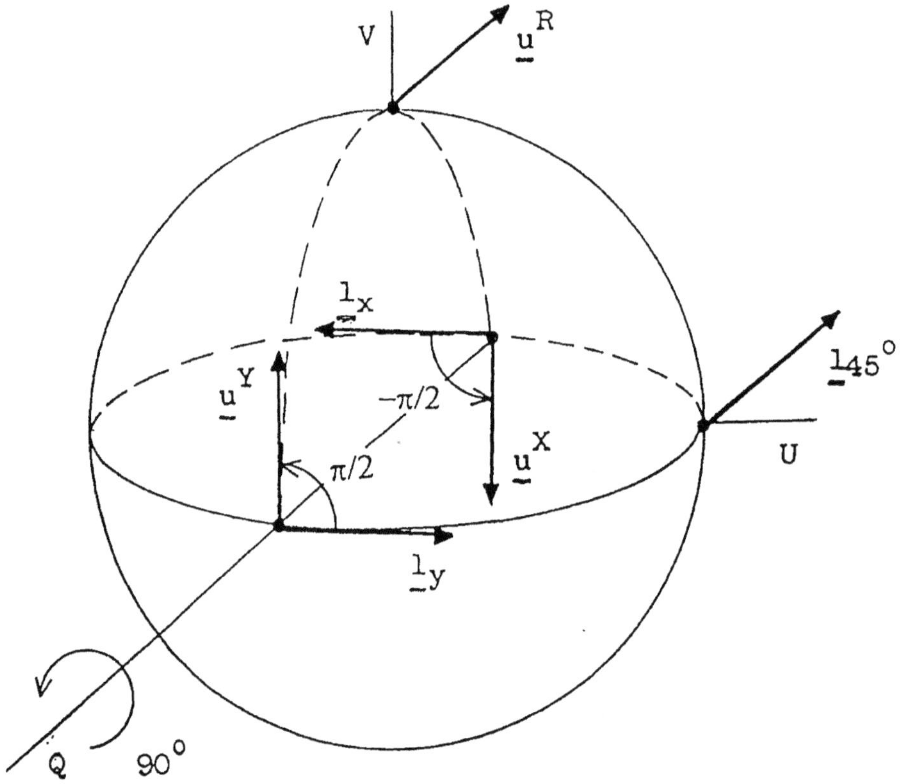

Fig. 4.2. Obtaining same rules for phasors' addition, independently of the PP sphere rotation (here, by 90° about the Q axis), requires change of each phasor's direction by the double value of its change of phase angle.

a)

$$E_{YBA} = \hat{A} + \hat{B} + 2\delta^A - 2\delta^B - \pi$$

b)

$$E_{RAB} = \hat{A}' + \hat{B}' + 2\beta^B - 2\beta^A - \pi$$

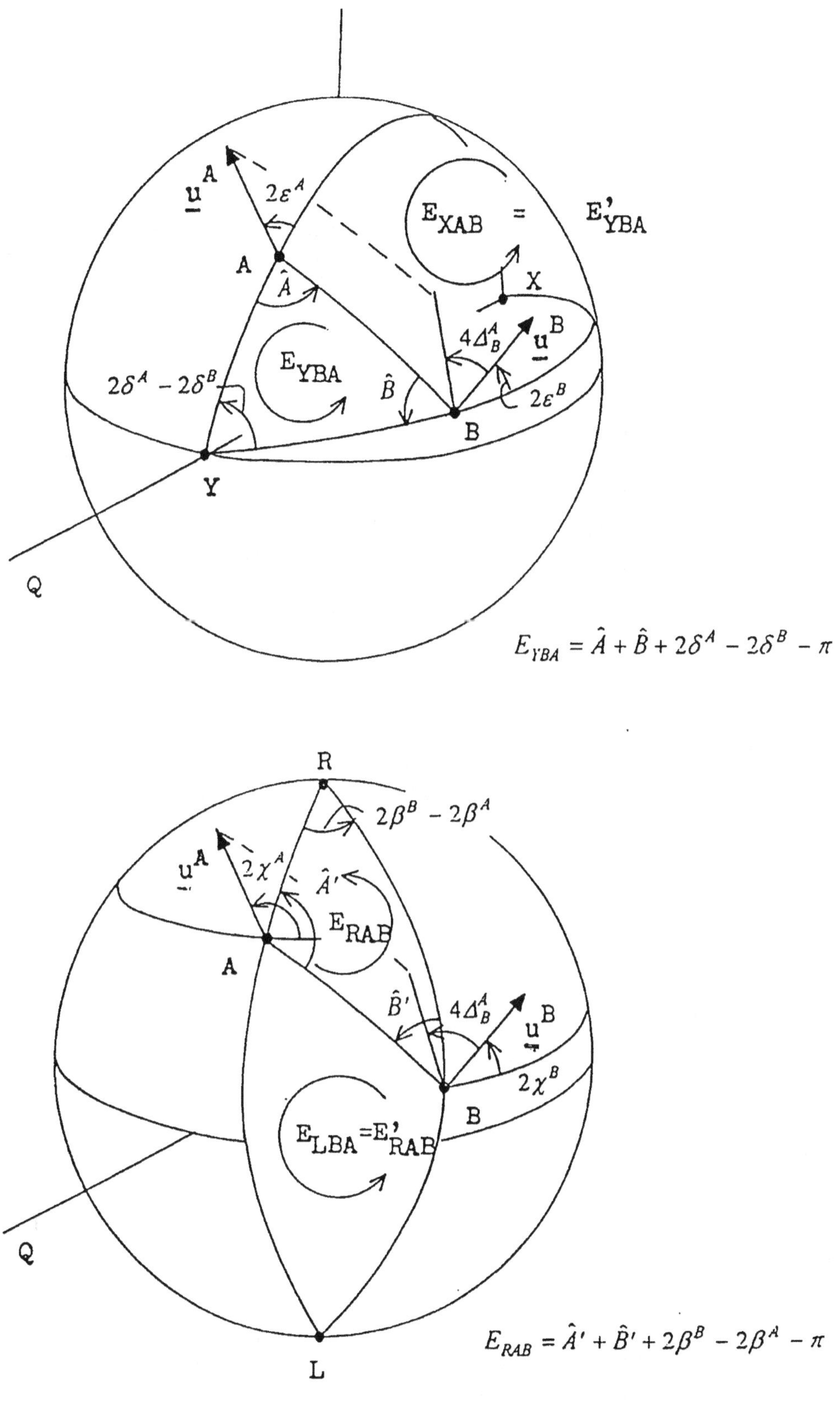

Fig. 4.3. To the evaluation of the spatial phase delay of phasor $\underline{u}^A$ versus $\underline{u}^B$, $2\Delta_B^A$, in terms of: a) analytical, and b) geometrical phasors' parameters.

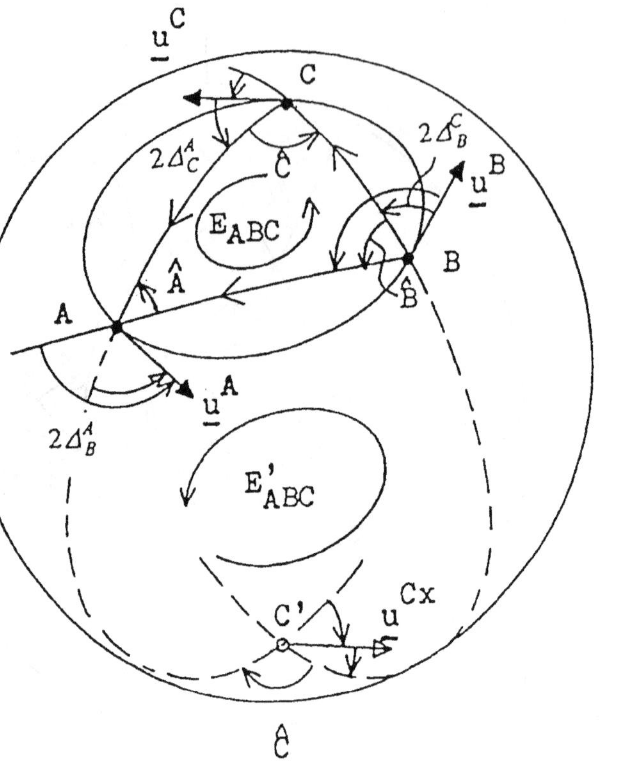

Fig. 4.4. To the explanation of the C point location; result of addition of C' (Cx) components of the two phasors being added must vanish.

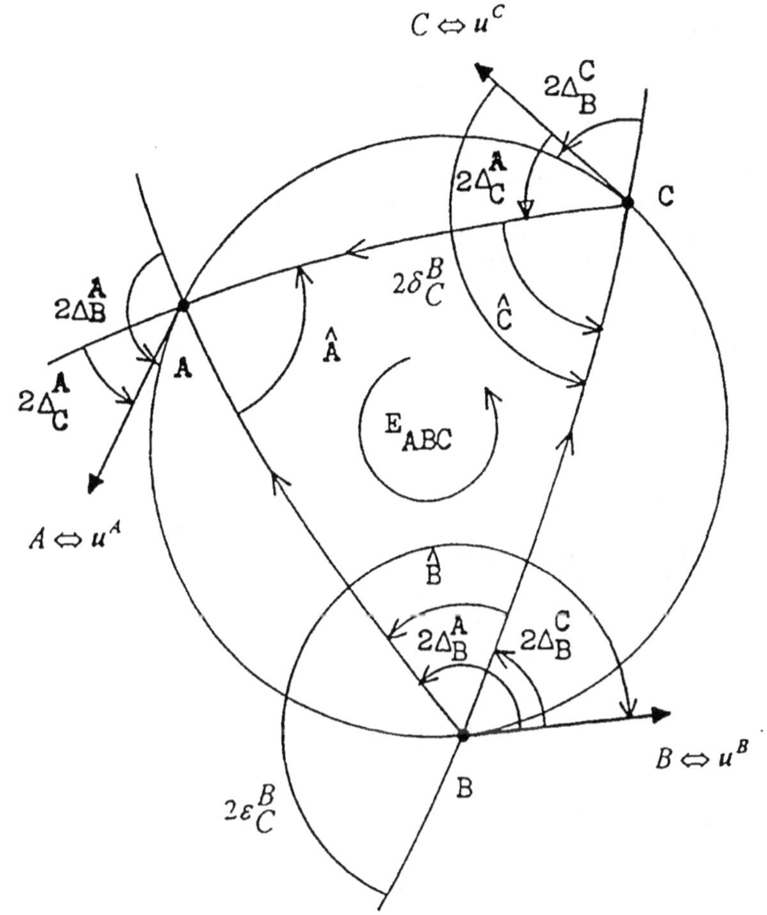

$$\left\{\begin{aligned} \hat{A} &= 2\Delta^A_B - 2\Delta^A_C \\ \hat{B} &= 2\Delta^A_B - 2\Delta^C_B \\ \hat{C} &= \pi - 2\Delta^A_C - 2\Delta^C_B \\ \hat{A} + \hat{B} + \hat{C} &= \pi + E_{ABC} \\ 4\Delta^B_C &= 2\delta^B_C + 2\varepsilon^B_C = -4\Delta^C_B \end{aligned}\right\} \Rightarrow \left\{\begin{aligned} 2\Delta^C_B + 2\Delta^A_C &= 4\Delta^A_B - \left(\hat{A} + \hat{B}\right) \\ &= \pi - \hat{C} \\ &= 2\Delta^A_B - \tfrac{1}{2} E_{ABC} \\ 2\Delta^A_B &= \pi - \hat{C} + \tfrac{1}{2} E_{ABC} \end{aligned}\right\}$$

Fig. 4.5. To the addition of phasors A and B resulting in the sum phasor C (phasors' magnitudes not shown). Mutual dependences between phasors' phase differences and angles of a spherical triangle ABC are presented.

5. Rotation Transformations on the PP Sphere

5. 1. The polarization phasor notation

Any vector is fully determined having components in a well-defined basis and established rules of transformation of those components under change of the basis.

When considering the PP vectors it is necessary to present both polarizations *and phases* of the basis vectors, otherwise their definition would be insufficient. The ONP polarization and phase basis discussed earlier will be applied here. Its first vector only determines the whole basis completely. That results from the uniqueness of the orthogonality transformation which applied to the first basis vector produces the second one unambiguously.

Having the PP tangential phasors determined, they will be used as an upper and lower indices of the PP column vectors. The upper index will represent the vector itself while the lower will denote the first basis phasor.

Also, to have a unified designation of phasors, the linear basis vectors will be denoted by $\boldsymbol{u}^H$ and $\boldsymbol{u}^{HX}$ instead of $\boldsymbol{1}_y$ and $\boldsymbol{1}_x$. The whole basis will be called now the $\boldsymbol{u}^H$ or, simply, the H basis. Formerly that basis was called the (y,x) basis of reversed order of its vectors.

In the proposed notation the $\boldsymbol{u}^A$ vector, represented by the A pohasor on the PP sphere, and the orthogonal $\boldsymbol{u}^{AX}$ vector can be written as follows:

$$\boldsymbol{u}^A = \left[\boldsymbol{u}^H \boldsymbol{u}^{H\mathrm{x}}\right] u_H^A \tag{5.1}$$

$$\boldsymbol{u}^{A\mathrm{x}} = \left[\boldsymbol{u}^H \boldsymbol{u}^{H\mathrm{x}}\right] u_H^{A\mathrm{x}} \tag{5.2}$$

where the column vectors expressed in terms of their analytical parameters are:

$$u_H^A = \begin{bmatrix} a \\ b \end{bmatrix}_H^A \equiv \begin{bmatrix} a_H^A \\ b_H^A \end{bmatrix} = \begin{bmatrix} \cos\gamma_H^A e^{-j(\delta_H^A+\varepsilon_H^A)} \\ \sin\gamma_H^A e^{+j(\delta_H^A-\varepsilon_H^A)} \end{bmatrix} \equiv \begin{bmatrix} \cos\gamma e^{-j(\delta+\varepsilon)} \\ \sin\gamma e^{+j(\delta-\varepsilon)} \end{bmatrix}_H^A \tag{5.3}$$

$$u_H^{A\mathrm{x}} = \begin{bmatrix} -b^* \\ a^* \end{bmatrix}_H^A \equiv \begin{bmatrix} -b_H^{A*} \\ a_H^{A*} \end{bmatrix} = \begin{bmatrix} -\sin\gamma_H^A e^{-j(\delta_H^A-\varepsilon_H^A)} \\ \cos\gamma_H^A e^{+j(\delta_H^A+\varepsilon_H^A)} \end{bmatrix} \equiv \begin{bmatrix} -\sin\gamma e^{-j(\delta-\varepsilon)} \\ \cos\gamma e^{+j(\delta+\varepsilon)} \end{bmatrix}_H^A \tag{5.4}$$

The phase delay of the phasor A over the phasor H will be denoted as

$$2\Delta_H^A = \delta_H^A + \varepsilon_H^A \tag{5.5}$$

what is numerically equal to ν_H^A. The two phasors in the angular and Stokes coordinate systems, according to the proposed polarization phasor notation, are shown in Fig.5.1.

5. 2. Rotation matrix and the change of basis rule for phasors

The rotation matrix can be defined when expressing the same PP vector in two different ONP polarization bases, H and B, and using the equalities (5.1) and (5.2):

$$\boldsymbol{u}^A = \left[\boldsymbol{u}^H \boldsymbol{u}^{H\mathrm{x}}\right] u_H^A \tag{5.6a}$$

$$= \left[\boldsymbol{u}^B \boldsymbol{u}^{B\mathrm{x}}\right] u_B^A \tag{5.6b}$$

$$= \left[\boldsymbol{u}^H \boldsymbol{u}^{H\mathrm{x}}\right]\left[u_H^B \quad u_H^{B\mathrm{x}}\right] u_B^A \tag{5.6c}$$

Comparison of (5.6c) with (5.6a) yields a change of basis rule for the PP column vectors:

$$u_H^A = \begin{bmatrix} u_H^B & u_H^{Bx} \end{bmatrix} u_B^A \tag{5.7}$$

Now, introducing the rotation matrix as

$$C_H^B = \begin{bmatrix} u_H^B & u_H^{Bx} \end{bmatrix} = \begin{bmatrix} a & -b^* \\ b & a^* \end{bmatrix}_H^B \tag{5.8}$$

one finally obtains the change of basis rule in a simple form

$$u_H^A = C_H^B \, u_B^A \tag{5.9}$$

from which the following rule of multiplication for rotation matrices results

$$C_H^A = C_H^B \, C_B^A \tag{5.10}$$

because

$$C_H^B \, C_B^A = C_H^B \begin{bmatrix} u_B^A \; u_B^{Ax} \end{bmatrix} = \begin{bmatrix} C_H^B u_B^A \;\; C_H^B u_B^{Ax} \end{bmatrix}$$
$$= \begin{bmatrix} u_H^A \; u_H^{Ax} \end{bmatrix}$$

Of course,

$$C_B^B = \begin{bmatrix} u_B^B \; u_B^{Bx} \end{bmatrix} = \begin{bmatrix} 1 & 0 \\ 0 & 1 \end{bmatrix}$$

It is immediately seen that the rotation matrix (5.8) is unitary:

$$C_H^B \tilde{C}_H^B{}^* = \begin{bmatrix} 1 & 0 \\ 0 & 1 \end{bmatrix} \tag{5.11}$$

and unimodular:

$$\det C_H^B = +1 \; . \tag{5.12}$$

The requirement about unimodularity is natural because three real parameters sufficiently determine any rotation of the polarization sphere of tangent phasors.

To have the engineering notation legible enough for engineers, the converse rotation matrix to that of (5.8) should be of the form

$$C_B^H = \begin{bmatrix} u_B^H & u_B^{Hx} \end{bmatrix} \tag{5.13}$$

with

$$C_B^H = \tilde{C}_H^B{}^* \tag{5.14}$$

or, according to (5.11),

$$C_H^B C_B^H = \begin{bmatrix} 1 & 0 \\ 0 & 1 \end{bmatrix} . \tag{5.15}$$

The last equality follows immediately from (5.10), but to obtain (5.13) the converse analytical parameters should satisfy the condition:

$$\Delta_B^H = -\Delta_H^B \tag{5.16}$$

where

$$\begin{aligned} 4\Delta_H^B &= 2\delta_H^B + 2\varepsilon_H^B \\ 4\Delta_B^H &= 2\delta_B^H + 2\varepsilon_B^H \end{aligned} \tag{5.17}$$

but with

$$2\gamma_B^H = 2\gamma_H^B \tag{5.18}$$

The following ranges of angular parameters are proposed for unambiguity reasons:

$$0 \le 2\gamma \le \pi$$

$$-\pi < 2\delta \le \pi$$

and

$$-2\pi \le 2\varepsilon \le 2\pi \tag{5.19}$$

or, instead,

$$-\pi \le 2\Delta \le \pi \tag{5.20}$$

5. 3. Change of polarization and spatial phase - a general formula

The active transformation rule can be found by inspection of the equalities

$$C_H^B C_A^H u_H^A = C_H^B u_A^A = \begin{bmatrix} u_H^B & u_H^{Bx} \end{bmatrix} \begin{bmatrix} 1 \\ 0 \end{bmatrix} = u_H^B \tag{5.21}$$

Denoting the rotation matrix C_A^B in the u^H basis as

$$C_{A,H}^B = C_H^B C_A^H \tag{5.22}$$

one obtains the desired active transformation presented by a simple formula

$$C_{A,H}^B u_H^A = u_H^B \tag{5.23}$$

Also, it is worth noticing that

$$C_A^B = C_{A,A}^B = C_{A,B}^B \tag{5.24}$$

Now, the general rule of the change of basis for rotation matrices can be found.

5. 4. Change of basis transformation for rotation matrices - a general rule

There is a very convenient and universal method which can be used to develop miscellaneous transformation formulae. It will be called the 'unit matrix insertion method'. Always, one should start with a transmission equation and apply a product of two matrices equal to the unit matrix. In the example under consideration the following 'unit matrices',

$$\begin{bmatrix} 1 & 0 \\ 0 & 1 \end{bmatrix} = \widetilde{C}_B^H \widetilde{C}_H^B = C_H^B * C_B^H * \tag{5.25}$$

will be inserted to the transmission equation representing a voltage, V_r, received from the incoming wave, the polarization and phase of which will be transformed, from the transmitted T to the scattered S, by a rotation matrix determined in the H basis:

$$\begin{aligned} V_r &= \widetilde{u}_H^R u_H^S * = \widetilde{u}_H^R C_{T,H}^S * u_H^T * \\ &= \widetilde{u}_H^R \left(\widetilde{C}_B^H \widetilde{C}_H^B\right) C_{T,H}^S * \left(C_H^B * C_B^H *\right) u_H^T * \\ &= \left(\widetilde{u}_H^R \widetilde{C}_B^H\right)\left(C_B^H * C_{T,H}^S * C_H^B *\right)\left(C_B^H * u_H^T *\right) \\ &= \widetilde{u}_B^R C_{T,B}^S * u_B^T * = \widetilde{u}_B^R u_B^S * \end{aligned} \tag{5.26}$$

What can be seen from the above formulae is the needed change of basis rule for the rotation matrix:

$$C^{S}_{T,B} = C^{H}_{B} C^{S}_{T,H} C^{B}_{H} \tag{5.27}$$

5. 5. Decomposition of the 3-parameter rotation matrices into products of 1-paramater rotation matrices

The same rotation matrix can be used for changing the vector's basis (passive transformation) or for changing the PP vector themselves when preserving its basis (active transformation). In both cases decomposition of the rotation matrix into product of three 1-parameter matrices is formally the same but corresponds to different relative movements of tangential phasors on the Poincare sphere surface. Directions of movement are reversed and routes changed. However, the decompositions are different when matrices are expressed in terms of analytical or geometrical parameters.

1. Analytical parameter case (Fig. 5.2)

Following the change of basis rule $K \to K' \to H''' \to H$:

$$u^{K}_{H} = C^{H'''}_{H} \underbrace{C^{K'}_{H'''} \underbrace{C^{K}_{K'} u^{K}_{K}}_{u^{K}_{K'}}}_{u^{K}_{H'''}} = C^{K}_{H} u^{K}_{K} \tag{5.28}$$

or the change of polarization and phase $H \to H' \to L' \to K$:

$$u^{K}_{H} = C^{K}_{L',H} \underbrace{C^{L'}_{H',H} \underbrace{C^{H'}_{H,H} u^{H}_{H}}_{u^{H'}_{H}}}_{u^{L'}_{H}} = C^{K}_{H,H} u^{H}_{H} \tag{5.29}$$

one obtains

$$C^{K}_{H} = \begin{bmatrix} \cos\gamma e^{-j(\delta+\varepsilon)} & -\sin\gamma e^{-j(\delta-\varepsilon)} \\ \sin\gamma e^{j(\delta-\varepsilon)} & \cos\gamma e^{j(\delta+\varepsilon)} \end{bmatrix}^{K}_{H}$$

$$= \underbrace{\begin{bmatrix} e^{-j\delta} & 0 \\ 0 & e^{j\delta} \end{bmatrix}}_{C^{H'''}_{H}=C^{K}_{L',H}} \underbrace{\begin{bmatrix} \cos\gamma & -\sin\gamma \\ \sin\gamma & \cos\gamma \end{bmatrix}}_{C^{K'}_{H'''}=C^{L'}_{H',H}} \underbrace{\begin{bmatrix} e^{-j\varepsilon} & 0 \\ 0 & e^{j\varepsilon} \end{bmatrix}}_{C^{K}_{K'}=C^{H'}_{H,H}} \tag{5.30}$$

2. Geometrical parameter case (Fig. 5.3)

Similarly, following the change of basis rule, $K \to K'' \to L \to H$:

$$u^{K}_{H} = C^{L}_{H} \underbrace{C^{K''}_{L} \underbrace{C^{K}_{K''} u^{K}_{K}}_{u^{K}_{K''}}}_{u^{K}_{L}} = C^{K}_{H} u^{K}_{K} \tag{5.31}$$

or the change of polarization and phase, $H \to H' \to P \to K$:

$$u_H^K = C_{P,H}^K C_{H'',H}^P \underbrace{\underbrace{C_{H,H}^{H''} u_H^H}_{u_H^{H''}}}_{u_H^P} = C_{H,H}^K u_H^H \quad (5.32)$$

one obtains

$$C_H^K = \begin{bmatrix} (\cos\alpha\cos\beta - j\sin\alpha\sin\beta)e^{-j\chi} & (-\cos\alpha\sin\beta + j\sin\alpha\cos\beta)e^{j\chi} \\ (\cos\alpha\sin\beta + j\sin\alpha\cos\beta)e^{-j\chi} & (\cos\alpha\cos\beta + j\sin\alpha\sin\beta)e^{j\chi} \end{bmatrix}_H^K$$

$$= \underbrace{\begin{bmatrix} \cos\beta & -\sin\beta \\ \sin\beta & \cos\beta \end{bmatrix}}_{C_H^L = C_{P,H}^K} \underbrace{\begin{bmatrix} \cos\alpha & j\sin\alpha \\ j\sin\alpha & \cos\alpha \end{bmatrix}}_{C_L^{K''} = C_{H'',H}^P} \underbrace{\begin{bmatrix} e^{-j\chi} & 0 \\ 0 & e^{j\chi} \end{bmatrix}}_{C_{K''}^K = C_{H,H}^{H''}} \quad (5.33)$$

5. 6. Geometrical form of the PP Sphere Rotation Matrix

The obtained formulae for the active transformation of the PP vectors and basis transformation of the rotation matrix can be applied to present rotation of the PP sphere about any OA axis by a 2ϕ angle as follows (Fig.5.4):

$$u_P^{P_1} = C_{P,A}^{P_1} u_A^P = \begin{bmatrix} e^{-j\phi} & 0 \\ 0 & e^{j\phi} \end{bmatrix} u_A^P \quad (5.34)$$

By inspection of the equation (5.30) one can see that the only changing parameter of the rotated phasor is

$$2\delta_A^{P_1} = 2\delta_A^P + 2\phi \quad (5.35)$$

Equation (5.34) is in the A basis. In any other B basis, the rotation matrix takes the form:

$$\begin{aligned} C_{P,B}^{P_1} &= C_B^A \begin{bmatrix} e^{-j\phi} & 0 \\ 0 & e^{j\phi} \end{bmatrix} C_A^B \\ &= \begin{bmatrix} \cos\phi - jn_1\sin\phi & (-n_3 - jn_2)\sin\phi \\ (n_3 - jn_2)\sin\phi & \cos\phi + jn_1\sin\phi \end{bmatrix}_{P,B}^{P_1} \\ &= C(\boldsymbol{n},2\phi)_{P,B}^{P_1} = -C(\boldsymbol{n},2\phi + 2\pi)_{P,B}^{P_1} \\ &\equiv C_B^{ROT}(\boldsymbol{n},2\phi) \end{aligned} \quad (5.36)$$

Here, the unit vector $\boldsymbol{n}$ is directed along the OA rotation axis. Its components in the B basis are

$$\begin{bmatrix} n_1 \\ n_2 \\ n_3 \end{bmatrix} = \begin{bmatrix} \mathrm{q} \\ \mathrm{u} \\ \mathrm{v} \end{bmatrix}_B^A = \begin{bmatrix} \cos 2\gamma \\ \sin 2\gamma\cos 2\delta \\ \sin 2\gamma\sin 2\delta \end{bmatrix}_B^A \equiv \begin{bmatrix} \cos 2\gamma_B^A \\ \sin 2\gamma_B^A \cos 2\delta_B^A \\ \sin 2\gamma_B^A \sin 2\delta_B^A \end{bmatrix} \quad (5.37)$$

The obtained formulae correspond to the earlier presented form of the rotation matrix as in (4.7).

5. 7. Comparison of Two Forms of Rotation Matrices

To obtain the axis, $\boldsymbol{n}$, and the angle of rotation, 2ϕ, realizing the change of polarization and phase, $B \to K$ (see Fig. 5.4), we have to solve the matrix equation

$$C_B^{ROT}(\boldsymbol{n},2\phi) = C_{B,B}^{K} = C_B^K = \begin{bmatrix} \cos\gamma e^{-j(\delta+\varepsilon)} & -\sin\gamma e^{-j(\delta-\varepsilon)} \\ \sin\gamma e^{+j(\delta-\varepsilon)} & \cos\gamma e^{+j(\delta+\varepsilon)} \end{bmatrix}_B^K \qquad (5.38)$$

Alternatively, rotation by 2ϕ about the $-\boldsymbol{n}$ vector realizes the change of basis transformation, $K \to B$, according to the equality

$$C_B^{ROT}(-\boldsymbol{n},2\phi) = C_K^B = \begin{bmatrix} \cos\gamma e^{-j(\delta+\varepsilon)} & -\sin\gamma e^{-j(\delta-\varepsilon)} \\ \sin\gamma e^{+j(\delta-\varepsilon)} & \cos\gamma e^{+j(\delta+\varepsilon)} \end{bmatrix}_K^B = \tilde{C}_B^K * \qquad (5.39)$$

Taking into account mutual relations (see Fig. 5.5)

$$\begin{array}{lll} \gamma_K^B = \gamma_B^K & & 0^0 \le \gamma_K^B \le 90^0 \\ \left.\begin{array}{l} \delta_K^B = \pm 90^0 - \varepsilon_B^K \\ \varepsilon_K^B = \mp 90^0 - \delta_B^K \end{array}\right\} \Rightarrow & \left\{\begin{array}{l} \delta_K^B + \varepsilon_K^B = -(\delta_B^K + \varepsilon_B^K) \\ \delta_K^B - \varepsilon_K^B = \pm 180^0 + (\delta_B^K - \varepsilon_B^K) \end{array}\right. & \left\{\begin{array}{l} -90^0 < \delta_K^B \le 90^0 \\ -180^0 \le \varepsilon_K^B \le 180^0 \end{array}\right. \end{array} \qquad (5.40)$$

one obtains the following (alternative) expressions which have to be computed successively

$$\begin{array}{ll} \cos\phi = \cos\gamma_B^K \cos(\delta_B^K + \varepsilon_B^K) = \cos\gamma_K^B \cos(\delta_K^B + \varepsilon_K^B); & 0^0 \le \phi < 180^0 \\ n_1 \sin\phi = \cos\gamma_B^K \sin(\delta_B^K + \varepsilon_B^K) = -\cos\gamma_K^B \sin(\delta_K^B + \varepsilon_K^B) & \\ n_2 \sin\phi = -\sin\gamma_B^K \sin(\delta_B^K - \varepsilon_B^K) = \sin\gamma_K^B \sin(\delta_K^B - \varepsilon_K^B) & \\ n_3 \sin\phi = \sin\gamma_B^K \cos(\delta_B^K - \varepsilon_B^K) = -\sin\gamma_K^B \cos(\delta_K^B - \varepsilon_K^B) & \end{array} \qquad (5.41)$$

The inverse formulae present angles which also have to be computed successively:

$$\begin{array}{ll} \gamma_B^K = arc\ \cos\sqrt{\cos^2\phi + n_1^2 \sin^2\phi}, & 0^0 \le \gamma_B^K \le 90^0 \\ \delta_B^K = arc\ \tan\left(\sin\phi \dfrac{n_1 \sin\gamma_B^K - n_2 \cos\gamma_B^K}{\cos\phi \sin\gamma_B^K + n_3 \sin\phi \cos\gamma_B^K}\right), & -90^0 \le \delta_B^K \le 90^0 \\ \varepsilon_B^K = arc\ \sin\dfrac{\sin\phi(n_1 \sin\gamma_B^K + n_2 \cos\gamma_B^K)}{\cos\delta_B^K \sin 2\gamma_B^K} & \\ \quad = arc\ \cos\dfrac{\sin\phi(n_1 \sin\gamma_B^K - n_2 \cos\gamma_B^K)}{\sin\delta_B^K \sin 2\gamma_B^K}, & -180^0 \le \varepsilon_B^K \le 180^0 \end{array} \qquad (5.42)$$

5. 8. Rotation matrix in the Stokes parameters space

Using standard procedure with Kronecker multiplication and applying the auxiliary U matrix, for the amplitude rotation matrices we find,

$$D_H^K = \tilde{U}^*\left(C_H^K \otimes C_H^{K*}\right)U = \frac{1}{2}\begin{bmatrix} 1 & 0 & 0 & 1 \\ 1 & 0 & 0 & -1 \\ 0 & 1 & 1 & 0 \\ 0 & j & -j & 0 \end{bmatrix}\left(\begin{bmatrix} a & -b^* \\ b & a^* \end{bmatrix} \otimes \begin{bmatrix} a^* & -b \\ b^* & a \end{bmatrix}\right)_H^K \begin{bmatrix} 1 & 1 & 0 & 0 \\ 0 & 0 & 1 & -j \\ 0 & 0 & 1 & j \\ 1 & -1 & 0 & 0 \end{bmatrix}$$

$$= \begin{bmatrix} 1 & 0 & 0 & 0 \\ 0 & aa^*-bb^* & -(ab+a^*b^*) & j(ab-a^*b^*) \\ 0 & ab^*+a^*b & (a^2+a^{*2}-b^2-b^{*2})/2 & j(-a^2+a^{*2}+b^2-b^{*2})/2 \\ 0 & j(ab^*-a^*b) & j(a^2-a^{*2}+b^2-b^{*2})/2 & (a^2+a^{*2}+b^2+b^{*2})/2 \end{bmatrix}_H^K \tag{5.43}$$

where, for the Cayley-Klein parameters:

$$a_H^K = \cos\gamma_H^K\, e^{-j(\delta_H^K+\varepsilon_H^K)}$$
$$b_H^K = \sin\gamma_H^K\, e^{j(\delta_H^K-\varepsilon_H^K)} \tag{5.44}$$

we finally obtain,

$$D_H^K = \begin{bmatrix} 1 & 0 & 0 & 0 \\ 0 & \cos 2\gamma & -\sin 2\gamma\cos 2\varepsilon & \sin 2\gamma \sin 2\varepsilon \\ 0 & \sin 2\gamma \cos 2\delta & \cos 2\gamma\cos 2\delta\cos 2\varepsilon - \sin 2\delta \sin 2\varepsilon & -\cos 2\gamma\cos 2\delta \sin 2\varepsilon - \sin 2\delta\cos 2\varepsilon \\ 0 & \sin 2\gamma \sin 2\delta & \cos 2\gamma \sin 2\delta \cos 2\varepsilon + \cos 2\delta \sin 2\varepsilon & -\cos 2\gamma\sin 2\delta\sin 2\varepsilon + \cos 2\delta \cos 2\varepsilon \end{bmatrix}_H^K \tag{5.45}$$

Similarly, in the geometrical form,the rotation matrix is

$$D_H^{ROT}(\boldsymbol{n},2\phi) = \begin{bmatrix} 1 & 0 & 0 & 0 \\ 0 & \cos 2\phi + 2n_1^2\sin^2\phi & -n_3\sin 2\phi + 2n_1n_2\sin^2\phi & n_2\sin 2\phi + 2n_1n_3\sin^2\phi \\ 0 & n_3\sin 2\phi + 2n_1n_2\sin^2\phi & \cos 2\phi + 2n_2^2\sin^2\phi & -n_1\sin 2\phi + 2n_2n_3\sin^2\phi \\ 0 & -n_2\sin 2\phi + 2n_1n_3\sin^2\phi & n_1\sin 2\phi + 2n_2n_3\sin^2\phi & \cos 2\phi + 2n_3^2\sin^2\phi \end{bmatrix}_H \tag{5.46}$$

5. 9. The Unitary and Con-Unitary Transformations

The change of basis formulae require some remarks. Let us look at the equation (5.27) rewritten with a slight modification

$$C_{T,B}^S = \left(C_H^B\right)^{-1} C_{T,H}^S C_H^B \tag{5.27a}$$

This is evidently a similarity transformation because of similar form of the two change of PP equations with the two mutually transformed rotation matrices:

$$u_H^S = C_{T,H}^S u_H^T$$

and (5.47)

$$u_B^S = C_{T,B}^S u_B^T$$

However, transformation matrices in (5.27) are also unitary. So, we have to call the (5.27) transformation, in that special case, the *unitary* transformation.

There is another transformation of the PP vectors that exhibit (other) similarity under change of basis. It is one which transforms, e.g., the Sinclair scattering matrix. There is an essential difference that can be observed when comparing these two kinds of transformations. The Sinclair transformation moves the PP vector of the illuminating wave to the conjugate complex space to represent the CA of scattered waves. This happens because the scattered wave propagates in the negative direction of the z axis of a local coordinate system of the receiving antenna. So its complex amplitude should be expressed by the conjugate value of the corresponding PP vector as in the equation (3.11b). In the proposed polarization phasor notation, the two analogous amplitude scattering equations in two different PP bases are:

$$\lambda^T u_H^S{}^* = A_H u_H^T$$

and (5.48)

$$\lambda^T u_B^S{}^* = A_B u_B^T$$

with λ^T as a positive real number depending on T only, and not on the polarization bases. The corresponding change of basis rule for the Sinclair matrices is

$$A_B = \tilde{C}_H^B A_H C_H^B \qquad (5.49)$$

It could be named the 'con-similarity' transformation, with the 'con-' prefix for the 'conjugate' space of the scattered PP vectors. However, the unitarity of transformation matrices again should be taken into account. For that reason the name of the '*con-unitary*' transformation seems to be more adequate.

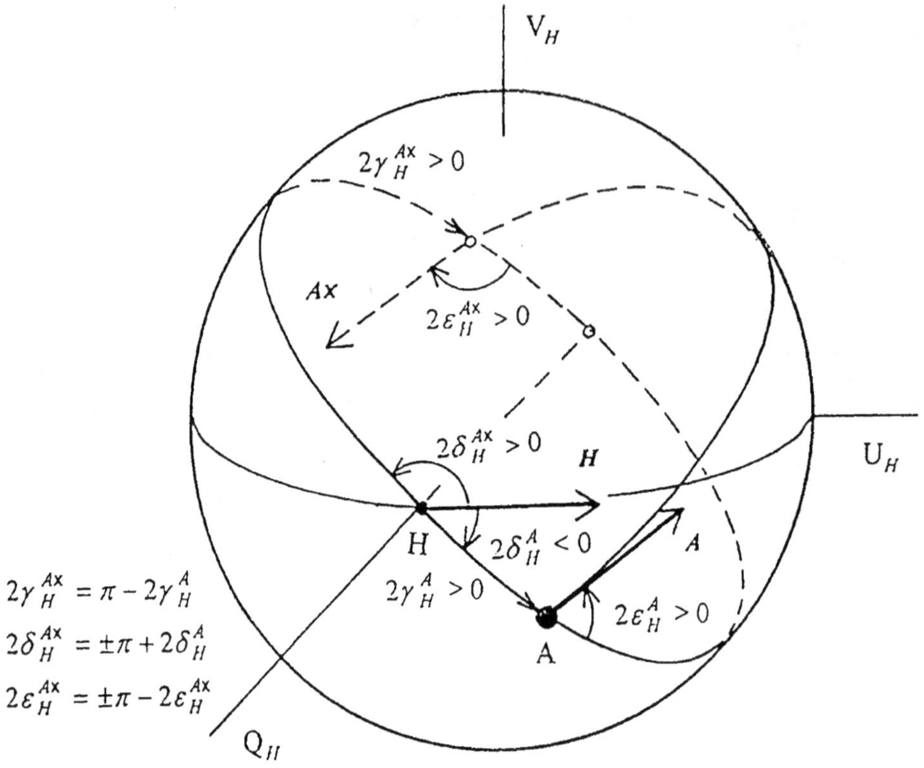

Fig. 5.1. Phasors of the ONP polarization basis A.

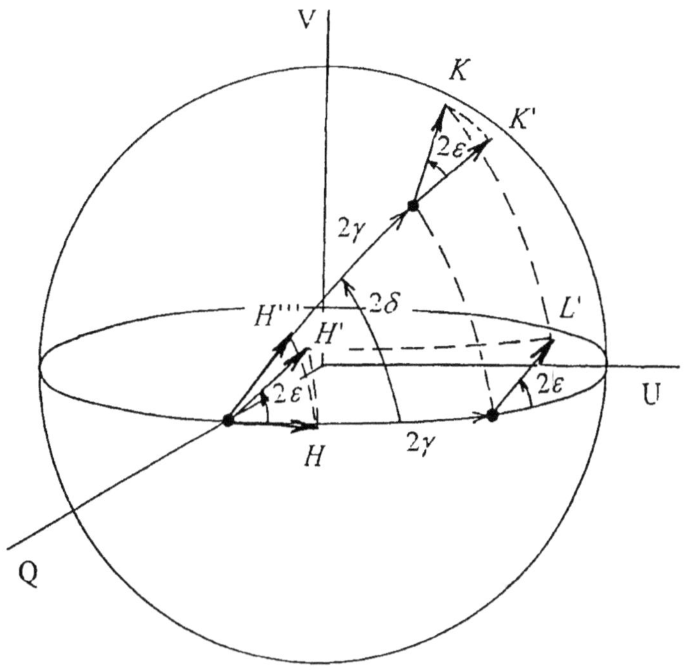

Fig. 5.2. To the rotation matrix decomposition
in terms of analytical parameters

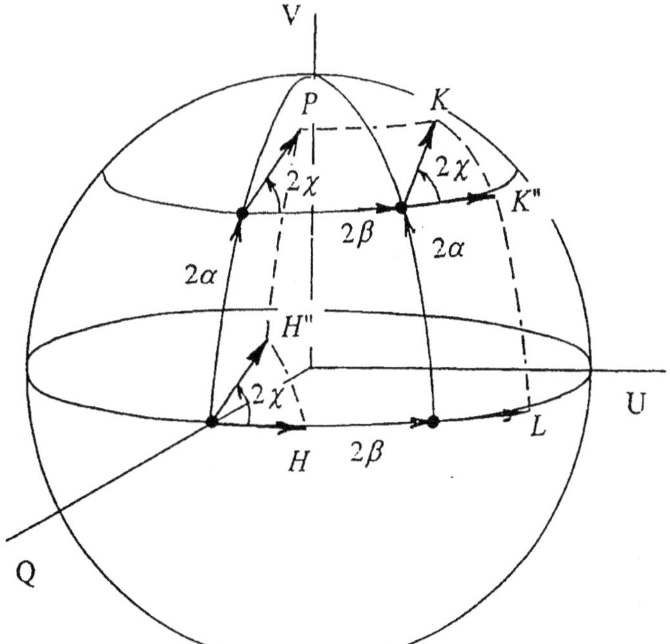

Fig. 5.3. To the rotation matrix decomposition
in terms of geometrical parameters

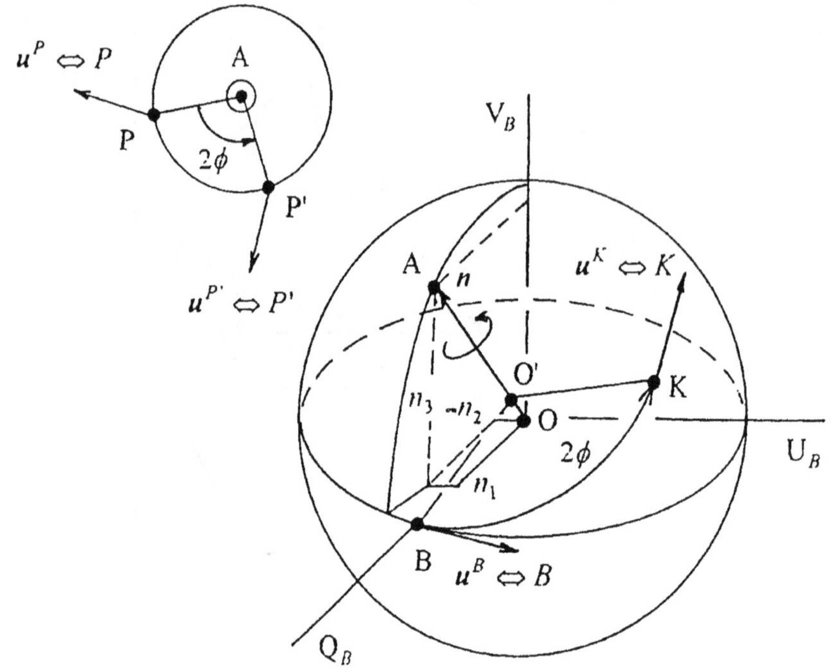

Fig. 5.4. The PP sphere rotation about the OA axis by the 2ϕ angle.

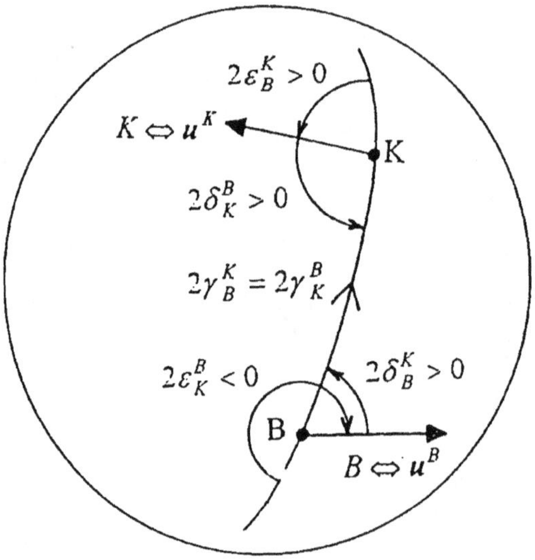

Fig. 5.5. An example of mutual dependence of two tangential phasors' angular parameters. The arrow on the grat circle arc segments indicate direction of the spatial phase lag.

6. Change of Phase, Orthogonality and Spatial Reversal Transformations on the PP Sphere

There are two fundamental transformations of the PP sphere of tangential phasors on itself which are independent of the ONP PP basis rotation: change of spatial phase and orthogonality transformations.

An essential difference between those two transformations is that they belong to two different classes. Change of spatial phase leaves PP vectors in the same complex C^2 space while orthogonality moves them to the conjugate space, similarly as reversal transformation does, but the latter is basis dependent.

However, there is another important common feature of the change of spatial phase and orthogonality. They both do not change the PP phasors' direction of rotation *in time*, unlike the reversal transformation.

Also, it should be mentioned here, that all these three transformations, as well as previously described rotation, have one common essential feature: they keep magnitudes of the PP vectors constant. That is why to all of them the '*C*' symbols will be ascribed with suitably differentiating indices.

One more *C*-type transformation could be considered but it will never be used. This would be a transformation which moves the CA vector into the conjugate space leaving its phasor unchanged, but rotating *in time* in the opposite direction.

The change of spatial phase transformation does not require any special description. It just rotate all tangential PP phasors by the same angle equal to the double phase angle of the PP vectors.

6. 1. The orthogonality transformation

The orthogonality transformation could be presented by the following matrix equation:

$$u_H^{Px} * = C_H^x u_H^P \tag{6.1}$$

with the orthogonality matrix in the H basis

$$C_H^x = \begin{bmatrix} 0 & -1 \\ 1 & 0 \end{bmatrix} \tag{6.2}$$

However, performing the change of basis suitable for transformations moving the PP vector into conjugate space:

$$\begin{aligned} C_B^x &= \tilde{C}_H^B C_H^x C_H^B \\ &= \begin{bmatrix} a & b \\ -b* & a* \end{bmatrix}_H^B \begin{bmatrix} 0 & -1 \\ 1 & 0 \end{bmatrix} \begin{bmatrix} a & -b* \\ b & a* \end{bmatrix}_H^B \\ &= \begin{bmatrix} -ab+ab & -aa*-bb* \\ aa*+bb* & a*b*-a*b* \end{bmatrix}_H^B = \begin{bmatrix} 0 & -1 \\ 1 & 0 \end{bmatrix} \end{aligned} \tag{6.3}$$

we can observe that the above transformation is basis independent and can be expressed by using of a *real* matrix

$$C^x = \begin{bmatrix} 0 & -1 \\ 1 & 0 \end{bmatrix} \tag{6.4}$$

Therefore, we will write the orthogonality transformation in the form

$$u_B^{Px} = C^x u_B^P *, \quad \text{and} \quad u_B^{Pxx} = C^x u_B^{Px} * = C^x C^x u_B^P = -u_B^P . \tag{6.5}$$

By moving conjugation to the right side of the equation, in comparison with (6.1), we stress that the orthogonal PP vector can be represented by the tangential phasor of the same kind (rotating in the same direction in time) as before the transformation, and of the same phase, as, e.g., $\boldsymbol{1}_y$ and $\boldsymbol{1}_x$ vectors are of the same (null) phase.

It can be shown that, in any ONP B basis, P and Px phasors are collinear and also form an ONP basis. In order to prove that we can start from the known PP change transformations:

$$u_H^P = C_{B,H}^P u_H^B \quad \text{and} \quad u_H^{Px} = C_{Bx,H}^{Px} u_H^{Bx} \tag{6.6}$$

It will be sufficient to demonstrate that the two PP sphere rotations used above are identical. Let us see first that

$$C_B^{Bx} = \left[u_B^{Bx} \; u_B^{Bxx} \right] = \begin{bmatrix} 0 & -1 \\ 1 & 0 \end{bmatrix} = C_P^{Px} = C^x \tag{6.7}$$

and

$$C_{Bx}^B = \widetilde{C}_B^{Bx} * = \widetilde{C}^x \tag{6.8}$$

what yields the expected equality:

$$\begin{aligned} C_{Bx,H}^{Px} &= C_H^{Px} C_{Bx}^H \\ &= \left(C_H^P C_P^{Px} \right) \left(C_{Bx}^B C_B^H \right) \\ &= C_H^P \begin{bmatrix} 0 & -1 \\ 1 & 0 \end{bmatrix} \begin{bmatrix} 0 & 1 \\ -1 & 0 \end{bmatrix} C_B^H \\ &= C_H^P C_B^H = C_{B,H}^P \end{aligned} \tag{6.9}$$

6. 2. The spatial reversal transformation and the direct (one-way) complex voltage transamission equation

Complex voltage received by an antenna from a wave can be presented by the following Hermitian product of two PP vectors expressed in the antenna local xyz coordinate system with the z-axis oriented out of the antenna:

$$V_r = \boldsymbol{u}^A \cdot \boldsymbol{u}^W * \tag{6.10}$$

Here constant coefficients of less importance for these considerations have been omitted. As usually, the upper indices denote phasors tangent to the Poincare sphere at points A and W, corresponding to the antenna and wave PP vectors respectively. Introduction of the Hermitian product of the PP vectors was necessary because of opposite orientations of the receiving antenna and the incoming wave.

In the similar form, of the Hermitian product of two PP vectors, the equation of transmission *between two antennas* could be presented. Unfortunately, however, it would suffer from one but essential inconvenience. It would not reflect the reciprocity of transmission. In order to resolve that difficulty, the PP vectors of the two antennas will be expressed in the PP bases B determined for their own local right-handed xyz coordinate systems, with the z-axes indicating, for both antennas, their directions of radiation/reception. The two antennas will be assumed 'looking' into each other with their local yz 'reference planes', called 'horizontal', coinciding. Then, the 'spatial reversal' matrix can be introduced that will reverse any of these two local coordinate systems by 180° rotation about its x-axis, called 'vertical'. As a result, complex amplitude vector of a wave radiated by one antenna, determined by the phasor T, can be obtained in the local coordinate system of the other antenna, with the PP vector then corresponding to the phasor To:

$$u_B^{To} * = C_B^o u_B^T \tag{6.11}$$

This is, alternatively, the conjugate value of the antenna's spatially *'reversed PP vector' by reversal of its local coordinate system*, or the conjugate value of the PP vector of the *'antenna reversed' versus its unchanged local coordinate system*. The conjugation tells us about −z direction of the antenna orientation, or the transmitted wave propagation, as has been proposed in (3.11b). Of course, the reversal matrix will be subject to a change-of-basis transformation. In the horizontal H basis its form is especially simple,

$$C_H^o = \begin{bmatrix} -1 & 0 \\ 0 & 1 \end{bmatrix}. \tag{6.12}$$

Now, the transmission equation can be obtained in the form compatible with the reciprocity principle (see also Fig. 6.1d):

$$V_r = \tilde{u}_B^R \, C_B^o \, u_B^T = \tilde{u}_B^R \, u_B^{To} * = \tilde{u}_B^{Ro} * \, u_B^T. \tag{6.13}$$

Applying to (6.12) the change-of-basis rule, we obtain

$$C_B^o = \tilde{C}_H^B C_H^o C_H^B = \begin{bmatrix} \text{-w} & \text{u} \\ \text{u} & \text{w}* \end{bmatrix}_H^B \tag{6.14}$$

with

$$\text{u}_H^B = \sin 2\gamma_H^B \cos 2\delta_H^B \tag{6.15}$$

and

$$\text{w}_H^B = \cos 2\gamma_H^B \cos 2\delta_H^B \cos 2\varepsilon_H^B - \sin 2\delta_H^B \sin 2\varepsilon_H^B - j\left(\cos 2\gamma_H^B \cos 2\delta_H^B \sin 2\varepsilon_H^B + \sin 2\delta_H^B \cos 2\varepsilon_H^B\right) \tag{6.16}$$

with

$$\det C_H^o = \det C_B^o = -\text{ww}*\text{-u}^2 = -1. \tag{6.17}$$

In any ONP polarization basis the spatial reversal matrix is symmetric and satisfies the equality

$$C_B^o C_B^o * = \begin{bmatrix} 1 & 0 \\ 0 & 1 \end{bmatrix}. \tag{6.18}$$

It may be interesting to observe that for the right-circular polarization, RC, defined by

$$2\delta_H^{RC} = 2\gamma_H^{RC} = -2\varepsilon_H^{RC} = 90° \tag{6.19}$$

the spatial reversal matrix remains unchanged:

$$C_{RC}^o = \begin{bmatrix} -1 & 0 \\ 0 & 1 \end{bmatrix} = C_H^o \tag{6.20}$$

similarly as for all in-phase, with H, basis phasors tangent to the polarization sphere at the great circle $2\delta = 90°$. These basis phasors, rotated by the $2\varepsilon = 180°$ angle, are eigenphasors of the (6.20) Sinclair scattering matrix. This is a special case of the general rule stating that, e.g., in the H basis, the spatial reversal operation changes analytical and geometrical parameters of the PP vector as follows:

$$2\delta_H^{To} = 180° - 2\delta_H^T, \qquad 2\gamma_H^{To} = 2\gamma_H^T, \qquad 2\varepsilon_H^{To} = 180° - 2\varepsilon_H^T \tag{6.21a}$$

$$2\alpha_H^{To} = 2\alpha_H^T, \qquad 2\beta_H^{To} = -2\beta_H^T, \qquad 2\chi_H^{To} = 360° - 2\chi_H^T \tag{6.21b}$$

and any PP vector with analytical parameters $2\delta_H^T = 2\varepsilon_H^T = \pm 90°$ will not change under reversal transformation. Figs. 6.1a,b present all those dependences for polarization phasors in the linear H basis (on the Poincare sphere in the corresponding to that basis the QUV coordinate system). In addition, Fig. 6.1b explains also the independence of the received voltage phase of the z-axis reversal by rotation of the xyz coordinate system.

6.3. The transformation changing the PP vector's propagation index

One more C transformation can be considered, changing for opposite one the direction of propagation without influence on the polarization and spatial phase. This is the only exception among transformations when the *domain and range* constitute the C^2 space of the PP vectors. If applied to the CA vectors it would cause a transfer to the conjugate C^2 space. This is the $C^{(\pm)}$ transformation which changes the PP vector's propagation index, leaving the value of the vector unchanged:

$$C^{(\pm)}\, u_B^{P(+)} = u_B^{P(-)}; \qquad C^{(\pm)} = \begin{bmatrix} 1 & 0 \\ 0 & 1 \end{bmatrix}, \qquad u_B^{P(+)} = u_B^{P(-)} = u_B^{P}. \tag{6.22}$$

As a result, the phasor of the transformed PP vector (if it belonges to a wave, not to an antenna) rotates *in time*, with the 2ω angular velocity, in the opposite direction.

Though the phasor of a wave only rotates, the propagation index of an antenna is also essential, but only when considering the sign of *temporal* phase of the voltage received by the antenna from a wave. That is for the reciprocity reasons. It is not important which phasor of the two under consideration represents the wave, and which one the antenna. The sign is positive (the voltage is delayed in time) if direction of rotation of the phasor with the minus propagation index to the phasor with the plus index is counter-clockwise (positive direction about the PP sphere radius to its piolarization point) because the two phasors tend to meet. The opposite direction means the voltage advanced in time, for the distance between the phasors is growing. So, the spatial and temporal phases are of opposite sign for phasors with minus propagation index.

Using the PP vectors with their propagation indices, instead of (3.11) and for z=0 one can write

$$\boldsymbol{E}^+(t,0) = E_0(\boldsymbol{u}^+ e^{j\omega t}) \tag{6.23a}$$

and

$$\boldsymbol{E}^-(t,0) = E_0\boldsymbol{u}^- * e^{j\omega t} = E_0(\boldsymbol{u}^- e^{-j\omega t})* \tag{6.23b}$$

The last equality just explains the opposite sense of rotation of the transformed polarization phasor in time. Such a phasor can be called of different 'kind', though its numerical value is the same for $t = z = 0$.

So, depending on its kind, the PP vector, or the polarization phasor, can be given a 'propagation index', plus or minus, indicating direction of wave propagation or antenna orientation along the z axis, as well as the sense of the wave's phasor rotation.

6. 4. Four types of scattering and propagation (transmission) matrices

Taking into account, according to (3.3), the dependence of the unit complex amplitude vector on the sense of the propagation z-coordinate axis (see Fig. 6.2a), and the following earlier presented form (5.48) of the amplitude scattering equation

$$\lambda^T u_B^S * = A_B u_B^T \tag{6.24a}$$

we will apply the unit matrix insertion (UMI) method by inserting the unit matrices (6.18) to different places in the amplitude two-way transmission equation

$$V_r = \tilde{u}_B^R A_B u_B^T \tag{6.24b}$$

with the regular Sinclair scattering matrix A_B (we shall call it the 'S1 type' matrix). By doing so we arrive at three other amplitude matrices and the corresponding scattering and transmission equations (see Fig. 6.2b-d):

the Jones propagation matrix of P1 type,

$$A_B^{\circ} = C_B^{\circ} * A_B; \qquad A_B^{\circ} u_B^T = \lambda^T u_B^{S\circ}, \qquad V_r = \tilde{u}_B^{R\circ} * A_B^{\circ}\, u_B^T \tag{6.25}$$

the Jones propagation matrix of P2 type,

$${}^{\circ}A_B = A_B C_B^{\circ} *; \qquad {}^{\circ}A_B u_B^{T\circ} * = \lambda^T u_B^S *, \qquad V_r = \tilde{u}_B^R\, {}^{\circ}A_B\, u_B^{T\circ} * \tag{6.26}$$

the Sinclair scattering matrix of S2 type,

$$ {}^{o}A_B^{o} = C_B^{o} * A_B C_B^{o} * \; ; \quad {}^{o}A_B^{o} u_B^{To} * = \lambda^T u_B^{So} , \quad V_r = \tilde{u}_B^{Ro} * {}^{o}A_B^{o} u_B^{To} * \tag{6.27} $$

These matrices and equations correspond to the reversed coordinate systems at the input, output, and on both sides of the scattering object, succesively. They will be especially useful when considering, e.g., the cascading connections of the polarimetric two-ports [14]. The reflectance and transmittance amplitude matrices of those two-ports can be of all the above mentioned types. However, in order to take full advantage of such representations, at first the Poincare sphere geometrical models of those matrices should be constructed, proceeded by their decomposition into product of matrices explaining successive transformations of. the PP vectors when scattering.

6. 5. Summary of advantages and envisaged applications resulting from introduction of the PP vectors

There are several essential advantages/applications of the waves' representation by the PP vectors (or tangential phasors), e.g.:

- The two Jones directional vectors (when applying Lueneburg's terminology [107]), i.e. complex amplitudes of waves or complex lengths of antennas oriented in opposite directions, can be determined by *one only PP vector* (its direct value or complex conjugate) what results in using *the same orthogonal null-phase (ONP) PP basis and its transformation or spatial basis reversal* (for the *xyz* coordinate system rotated by 180° about an axis perpendicular to the *z*-axis) *independently of direction of wave's propagation*. It simplifies considerably presentation of transmission equations.
- Sum and product of phasors can be used to present, on the PP sphere, results of waves' *interference* and values of *complex voltages received* by an antenna from a wave.
- When considering scattering by targets, the polarization and spatial phase of both the incident and scattered wave, and of the receiving antenna, all can be presented by phasors *on the same PP sphere*, what will allow for a geometrical interpretation of scattering and two-way transmission on that sphere as its phasors' inversion, rotation, and change of phase, followed by multiplication of results by the receiving antenna phasor.
- Not only special polarization points, but also special polarization phasors, e.g. eigenphasors, can be found for Sinclair matrices given.
- Convenient notation can be applied which uses *phasors* of PP vectors and PP bases as *upper and lower indices*, thus obtaining simple comprehensive, and easy to remember formulae for transmission equations, the spatial coordinate system reversal, and for the ONP PP basis or the PP sphere rotation.
- Changing phase of the first phasor of the ONP PP basis (together with automatic opposite change of phase of the second phasor in order to keep the basis null-phase) will be of fundamental importance when studying polarization properties of scattering matrices. It will allow one to arrive at *canonical forms of bistatic (in general) scattering matrices* with a minimum but sufficient number of parameters necessary for polarimetric considerations, e.g., for classification of targets.

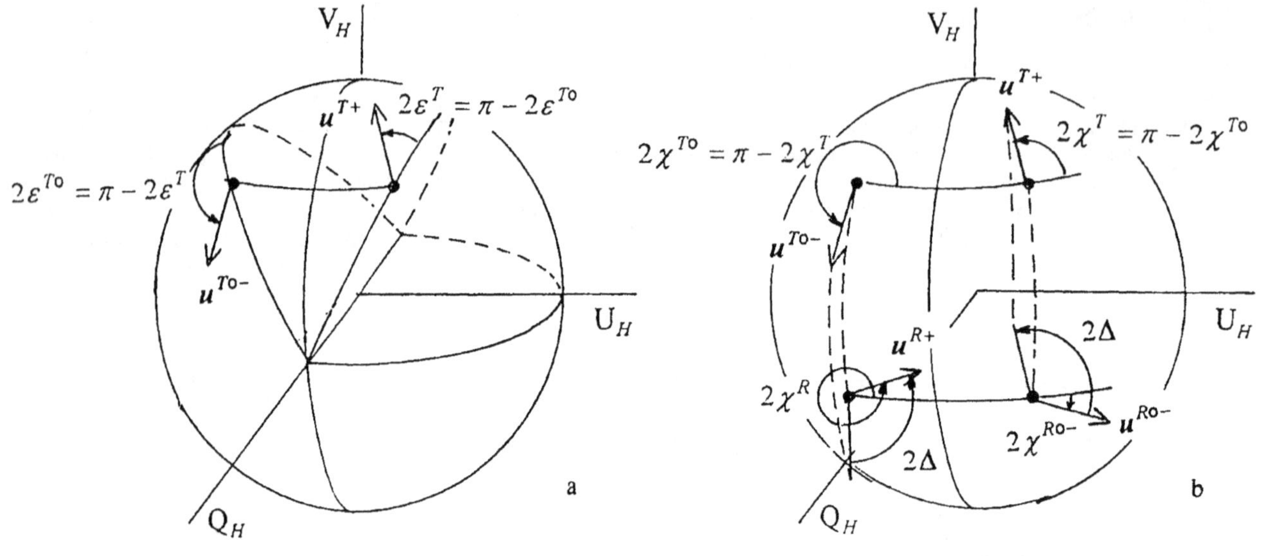

Compare: $u_B^{xx} = C^x C^x u_B = -u_B$ with $u_B^{oo} = C_B^o * C_B^o u_B = +u_B$

Fig. 6.1a,b. Mutual dependence between phase parameters: (a) analytical, ε, and (b) geometrical, χ, of the $\boldsymbol{u}$ and $\boldsymbol{u}^{o}$ vectors.
In (b): Phase delay Δ of the received voltage does not depend on the z-axis reversal by the coordinate system rotation. (See also Fig. 6.1c. for the sign of Δ dependence on directional upper indices of polarization phasors)

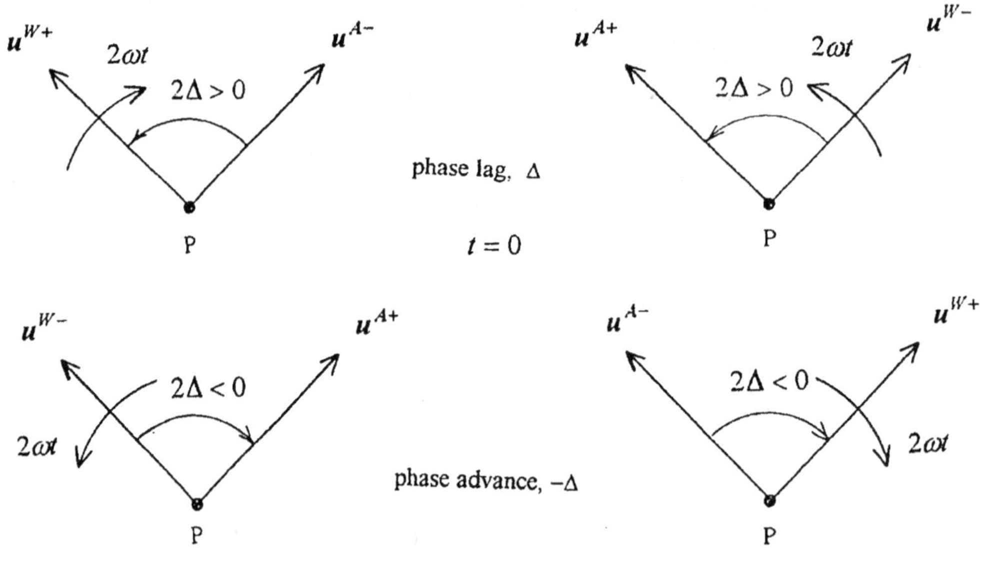

Fig. 6.1c. Determination for the received voltage its phase lag, Δ, or phase advance, $-\Delta$, by inspection of the polarization phasors shifted parallel to a common polarization point P.

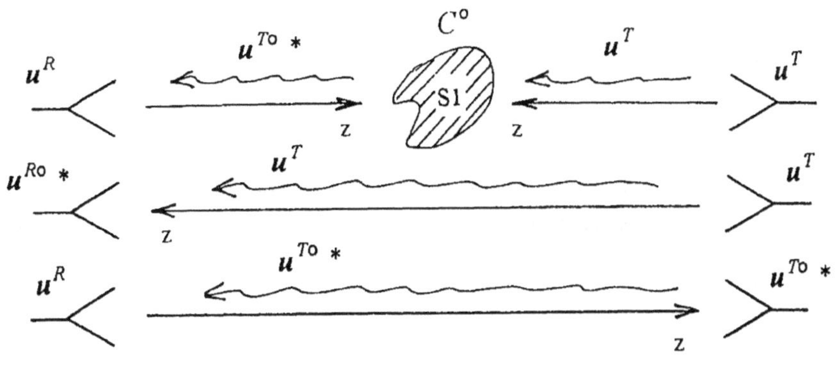

Fig. 6.1d. To the direct transmission equations (6.13).

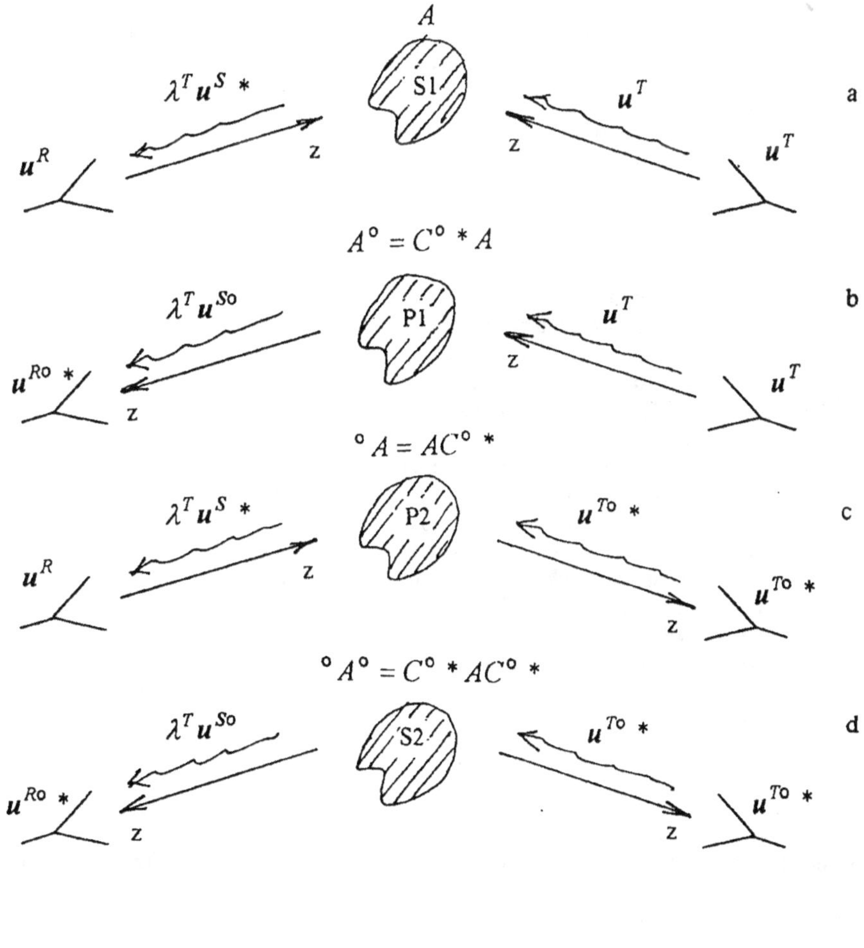

Fig. 6.2a-d. Complex amplitudes (directive Jones vectors) in four alignments: two 'scattering', of S1 and S2 type, and two 'propagation', of P1 and P2 type.

7. Scattering and Propagation Matrices

The same scattering phenomenon can be described from two points of view. Historically the first one, corresponding to the so called forward scattering alignment (FSA) or 'wave coordinate system', is usually being applied in optics. In radar transmission it is being used mostly to determine the change of polarization and phase of a wave propagating through the surrounding medium to and from an observed target. Polarimetric properties of that medium are determined in the FSA by the Jones or Mueller matrices. Another point of view corresponds to the so called backward scattering alignment (BSA) or 'antenna coordinate system'. It is especially useful when analyzing the change of polarization by an observed object itself. Sinclair and Kennaugh matrices are then in use to describe polarimetric properties of that target. Though scattering in the two alignments is the same, the two points of view impose different interpretations of polarimetric signatures obtained for matrices in the two alignments. On the other hand, because of dealing with the same scattering phenomenon, an exact relation between the Jones and Sinclair matrices exists and will be presented. Obtaining of such relation is only possible when applying the spatial reversal transformation joining matrices in the two alignments.

In order to gain geometrical representation of scattering, vectors of incident and scattered waves, as well as the receiving antenna vector, should be presented on one Poincare sphere. For that reason all matrices and vectors used in one transmission equation will be expressed in the same precisely determined orthogonal null-phase (ONP) polarimetric basis. Such a basis will be represented on the Poincare sphere by two collinear polarization phasors tangent to the sphere at its antipodal points. And, what should be stressed, that basis will be exactly the same for illuminating and scattered waves independently of the alignment used, FSA or BSA.

To arrive at such representation of scattering, a concept of polarization and spatial phase (PP) vectors for antennas and waves, different from complex antenna heights (CH) and complex wave amplitudes (CA), will be applied. It will be based on time reversal symmetry of Maxwell equations and their solutions, and on precise relation of these PP vectors versus each local spatial coordinate system: for illuminating waves, scattered waves, and for a receiving antenna.

In case of using the Sinclair scattering matrix, the last two spatial coordinate systems are identical with the first one. When applying the Jones propagation matrix, they are reversed by 180^0 rotation about an axis perpendicular to direction of propagation and to a chosen 'horizontal' reference axis. Therefore, the corresponding PP vectors depend on the reversal (by rotation) of a local spatial coordinate system and such reversal is the only reason for a difference between Sinclair and Jones matrices.

The PP vectors of antennas and waves also will be represented by tangential phasors on the Poincare sphere. Mutual locations and orientations of the antenna and wave phasor with respect to a basis phasor define elements of the corresponding antenna/wave PP column vector.

Spatial reversal matrices determining relations between Jones and Sinclair matrices as well as between Mueller and Kennaugh matrices depend, what has been stressed, on the orthogonal polarimetric bases used. It has been observed that different kinds of orthogonal bases are in use for different applications. Usually the ONP PP bases present the best choice. However, bases with their both phasors rotated by 90^0 seem to be more adequate for meteorological applications. For instance, they have been used in the McCormick's works. Here, derivation of 'reversal' transformations joining vectors and matrices in the two alignments has been presented based on the radar transmission equations.

For Sinclair and Jones matrices, describing the same physical phenomenon of nondepolarizing scattering, two different Poincare sphere models are proposed. The Sinclair matrix has been decomposed into a product of two operations of inversion and rotation of the Poincare sphere. For the Jones matrix model another Sinclair matrix has been applied followed by additional orthogonality transformation. Such difference in construction of the two models can best explain diverse physical interpretations of their operation.

Limited amount of simple formulae and the Poincare sphere geometrical models of scattering matrices can be obtained when applying to radar polarimetry the PP vector approach, based on the time-symmetry of Maxwell equations and followed by the polarization phasor notation which uses phasors as upper and lower indices for vectors and matrices.

Before approaching presentation of scattering and propagation matrices short summary of fundamental transformation matrices will be attached and their meanings explained. Then the two 'elementary' transformations will be presented by which, or by their special forms, all others can be expressed.

7. 1. Setting-up of fundamental transformations

The following set of four transformation equations completes the u vector definition. A general rule is that the domain and range of all these transformations are CA vectors expressed by the corresponding PP vectors. In such a sense we can speak about the following transformations of the PP vectors:

- *Scattering* transformation in the BSA, with scattering (Sinclair) matrix A (of the so called S1 type), transforms the PP(+) unit vector u of an incident wave into another, scattered wave PP(-) unit vector u' with real nonnegative coefficient λ:

$$Au = \lambda u'^*, \quad \text{alias} \quad A \underbrace{\overbrace{u^+}^{PP(+)}}_{CA(+)} = \lambda \underbrace{\overbrace{u'^-}^{PP(-)} *}_{CA(-)}; \qquad \lambda \in R^+ \tag{7.I}$$

The A matrix should be determined for local spatial coordinate systems of the two PP vectors, u and u', of the incident and scattered waves. Both the A matrix and u vector should be expressed in a common ONP PP basis, what results in the u' vector also in that basis, i.e. the basis identically determined, like the u vector, versus its local spatial coordinate system.
Conjugate value of the scattered unit PP vector means the unit CA vector of the scattered wave moving in the negative direction of the propagation axis in the BSA.

- *Orthogonality* transformation of the PP vector, $u \to u^\times$, also results in a conjugate value of the orthogonal PP vector:

$$C^\times u = u^\times *, \quad \text{alias} \quad \begin{cases} C^\times \underbrace{\overbrace{u^+}^{PP(+)}}_{CA(+)} = \underbrace{\overbrace{u^{\times-}}^{PP(-)} *}_{CA(-)} \\ \text{or} \\ C^\times \underbrace{\overbrace{u^-}^{PP(-)} *}_{CA(-)} = \underbrace{\overbrace{u^{\times+}}^{PP(+)}}_{CA(+)} \end{cases}, \quad C^\times = \begin{bmatrix} 0 & -1 \\ 1 & 0 \end{bmatrix}, \quad u^\times = \begin{bmatrix} -b^* \\ a^* \end{bmatrix}, \quad \tilde{u} u^\times * = 0. \tag{7.II}$$

This is compatible with the statement for scattering in the BSA because the received voltage expressed by the equation like (3.8), for mutually orthogonal receiving antenna and incoming wave, will vanish only for their opposite 'orientations', accounted for in the Hermitian product. There is no change of basis under the orthogonality transformation.

- *Rotation* transformation in the PP (C^2) space will be defined for three different (in general) PP column vectors u, u', and u'' as follows:

$$Cu' = u''; \qquad C = \begin{bmatrix} u & u^\times \end{bmatrix} = \begin{bmatrix} a & -b^* \\ b & a^* \end{bmatrix}, \quad C\tilde{C}^* = \begin{bmatrix} 1 & 0 \\ 0 & 1 \end{bmatrix}, \qquad \det C = +1. \tag{7.III}$$

No conjugation of PP vectors is experienced because rotation in the PP space does not change the direction of wave's propagation.

- *Spatial reversal* transformation rotates the antenna/wave or, *equivalently*, the spatial coordinate system by 180^0 in the R^3 space about an axis perpendicular to direction of propagation. In both cases it changes the PP vector of the antenna/wave which becomes oriented in the opposite direction along the propagation Oz axis:

$$C^\circ u = u^\circ *, \quad \text{alias} \quad \begin{cases} C^\circ \underbrace{\overbrace{u^+}^{PP(+)}}_{CA(+)} = \underbrace{\overbrace{u^{\circ-}}^{PP(-)} *}_{CA(-)} \\ \text{or} \\ C^\circ * \underbrace{\overbrace{u^-}^{PP(-)} *}_{CA(-)} = \underbrace{\overbrace{u^{\circ+}}^{PP(+)}}_{CA(+)} \end{cases} \tag{7.IV}$$

with

$$\det C^\circ = -1, \qquad C^\circ = \tilde{C}^\circ = \tilde{C} \begin{bmatrix} -1 & 0 \\ 0 & 1 \end{bmatrix} C, \quad C^\circ C^\circ * = \begin{bmatrix} 1 & 0 \\ 0 & 1 \end{bmatrix}$$

Here C^{o} is a complex matrix defined in any ONP PP basis to which it is transformed, by using the rotation matrix *C*, from the simplest *real* form in the horizontal/vertical linear polarization basis.

The following essential feature of definition of the reversal transformation should be observed: the PP vector of an antenna changes when its spatial local coordinate system reverses (by rotation) against the antenna. This is because the polarization and phase is always determined in its local spatial coordinate system, and never changes when antenna rotates together with that system. So, no matter what is reversing, the antenna or the coordinate system, the PP vector of the antenna always undergoes the same change.

It is worth noticing the following dependencies between matrices C^{x}, C, and C^{o}:

$$C^{x}C = C * C^{x} \qquad \text{and} \qquad C^{x}C^{o} = C^{o} * \tilde{C}^{x} \tag{7.1}$$

The four transformations, (I) - (IV), represented by the matrices A, C^{x}, C, and C^{o}, form a corner stone for the PP vector approach to the theory of radar polarimetry together with an 'polarization phasor notation' explained in Chapter 5. All other transformation formulae can be obtained by using those four only.

7. 2. Two Elementary Transformations

For purposes of representation of scattering on the Poincare sphere, the normalized inversion matrix A_{0n}^{INV} can be introduced instead of A, more elementary than the Sinclair matrix, realizing the following inversion transformation:

$$A_{0n}^{INV} u = \lambda u^{INV} * \tag{7.1'}$$

where

$$A_{0n}^{INV} = \begin{bmatrix} -U - jV & 1+Q \\ -1+Q & U - jV \end{bmatrix}; \qquad \det A_{0n}^{INV} = 1 - (Q^2 + U^2 + V^2) \le 1 \tag{7.2}$$

in which the Q, U, V values are coordinates of the 'inversion point' inside the Poincare sphere of unit radius. The Sinclair scattering matrix can then be presented in the form:

$$A = e^{j\xi} C * \frac{\sqrt{\sigma_0}}{2} A_{0n}^{INV} \tag{7.3}$$

Here

$$\xi = \frac{1}{2} \arg \det A, \quad \sigma_0 = SpanA + 2|\det A|, \tag{7.4}$$

and $\frac{\sqrt{\sigma_0}}{2}$ denotes the radius of the Poincare sphere model of the scattering matrix A. Of course, the inversion matrix depends on the ONP PP basis, contrary to the othogonality matrix, the transposed version of which presents its special case for the inversion point in the center of the sphere.

Also the spatial reversal matrix can be presented as a product of simpler matrices: of conjugate rotation and transposed orthogonality multiplied by *-j* phase factor. In that form it resembles the Sinclair scattering matrix and can be considered as that matrix for the free space. Its inversion matrix is the inverse (the transposed version) of the orthogonality matrix:

$$C^{o} = \begin{bmatrix} -1 & 0 \\ 0 & 1 \end{bmatrix} = -j \begin{bmatrix} 0 & j \\ j & 0 \end{bmatrix} \begin{bmatrix} 0 & 1 \\ -1 & 0 \end{bmatrix} = -jC^{ROT} * (I_{U}, \pi) A^{INV}(C^{o}) \Rightarrow A^{INV}(C^{o}) = \tilde{C}^{x} \tag{7.3'}$$

Taking all that into account one can see that all polarimetric matrices can be expressed by a product of only two *'elementary'* matrices of *rotation* and *inversion* with a complex number factor.

What can be seen from the above approach to the theory of radar polarimetry is that an efficient reduction in number of basic formulae has been obtained owing to expression of complex wave's amplitude and complex antenna height vectors, equipped with their directional indices, by the PP vectors numerically identical for both directional indices.

7. 3. The Poincare sphere of tangential phasors and the ONP PP basis

In order to construct the Poincare sphere model of a Sinclair matrix for bistatic scattering, it is necessary to present that matrix in its characteristic coordinate system (CCS) in Stokes' parameters space, corresponding to some characteristic ONP PP basis.

Any ONP PP basis is formed by two orthogonal PP vectors mutually related by the orthogonality transformation (7.II). In the polarization phasor representation they will be shown on the Poincare sphere as the ordered pair of two collinear phasors tangent to the sphere in two antipodal points [3]. It is evident that only the first vector of the basis should be determined because the second one can be found using formula (7.II). Let those phasors and the corresponding unit PP vectors be denoted by:

$$B \Leftrightarrow \boldsymbol{u}^{B} \quad \text{and} \quad B\text{x} \Leftrightarrow \boldsymbol{u}^{B\text{x}} \tag{7.5}$$

and let such basis be denoted by its first phasor B. Then, any other PP vector, corresponding to, say, P phasor, can be expressed by its unit column PP vector in the ONP PP basis B in a matrix form as follows:

$$\boldsymbol{u}^{P} = [\boldsymbol{u}^{B}\ \boldsymbol{u}^{B\text{x}}]u_{B}^{P} \tag{7.6}$$

where the unit column PP vector itself will be determined by Cayley-Klein parameters a and b, expressed by halves of Euler angles between tangential phasors on the Poincare sphere: of that PP vector, and of the first vector of its ONP PP basis (Fig. 7.1):

$$u_{B}^{P} = \begin{bmatrix} a \\ b \end{bmatrix}_{B}^{P} = \begin{bmatrix} \cos\gamma e^{-j(\delta+\varepsilon)} \\ \sin\gamma e^{+j(\delta-\varepsilon)} \end{bmatrix}_{B}^{P} \tag{7.7}$$

7. 4. The Stokes' four-vector of complete polarization

The corresponding unit Stokes four-vector will be found according to (2.4) and (3.31) or (5.37) as

$$\mathsf{P}_{B}^{P} = \widetilde{\mathsf{U}} * (u_{B}^{P} \otimes u_{B}^{P} *) = \frac{1}{\sqrt{2}} \begin{bmatrix} 1 \\ \cos 2\gamma \\ \sin 2\gamma \cos 2\delta \\ \sin 2\gamma \sin 2\delta \end{bmatrix}_{B}^{P} \tag{7.8}$$

thus presenting rectangular coordinates of the polarization P point in the four Stokes' parameters space, on the polarization four-sphere of unit radius.

7. 5. About different orthogonal PP bases - The collinear and parallel phasor bases

The ONP PP bases can be called the bases of collinear phasors. Another often used is the basis of parallel phasors. This is a different class basis which can be obtained from the ONP PP basis not by its rotation but by multiplication of, e. g., its second (orthogonal) vector by $+j$ or $-j$ factor (the corresponding basis transformation matrix is of determinant equal to plus or minus j, accordingly). There is, however, a problem about location of sum of the new basis phasors. There was no such problem for the ONP PP basis, for which the sum of its phasors was also collinear and always shown by the arrow of its first phasor. Simple rule can be proposed for the parallel basis phasors: if the second vector of the original ONP basis has been multiplied by $-j$ ($+j$), then the sum of the new basis phasors is on the right (left) site of the first phasor of that basis, e. g. (Fig.7.2),

$$\begin{bmatrix} u_H^R & u_H^{Rx} \end{bmatrix} = \underbrace{\frac{1}{\sqrt{2}}\begin{bmatrix} 1 & j \\ j & 1 \end{bmatrix}}_{\substack{\text{vectors of} \\ \text{collinear} \\ \text{phasors}}} \rightarrow \begin{bmatrix} u_H^R & -ju_H^{Rx} \end{bmatrix} = \underbrace{\frac{1}{\sqrt{2}}\begin{bmatrix} 1 & 1 \\ j & -j \end{bmatrix}}_{\substack{\text{vectors of} \\ \text{parallel} \\ \text{phasors}}} \Rightarrow \underbrace{\Sigma = \sqrt{2}\begin{bmatrix} 1 \\ 0 \end{bmatrix} = \sqrt{2}\, u_H^H}_{\substack{\text{vector of the sum} \\ \text{of basis phasors}}} . \qquad (7.9)$$

7. 6. Opposite orders of the (H,V) or (V,H) linear bases

When using the (H,V) or (V,H) ONP PP bases (H - horizontal linear, V - vertical linear) there is always a doubt about the order of basis vectors. Altogether four possibilities exist for choosing the horizontal/vertical linear polarization basis:

$$\begin{bmatrix} E_H \\ E_V \end{bmatrix} = \begin{bmatrix} E_x \\ E_y \end{bmatrix}, \quad \begin{bmatrix} E_H \\ E_V \end{bmatrix} = \begin{bmatrix} E_y \\ E_x \end{bmatrix}, \quad \begin{bmatrix} E_V \\ E_H \end{bmatrix} = \begin{bmatrix} E_x \\ E_y \end{bmatrix}, \quad \text{and} \quad \begin{bmatrix} E_V \\ E_H \end{bmatrix} = \begin{bmatrix} E_y \\ E_x \end{bmatrix}. \qquad (7.10)$$

The above Jones or PP column vectors may correspond to the following linear bases:

$$(H,V) \Leftrightarrow (x,y), \quad (H,V) \Leftrightarrow (y,x), \quad (V,H) \Leftrightarrow (x,y), \quad \text{and} \quad (V,H) \Leftrightarrow (y,x). \qquad (7.11)$$

Recognition of the order of basis used by an author is necessary to identify handedness of polarization on the Poincare sphere.

7. 7. Phasors in bases of the opposite order

Basis phasors, and all other phasors, if referred to the basis of the opposite order, should be represented on the Poincare sphere as oppositely oriented.

If the original (x,y), alias X, ONP PP basis will be rotated to the position of the (a,b), alias A, basis, then bases of the opposite order, called (y,x), alias Y'', and (b,a), alias B'', will be represented by their first phasors, Y'' and B'', oriented oppositely to Y and B. That is in agreement with the known relations between basis-dependent polarization parameters for mutually reversed basis order.

There are two main transformations governing the PP column vectors expressed in the orthogonal bases of opposite order (not necessarily the ONP PP bases):

$$\begin{aligned} u_A^P &\rightarrow u_{B''}^{P''} = \begin{bmatrix} 0 & 1 \\ 1 & 0 \end{bmatrix} u_A^P \\ u_A^P &\rightarrow u_{A''}^{P''} = \begin{bmatrix} 1 & 0 \\ 0 & -1 \end{bmatrix} u_A^P \end{aligned} \qquad (7.12)$$

valid also in the 'opposite direction' (all double-primed phasors can be exchanged for non-double-primed and vice versa).

7. 8. Amplitude Equations of Scattering and the Two-Way Transmission

The scattering transformation in the, e.g., horizontal/vertical H basis can now be written according to (I) as

$$A_H u_H^T = \lambda^T u_H^S * \qquad (7.13)$$

and the received voltage, after (3.8) and (7.13), can be given by a Hermitian product

$$V_r = \lambda^T \tilde{u}_H^R u_H^S * = \tilde{u}_H^R A_H u_H^T \qquad (7.14)$$

This is a simplified two-way (radar) transmission equation presented in terms of polarimetrically essential parameters only.

7. 9. Change of Basis Transformation

Let the characteristic basis of the scattering matrix be denoted by phasor K symbol. In order to transform equation (7.14) to that basis the following rotation matrix will be used

$$C_H^K = \begin{bmatrix} u_H^K & u_H^{Kx} \end{bmatrix} = \begin{bmatrix} a & -b^* \\ b & a^* \end{bmatrix}_H^K, \quad \text{with } C_K^H = \widetilde{C}_H^K{}^* \text{ and } C_H^K C_K^H = \begin{bmatrix} 1 & 0 \\ 0 & 1 \end{bmatrix} \quad (7.15)$$

Change of basis rules with that rotation matrix are [2]:

$$C_K^H u_H^S = u_K^S \qquad \text{and} \qquad A_K = \widetilde{C}_H^K A_H C_H^K \quad (7.16)$$

So, for transformation to the K basis, one obtains

$$V_r = \lambda^T \widetilde{u}_H^R u_H^S{}^* = \lambda^T \widetilde{u}_H^R \widetilde{C}_K^H C_K^H{}^* u_H^S{}^* = \lambda^T \widetilde{u}_K^R u_K^S{}^* \quad (7.17)$$

or

$$V_r = \widetilde{u}_H^R A_H u_H^T = \widetilde{u}_H^R \widetilde{C}_H^K \widetilde{C}_K^H A_H C_H^K C_K^H u_H^T = \widetilde{u}_K^R A_K u_K^T \quad (7.18)$$

7. 10. Sinclair to Kennaugh matrix transformation

The received power can be found in terms of a Kennaugh matrix, K_K, in a K basis, say, determined as follows:

$$\begin{aligned} P_r &= V_r \otimes V_r{}^* = (\widetilde{u}_K^R A_K u_K^T) \otimes (\widetilde{u}_K^R A_K u_K^T)^* \\ &= (\widetilde{u}_K^R \otimes \widetilde{u}_K^R{}^*) \mathsf{U}^* \underbrace{\widetilde{\mathsf{U}}(A_K \otimes A_K{}^*) \mathsf{U}}_{\mathsf{K}_K} \widetilde{\mathsf{U}}^* (u_K^T \otimes u_K^T{}^*) \\ &= \widetilde{\mathsf{P}}_K^R \mathsf{K}_K \mathsf{P}_K^T \\ &= \sigma^T \widetilde{\mathsf{P}}_K^R \mathsf{P}_K^S \quad \text{with } \sigma^T = (\lambda^T)^2 \text{ and } \mathsf{P}_K^S = \widetilde{\mathsf{U}}^* (u_K^S \otimes u_K^S{}^*) \end{aligned} \quad (7.19)$$

Elements of the Kennaugh matrix will be denoted in a way used by Perrin [126] and van de Hulst [83], but with a slight modification resulting from another alignment (BSA), in comparison with the FSA applied by those authors. In effect, the following forms will be obtained for Kennaugh matrices in the polarization phasor notation, in the H (horizontal/vertical linear) and K (characteristic) bases:

$$\mathsf{K}_H = \begin{bmatrix} a_1 & b_1 & b_3 & b_5 \\ c_1 & a_2 & b_4 & b_6 \\ c_3 & c_4 & a_3 & b_2 \\ c_5 & c_6 & c_2 & a_4 \end{bmatrix}_H ; \qquad \mathsf{K}_K = \begin{bmatrix} a_1 & b_1 & b_3 & b_5 \\ b_1 & a_2 & b_4 & b_6 \\ -b_3 & -b_4 & a_3 & 0 \\ -b_5 & -b_6 & 0 & a_4 \end{bmatrix}_K \quad (7.20a,b)$$

Elements of those matrices will be presented in terms of elements of the following Sinclair A matrices in both bases (exact relations between A_H and A_K matrices can be found, e.g., in [53]):

$$A_H = \begin{bmatrix} A_2 & A_3 \\ A_4 & A_1 \end{bmatrix}_H ; \qquad A_K = \begin{bmatrix} A_2 & B_1 + jB_2 \\ -B_1 - jB_2 & A_1 \end{bmatrix}_{CCS} e^{j\mu}, \quad (7.21a,b)$$

Here: $A_{iH} \in C^1$, A_{iCCS} and $B_{iCCS} \in R^1$, and in the CCS: $A_2 \geq A_1 \geq 0$, $B_2 \geq 0$, and $B_1 \geq 0$ but only if $B_2 = 0$ (for unambiguity of the characteristic basis, K). In the H basis, the one digit lower indices of the scattering matrix elements are taken after van de Hulst [6]. It has been done for simplifying the notation, but also with one essential difference. In the proposal offered by van de Hulst his transformation matrix was designed for the FSA. Here the BSA has beenchosen, which is more convenient for symmetry reasons in the case of backscattering.

Relations between elements of the (20a) and (21a) matrices, in the H or in any other ONP PP basis, will be presented in terms of the following real valued expressions (also after van de Hulst), for i , k = 1, 2, 3, and 4:

$$\begin{aligned} M_k &= A_k A_k{}^* \\ S_{ki} = S_{ik} &= \frac{1}{2}(A_i A_k{}^* + A_k A_i{}^*) \\ -D_{ki} = D_{ik} &= \frac{j}{2}(A_i A_k{}^* - A_k A_i{}^*) \end{aligned} \tag{7.22}$$

In terms of those expressions, the K_H matrix as in (20a) takes the form (compare [6], [1], and [5]):

$$K_H = \begin{bmatrix} \frac{1}{2}(M_2+M_3+M_4+M_1) & \frac{1}{2}(M_2-M_3+M_4-M_1) & S_{32}+S_{14} & D_{32}+D_{14} \\ \frac{1}{2}(M_2+M_3-M_4-M_1) & \frac{1}{2}(M_2-M_3-M_4+M_1) & S_{32}-S_{14} & D_{32}-D_{14} \\ S_{42}+S_{13} & S_{42}-S_{13} & S_{34}+S_{12} & D_{12}+D_{34} \\ D_{42}+D_{13} & D_{42}-D_{13} & D_{12}-D_{34} & S_{34}-S_{12} \end{bmatrix}_H \tag{7.23}$$

Here, however, a caution is advisable: the similar van de Hulst's formula is for the bistatic propagation, P1-type, or Mueller matrix obtained for the corresponding Jones matrix.

In case of symmetrical Kennaugh matrices, its diagonal elements, denoted as in (7.20a), satisfy the linear equation

$$a_1 = a_2 + a_3 + a_4 \tag{7.24}$$

Elements of the K_K matrix (7.20b), determined in the characteristic coordinate system (CCS) corresponding to the characteristic ONP PP basis K, will be given directly by the (7.21b) matrix elements:

with nine different elements:

$$\begin{aligned} &a_1 = \tfrac{1}{2}(A_2^2 + A_1^2) + B_1^2 + B_2^2 \geq 0, && b_3 = B_1(A_2 - A_1), \\ &a_2 = \tfrac{1}{2}(A_2^2 + A_1^2) - B_1^2 - B_2^2, && b_4 = B_1(A_2 + A_1), \\ &a_3 = A_1 A_2 - B_1^2 - B_2^2, && b_5 = -B_2(A_2 + A_1) \leq 0, \\ &a_4 = -A_1 A_2 - B_1^2 - B_2^2 \leq 0, && b_6 = -B_2(A_2 - A_1) \leq 0. \\ &b_1 = \tfrac{1}{2}(A_2^2 - A_1^2) \geq 0, && b_2 = 0 \end{aligned} \tag{7.25}$$

satisfying the linear equation

$$a_2 = a_1 + a_3 + a_4 \tag{7.26}$$

They are mutually related by the folloeing equation known as conditions for preservation of complete polarization of the scattered wave:

$$\begin{aligned} a_1 a_2 + a_3 a_4 &= b_1^2 \\ a_1 a_3 + a_2 a_4 &= -b_3^2 - b_6^2 \\ a_1 a_4 + a_2 a_3 &= -b_5^2 - b_4^2 \end{aligned} \tag{7.27a}$$

$$a_1^2 - a_2^2 = b_3^2 + b_5^2 + b_4^2 + b_6^2$$
$$a_1^2 - a_3^2 = b_1^2 + b_5^2 + b_4^2 \qquad (7.27b)$$
$$a_1^2 - a_4^2 = b_1^2 + b_3^2 + b_6^2$$

$$b_1(a_1 - a_2) = b_3 b_4 + b_5 b_6$$
$$0 = b_3 b_5 - b_4 b_6$$
$$b_1(a_1 + a_3) = b_1 b_4$$
$$b_4(a_1 + a_4) = b_1 b_3 \qquad (7.27c)$$
$$b_5(a_1 + a_4) = b_1 b_6$$
$$b_6(a_1 + a_3) = b_1 b_5$$

An additional remark: The above presented and commonly accepted form of the Kennaugh matrix is not the only one possible. Its another version was recommended by Kennaugh himself in his Reports [95]. Istead of (7.19), an equivalent procedure can be applied for expressing the received power:

$$\begin{aligned} P_r &= V_r \otimes V_r{}^* = (\tilde{u}_K^R A_K u_K^T) \otimes (\tilde{u}_K^R A_K u_K^T)^* \\ &= (\tilde{u}_K^R \otimes \tilde{u}_K^R{}^*) \mathrm{U} \underbrace{\tilde{\mathrm{U}}^* (A_K \otimes A_K{}^*) \mathrm{U}^*}_{\mathrm{K}'_K} \tilde{\mathrm{U}} (u_K^T \otimes u_K^T{}^*) \\ &= \tilde{\mathrm{P}}'^R_K \mathrm{K}'_K \mathrm{P}'^T_K \\ &= \sigma^T \tilde{\mathrm{P}}'^R_K \mathrm{P}'^S_K \quad \text{with} \quad \sigma^T = (\lambda^T)^2 \quad \text{and} \quad \mathrm{P}'^S_K = \tilde{\mathrm{U}}(u_K^S \otimes u_K^S{}^*) \end{aligned} \qquad (7.19')$$

The unacceptable consequence of such approach are negative values of 2δ and 2α angles for the positive fourth V component of the new Stokes four-vector because, customarily, positive V values correspond to the upper part of the Poincare sphere. Of course, the right-handed circular polarization would be obtained for the upper pole of the sphere when applying the 'natural' (x,y) basis, what probably was the purpose of the Author. Also, in the new (precisely: original) Kennaugh matrix the following elements of it will change their signs: b_5, b_6, b_2, c_5, c_6, c_2 (or b-elements only for symmetrical matrices, because in his reports Kennaugh considered symmetrical matrices).

7. 11. Change of Alignment: from the BSA to FSA

When applying the reversal transformation to the transmission equation (7.18), the Jones propagation matrix A°, in the FSA, can be immediately determined as dependent on the Sinclair A matrix:

$$V_r = \tilde{u}_K^R A_K u_K^T = \underbrace{\tilde{u}_K^R (C_K^{\circ}}_{\tilde{u}_K^{R\circ *}} \underbrace{C_K^{\circ *}) A_K}_{A_K^{\circ}} u_K^T = \tilde{u}_K^{R\circ} {}^* A_K^{\circ} u_K^T \qquad (7.28)$$

where

$$A_K^{\circ} = C_K^{\circ} {}^* A_K \quad \text{and} \quad C_K^{\circ} = \tilde{C}_H^K C_H^{\circ} C_H^K \qquad (7.29)$$

Also the Mueller matrix, K°, can be found when using procedure similar to that of (7.19):

$$\begin{aligned} P_r &= V_r \otimes V_r{}^* = (\tilde{u}_K^{R\circ} {}^* A_K^{\circ} u_K^T) \otimes (\tilde{u}_K^{R\circ} {}^* A_K^{\circ} u_K^T)^* \\ &= (\tilde{u}_K^{R\circ} {}^* \otimes \tilde{u}_K^{R\circ}) \mathrm{U} \underbrace{\tilde{\mathrm{U}}^* (A_K^{\circ} \otimes A_K^{\circ}{}^*) \mathrm{U}}_{\mathrm{K}_K^{\circ}} \tilde{\mathrm{U}}^* (u_K^T \otimes u_K^T{}^*) \\ &= \tilde{\mathrm{P}}_K^{R\circ} \mathrm{K}_K^{\circ} \mathrm{P}_K^T \end{aligned} \qquad (7.30)$$

with

$$\mathsf{K}_K^o = \widetilde{\mathsf{U}}^*(A_K^o \otimes A_K^o{}^*)\mathsf{U} = \mathsf{D}_K^o \mathsf{K}_K \quad \text{where} \quad \mathsf{D}_K^o = \widetilde{\mathsf{U}}(C_K^o \otimes C_K^o{}^*)\mathsf{U} \tag{7.31}$$

In the H basis,

$$C_H^o = \begin{bmatrix} -1 & 0 \\ 0 & 1 \end{bmatrix}, \quad \text{and} \quad \mathsf{D}_H^o = \widetilde{\mathsf{U}}(C_H^o \otimes C_H^o{}^*)\mathsf{U} = \begin{bmatrix} 1 & & & \\ & 1 & & \\ & & -1 & \\ & & & 1 \end{bmatrix} \tag{7.32}$$

So, from (7.20b) we obtain

$$\mathsf{K}_H^o = \mathsf{D}_H^o \mathsf{K}_H = \begin{bmatrix} a_1 & b_1 & b_3 & b_5 \\ c_1 & a_2 & b_4 & b_6 \\ -c_3 & -c_4 & -a_3 & -b_2 \\ c_5 & c_6 & c_2 & a_4 \end{bmatrix}_H \tag{7.33}$$

In any other ONP PP basis B, the spatial reversal transformation matrix in the Stokes parameters space becomes

$$D_B^o = \widetilde{U}(C_B^o \otimes C_B^o{}^*)U$$

$$= \frac{1}{2}\begin{bmatrix} 1 & 0 & 0 & 1 \\ 1 & 0 & 0 & -1 \\ 0 & 1 & 1 & 0 \\ 0 & -j & j & 0 \end{bmatrix} \left(\begin{bmatrix} -w & u \\ u & w^* \end{bmatrix} \otimes \begin{bmatrix} -w^* & u \\ u & w \end{bmatrix} \right)_H^B \begin{bmatrix} 1 & 1 & 0 & 0 \\ 0 & 0 & 1 & -j \\ 0 & 0 & 1 & j \\ 1 & -1 & 0 & 0 \end{bmatrix} \tag{7.34}$$

and finally

$$\mathsf{D}_B^o = \begin{bmatrix} 1 & 0 & 0 & 0 \\ 0 & 1-2u^2 & -u(w+w^*) & ju(w-w^*) \\ 0 & -u(w+w^*) & u^2 - \frac{1}{2}(w^2+w^{*2}) & \frac{j}{2}(w^2-w^{*2}) \\ 0 & ju(w-w^*) & \frac{j}{2}(w^2-w^{*2}) & u^2+\frac{1}{2}(w^2+w^{*2}) \end{bmatrix}_H^B . \tag{7.35}$$

In special cases of $B = H$ or $B = RC$ (right-circular and co-phased with H, represented by the parallel phasor), when $u = 0$ and $w = 1$, we obtain the previously presented form,

$$D_{H\,or\,C}^0 = \begin{bmatrix} 1 & 0 & 0 & 0 \\ 0 & 1 & 0 & 0 \\ 0 & 0 & -1 & 0 \\ 0 & 0 & 0 & 1 \end{bmatrix}. \tag{7.36}$$

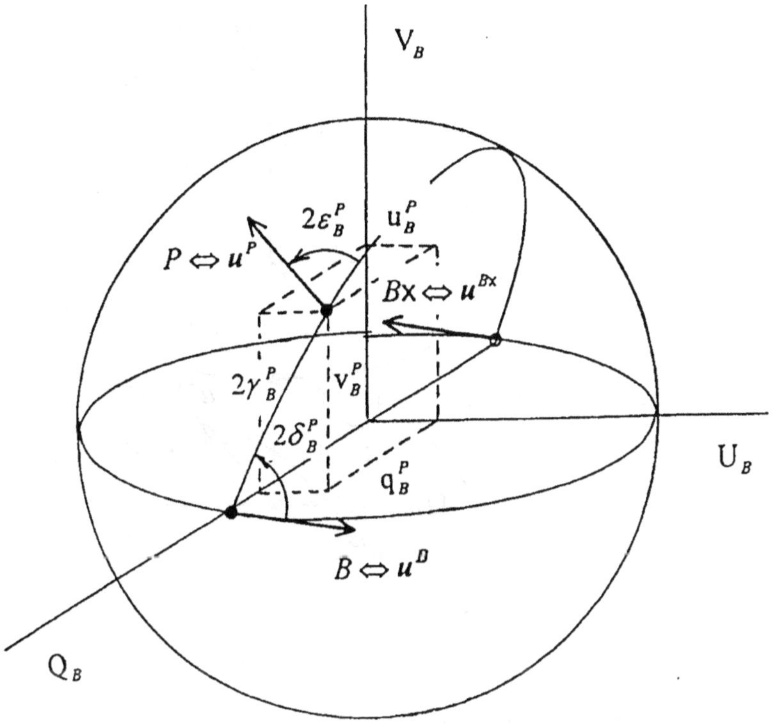

Fig. 7.1. Tangent phasor P, its angular coordinates in the ONP polarization basis B, and the corresponding rectangular coordinate system $Q_B\ U_B\ V_B$

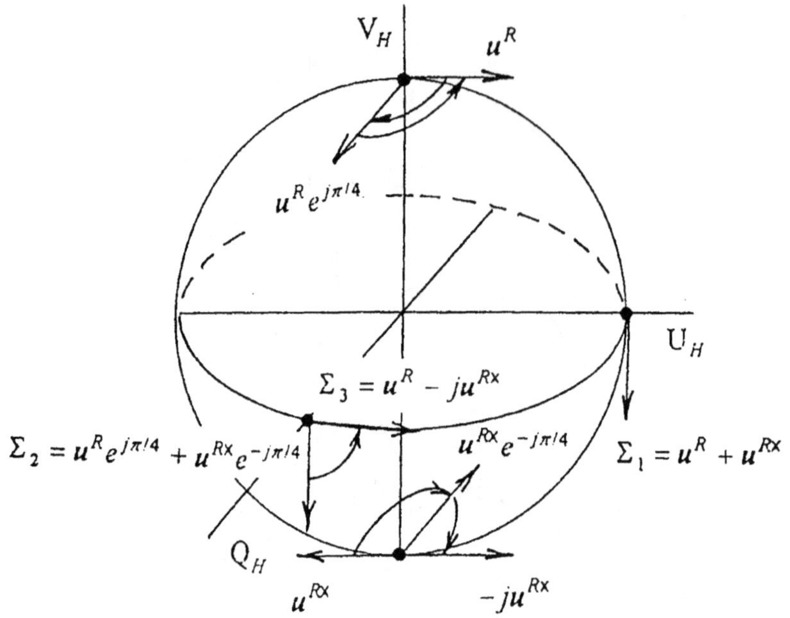

Fig. 7.2. An example of formation of the parallel phasor basis.

8. The Poincare Sphere Analysis

8. 1. Scattering as inversion, rotation, and change of phase

In the H basis, the expression for the bistatic scattering Sinclair matrix can be presented in the form

$$A_H = \begin{bmatrix} A_{2H} & A_{3H} \\ A_{4H} & A_{1H} \end{bmatrix} \equiv \begin{bmatrix} A_2 & A_3 \\ A_4 & A_1 \end{bmatrix}_H \tag{8.1}$$

where elements of the matrix are numbered according to the notation proposed by van de Hulst [83].

The Sinclair matrix can be decomposed into a product of the inversion (dephased and normalized) and rotation matrices, and of two scalar factors: the radius k of the Poincare sphere model of the matrix, and the exponential absolute phase factor:

$$A_H = k\, C_{P,H}^{K} * A_{0nH}^{INV} e^{j\xi} \tag{8.2}$$

where the subscript '$0n$' means dephased and normalized value of the inversion matrix corresponding to the unit radius of the Poincare sphere model of the matrix. The radius of the model before normalization is

$$k = r_0 = \frac{\sqrt{\sigma_0}}{2} \tag{8.3}$$

It is called also the 'stress' of the matrix, after Krogager [102]. The square of the Poincare sphere diameter (before normalization) can be defined as

$$\sigma_0 = SpanA + 2|\det A| \tag{8.4}$$

and the absolute phase of the matrix is

$$\xi = \frac{1}{2} \arg \det A \tag{8.5}$$

Span and determinant of the matrix are both independent of the PP basis. Therefore the name of the basis, usually presented by the lower index, has been omitted here.

General idea of such decomposition for nonsymmetrical scattering matrices has been propesed by Kennaugh in [96].

The matrix $C_{P,H}^{K}$ means rotation of the PP sphere placing some P phasor to cover with the characteristic basis K phasor , the rotation being expressed in the H basis. However, such a matrix describes rotation all tangential phasors on the Poincare sphere by an angle 2ϕ about an axis along the sphere unit vector $\boldsymbol{n}$. Therefore, also another symbol for that matrix can be used,

$$C_H^{ROT}(\boldsymbol{n}, 2\phi) \equiv C_H^{ROT} = C_{P,H}^{K} \tag{8.6}$$

and the Sinclair scattering matrix can take the form:

$$A_H = C_H^{ROT} * A_{0H}^{INV} e^{j\xi}, \tag{8.7}$$

Inversion always reverses direction of propagation versus the z axis and therefore produces the conjugate inverted PP vector. That requires an application of a conjugate rotation matrix which is:

$$\begin{aligned} C_H^{ROT}* &\equiv C_H^{ROT}*(\boldsymbol{n}, 2\phi) \\ &= \begin{bmatrix} \cos\phi + jn_1 \sin\phi & (-n_3 + jn_2)\sin\phi \\ (n_3 + jn_2)\sin\phi & \cos\phi - jn_1 \sin\phi \end{bmatrix}_H . \end{aligned} \tag{8.8}$$

Here ϕ is one half of the rotation after inversion angle, and components of the unit n vector of the rotation axis are along the Q_H , U_H, and V_H axes. The form of the above expression remains unchanged for any other ONP PP basis B.

That conjugate rotation matrix can be presented in terms of elements of the Sinclair matrix A (in any ONP PP basis). The derivation starts with determination of the dephased matrix $Ae^{-j\xi}$ and gives the following result:

$$C^{ROT}* = \frac{1}{\sqrt{\sigma_0}}\left\{\left(Ae^{-j\xi}\right)C^{\times} + C^{\times}\left(A*e^{j\xi}\right)\right\} = \frac{1}{\sqrt{\sigma_0}}\begin{bmatrix} A_3e^{-j\xi} - A_4*e^{j\xi} & -A_2e^{-j\xi} - A_1*e^{j\xi} \\ A_1e^{-j\xi} + A_2*e^{j\xi} & -A_4e^{-j\xi} + A_3*e^{j\xi} \end{bmatrix} \tag{8.9}$$

One can easily check that this is really a rotation matrix by just comparing mutual dependence of its elements with those defined in (III). Then, the inversion matrix can be found immediately when using the equation (8.7):

$$\begin{aligned} A_0^{INV} &= e^{-j\xi}\tilde{C}^{ROT}A \\ &= \frac{1}{\sqrt{\sigma_0}}\left\{\tilde{C}^{\times}\tilde{A}* + \tilde{A}\tilde{C}^{\times}e^{-2j\xi}\right\}A = \frac{1}{\sqrt{\sigma_0}}\begin{bmatrix} A_2A_3* + A_4A_1* & M_3 + M_1 + |\det A| \\ -M_2 - M_4 - |\det A| & -A_3A_2* - A_1A_4* \end{bmatrix} \end{aligned} \tag{8.10}$$

And again one can check that this is an inversion matrix because for the inversion I point coordinates inside the sphere of diameter $\sqrt{\sigma_0}$ are:

$$\begin{bmatrix} Q \\ U \\ V \end{bmatrix}^{I}_{2r_0=\sqrt{\sigma_0}} = \frac{-1}{\sqrt{\sigma_0}}\begin{bmatrix} b_1 \\ b_3 \\ b_5 \end{bmatrix}. \tag{8.11}$$

Taking b_i values for i = 1, 2, 3, from expressions (7.20a), (7.22) and (7.23), one can find that

$$A_0^{INV} = \begin{bmatrix} -U - jV & \frac{\sqrt{\sigma_0}}{2} + Q \\ -\frac{\sqrt{\sigma_0}}{2} + Q & U - jV \end{bmatrix}^{I}_{2r_0=\sqrt{\sigma_0}} \tag{8.12}$$

is exactly equal to (8.10).

8. 2. Propagation as Lorentz transformation, rotation, and change of phase

Similar considerations conducted for the Jones propagation matrix of the form

$$A^{o} = C^{o}*A = e^{j\xi^{o}}C'^{ROT}A_0^{LOR} \tag{8.13}$$

with (in any ONP PP B basis):

$$C_B^{o} = \tilde{C}_H^{B}\begin{bmatrix} -1 & 0 \\ 0 & 1 \end{bmatrix}C_H^{B} \tag{8.14}$$

$$A_0^{LOR} = \frac{\sqrt{\sigma_0}}{2}A_{0n}^{LOR}, \qquad \xi^{o} = \frac{1}{2}\arg\det A^{o} = \xi - \frac{\pi}{2} \tag{8.15}$$

lead to the following results:

$$C'^{ROT} = \frac{1}{\sqrt{\sigma_0}}\left\{\left(A^{\circ}e^{-j\xi^{\circ}}\right) + C^{\times}\left(A^{\circ} * e^{j\xi^{\circ}}\right)\tilde{C}^{\times}\right\}$$

$$= \frac{1}{\sqrt{\sigma_0}}\begin{bmatrix} A_2^{\circ}e^{-j\xi^{\circ}} + A_1^{\circ} * e^{j\xi^{\circ}} & A_3^{\circ}e^{-j\xi^{\circ}} - A_4^{\circ} * e^{j\xi^{\circ}} \\ A_4^{\circ}e^{-j\xi^{\circ}} - A_3^{\circ} * e^{j\xi^{\circ}} & A_1^{\circ}e^{-j\xi^{\circ}} + A_2^{\circ} * e^{j\xi^{\circ}} \end{bmatrix} \tag{8.16}$$

$$= jC^{\times}C^{\circ}C^{ROT}$$

and

$$A_0^{LOR} = e^{-j\xi^{\circ}}\tilde{C}'^{ROT} * A^{\circ}$$

$$= \frac{1}{\sqrt{\sigma_0}}\left\{\tilde{A}^{\circ} * + C^{\times}\tilde{A}^{\circ}\tilde{C}^{\times}e^{-2j\xi^{\circ}}\right\}A^{\circ}$$

$$= \begin{bmatrix} M_2^{\circ} + M_4^{\circ} + |\det A| & A_3^{\circ}A_2^{\circ} * + A_1^{\circ}A_4^{\circ} * \\ A_2^{\circ}A_3^{\circ} * + A_4^{\circ}A_1^{\circ} * & M_3^{\circ} + M_1^{\circ} + |\det A| \end{bmatrix} \tag{8.17}$$

$$= C^{\times}A_0^{INV}$$

$$= \begin{bmatrix} \frac{1}{2}\sqrt{\sigma_0} - \mathrm{Q} & -\mathrm{U} + j\mathrm{V} \\ -\mathrm{U} - j\mathrm{V} & \frac{1}{2}\sqrt{\sigma_0} + \mathrm{Q} \end{bmatrix}^{\mathrm{I}}_{2r_0 = \sqrt{\sigma_0}} \tag{8.18}$$

8. 3. Jones matrix in the horizontal/vertical linear basis

When changing the alignment from BSA to FSA in the H ONP PP basis, the obtained Jones matrix, called also the propagation amplitude matrix, is of the form

$$A_H^{\circ} = C^{\times}C'^{ROT}_H * A_{0H}^{INV}e^{j(\xi - \pi/2)} \tag{8.19}$$

with

$$A_{0H}^{INV} = \frac{\sqrt{\sigma_0}}{2}A_{0nH}^{INV}$$

and with another rotation matrix

$$C_H^{ROT'} * = C_H^{ROT'} * (\boldsymbol{n}', 2\phi') \tag{8.20}$$

preceded by the orthogonality matrix and followed by *the same inversion matrix.*

Simple derivation of that dependence can be based on the original definition of the Jones matrix:

$$\begin{aligned} A_H^{\circ} &= C_H^{\circ}A_H = C_H^{\circ}\tilde{C}^{\times}C^{\times}A_H = C^{\times}C_H^{\circ}C^{\times}A_H \\ &= -jC^{\times}(jC_H^{\circ}C^{\times}C_{P,H}^{K} *)A_{0H}^{INV}e^{j\xi} = C^{\times}A_H^{\times\circ} \end{aligned} \tag{8.21}$$

Above have been determined: the new rotation matrix, and the new absolute phase angle, $\xi^{\circ} = \xi - \pi/2$; *the inversion matrix* of the newly defined another Sinclair scattering matrix, $A_H^{\times\circ}$, *has not been changed.*

The two Sinclair matrices have different characteristic bases, K and K°, and different angles and axes of rotation after inversion:

$$\cos\phi' = -n_{2H}\sin\phi\,; \qquad 0 \le 2\phi' \le 2\pi\,, \tag{8.22}$$

$$n'_H = \begin{bmatrix} n'_1 \\ n'_2 \\ n'_3 \end{bmatrix}_H = \frac{1}{\sqrt{1-n_2^2\sin^2\phi}} \begin{bmatrix} n_3 \sin\phi \\ \cos\phi \\ -n_1 \sin\phi \end{bmatrix} \tag{8.23}$$

The new rotation matrix differs from the previous one by an additional rotation by 180^0 about the U_H axis.

8. 4. Decomposition of the Sinclair matrix into product of matrices in the characteristic ONP PP basis

Especially simple formulae for the rotation and inversion matrices of the Sinclair matrix decomposition in the characteristic K basis result from expressions (7.20b), (7.21b), and (7.25), from which one obtains:

$$A_K = C_K^{ROT} * A_{0K}^{INV} e^{j\xi}; \qquad A_{0K}^{INV} = \frac{\sqrt{\sigma_0}}{2} A_{0nK}^{INV} \tag{8.24}$$

with

$$\xi = \xi_0 + \mu; \qquad \xi_0 = \frac{1}{2}\arg(A_2 A_1 + B_1^2 - B_2^2 + j2B_1B_2)_{CCS} \ , \tag{8.25}$$

$$\sigma_0 = \left\{ A_2^2 + A_1^2 + 2(B_1^2 + B_2^2) + 2\sqrt{\left(A_2A_1 + B_1^2 - B_2^2\right)^2 + 4B_1^2B_2^2} \right\}_{CCS} \ , \tag{8.26}$$

$$C_K^{ROT} * = C_K^{ROT} * (\boldsymbol{n}, 2\phi) = \begin{bmatrix} \cos\phi & (-n_3 + jn_2)\sin\phi \\ (n_3 + jn_2)\sin\phi & \cos\phi \end{bmatrix}_K$$

$$= \frac{1}{\sqrt{\sigma_0}} \begin{bmatrix} 2B_1\cos\xi_0 + 2B_2\sin\xi_0 & -(A_2 + A_1)\cos\xi_0 + j(A_2 - A_1)\sin\xi_0 \\ (A_2 + A_1)\cos\xi_0 + j(A_2 - A_1)\sin\xi_0 & 2B_1\cos\xi_0 + 2B_2\sin\xi_0 \end{bmatrix}_{CCS}$$

and

$$A_{0K}^{INV} = \frac{1}{\sqrt{\sigma_0}} \begin{bmatrix} B_1(A_2 - A_1) - jB_2(A_2 + A_1) & \frac{\sigma_0}{2} - \frac{1}{2}(A_2^2 - A_1^2) \\ -\frac{\sigma_0}{2} - \frac{1}{2}(A_2^2 - A_1^2) & -B_1(A_2 - A_1) - jB_2(A_2 + A_1) \end{bmatrix}_{CCS} \tag{8.27, 8.28}$$

$$= \frac{1}{\sqrt{\sigma_0}} \begin{bmatrix} b_3 + jb_5 & \frac{\sigma_0}{2} - b_1 \\ -\frac{\sigma_0}{2} - b_1 & -b_3 + jb_5 \end{bmatrix}_K = \begin{bmatrix} -\mathrm{U} - j\mathrm{V} & \frac{1}{2}\sqrt{\sigma_0} + \mathrm{Q} \\ -\frac{1}{2}\sqrt{\sigma_0} + \mathrm{Q} & \mathrm{U} - j\mathrm{V} \end{bmatrix}^{\mathrm{I}}_{K;\ 2r_0 = \sqrt{\sigma_0}} \tag{8.29, 8.30}$$

Geometrically, the rule of inversion can be best explained when considering the operation in the equatorial plane of linear polarizations (see Fig. 8.1). That rule remains unchanged after any rotation of the ONP basis, similarly as the form of equation (8.30) or (8.12) does. Fig. 8.2 presents an example of inversion and rotation in the simplest case, of monostatic scattering.

As seen from the above formulae, the rotation after inversion axis in the CCS is in the Q_K=0 plane. The rotation axis components are n_1=0, and $n_3 \geq 0$. The rotation after inversion angles are:

$$0 \leq 2\phi \leq 180^\circ \quad \text{for} \quad \mathrm{U}_K^{\mathrm{I}} \leq 0, \quad \text{and} \quad 180^\circ < 2\phi \leq 360^\circ \quad \text{for} \quad \mathrm{U}_K^{\mathrm{I}} > 0. \tag{8.31}$$

In the Stokes parameter space the corresponding rotation and inversion matrices are:

$$\widetilde{D}_K^P = \frac{1}{a_0(a_1+a_0)}\begin{bmatrix} a_0(a_1+a_0) & 0 & 0 & 0 \\ 0 & b_1^2 - a_2(a_1+a_0) & b_4(a_4-a_0) & b_6(a_3-a_0) \\ 0 & -b_4(a_4-a_0) & -b_3^2 - a_3(a_1+a_0) & -b_3b_5 \\ 0 & -b_6(a_3-a_0) & -b_3b_5 & -b_5^2 - a_4(a_1+a_0) \end{bmatrix}_K \tag{8.32}$$

$$K_K^{INV} = \frac{1}{(a_1+a_0)}\begin{bmatrix} a_1(a_1+a_0) & b_1(a_1+a_0) & b_3(a_1+a_0) & b_5(a_1+a_0) \\ -b_1(a_1+a_0) & -b_1^2 - a_0(a_1+a_0) & -b_1b_3 & -b_1b_5 \\ -b_3(a_1+a_0) & -b_1b_3 & -b_3^2 - a_0(a_1+a_0) & -b_3b_5 \\ -b_5(a_1+a_0) & -b_1b_5 & -b_3b_5 & -b_5^2 - a_0(a_1+a_0) \end{bmatrix}_K \tag{8.33}$$

The amplitude and power expression for the inversion matrix in terms of Kennaugh matrix elements or I point coordinates are true for any ONP PP basis, not only for the characteristc basis K . Therefore, e.g., for coordinates of the inversion point in the linear H basis

$$\begin{bmatrix} Q \\ U \\ V \end{bmatrix}^{I}_{H;\, 2r_0=\sqrt{\sigma_0}} = \frac{-1}{\sqrt{\sigma_0}}\begin{bmatrix} b_1 \\ b_3 \\ b_5 \end{bmatrix}_H \tag{8.34}$$

we obtain

$$K_H^{INV} = $$

$$= \begin{bmatrix} Q^2 + U^2 + V^2 + \frac{\sigma_0}{4} & -\sqrt{\sigma_0}Q & -\sqrt{\sigma_0}U & -\sqrt{\sigma_0}V \\ \sqrt{\sigma_0}Q & -Q^2 + U^2 + V^2 - \frac{\sigma_0}{4} & -2QU & -2QV \\ \sqrt{\sigma_0}U & -2QU & Q^2 - U^2 + V^2 - \frac{\sigma_0}{4} & -2UV \\ \sqrt{\sigma_0}V & -2QV & -2UV & Q^2 + U^2 - V^2 - \frac{\sigma_0}{4} \end{bmatrix}_H \tag{8.35}$$

Having expressions for rotation and inversion matrices in the Stokes parameter space, we can present also in that space similar decomposition of the (power) scattering matrix. In any ONP PP basis B we obtain:

$$K_B = \widetilde{D}_{K,B}^P K_B^{INV} \tag{8.36}$$

with

$$\widetilde{D}_{K,B}^P = \widetilde{D}_K^B \widetilde{D}_B^P = \widetilde{D}_K^B \widetilde{D}_K^P \widetilde{D}_B^K \tag{8.37}$$

All special polarization points are symmetrically located versus the CCS coordinate planes and axes (see Chapter 9 and, e.g., [45], [77]). Of special interest is the eigencircle on the Poincare sphere model of the scattering matrix. It is determined by the crossection of the sphere with a plane through the inversion I point and perpendicular to the rotation after inversion axis. It contains two eigenpolarizations: one of them, situated closer to the inversion point, is repelling, whereas another one is attracting the scattered polarization point. Therefore, the eigenpolarization point which is located farther from the I point is polarimetrically more stable and also corresponds to a greater scattered intensity.

8. 5. Comparison of polarimetric transformations for the two alignments

Polarimetric signatures for SAR objects can be considered as various dependencies of the received part of scattered power on transmit and receive polarizations. Therefore, special polarization points of scattering matrices in the two alignments should be examined.

Change of the alignment has no influence upon the scattered polarization, though components of the PP vector in the reversed spatial coordinate system, being always related to that system, are different. Also the scattered power cannot change with the alignment. This is in agreement with the equality of the inversion matrices for the two alignments in the same ONP PP basis. They depend on the coordinates of the same inversion point and the scattered power is proportional to the square of the distance between the incoming polarization point on the Poincare sphere and the inversion point inside that sphere.

However, *in two different characteristic bases,* corresponding to two Sinclair matrices for the two alignments: A, and A^{xo} as in (8.21), *there are two different inversion matrices* (not only the rotation matrices). It results in two sets of special polarization points of the two Sinclair matrices which differ in mutual positions. Moreover, because of the extra orthogonality transformation that follows the Sinclair scattering in the Jones propagation transformation, physical interpretation of those polarizations should be changed. And so: the CO-POL NULLs become eigenpolarizations, as their antipodal points, which therefore do always exist in the FSA (contrary to what is being observed in the BSA). Also X-POL NULLs (eigen-polarizations) of the A^{xo} matrix in its BSA become cross-polarized versus the incident polarizations in the FSA. And these polarizations, similarly as X-POL NULLs in the BSA, do not always exist in the FSA.

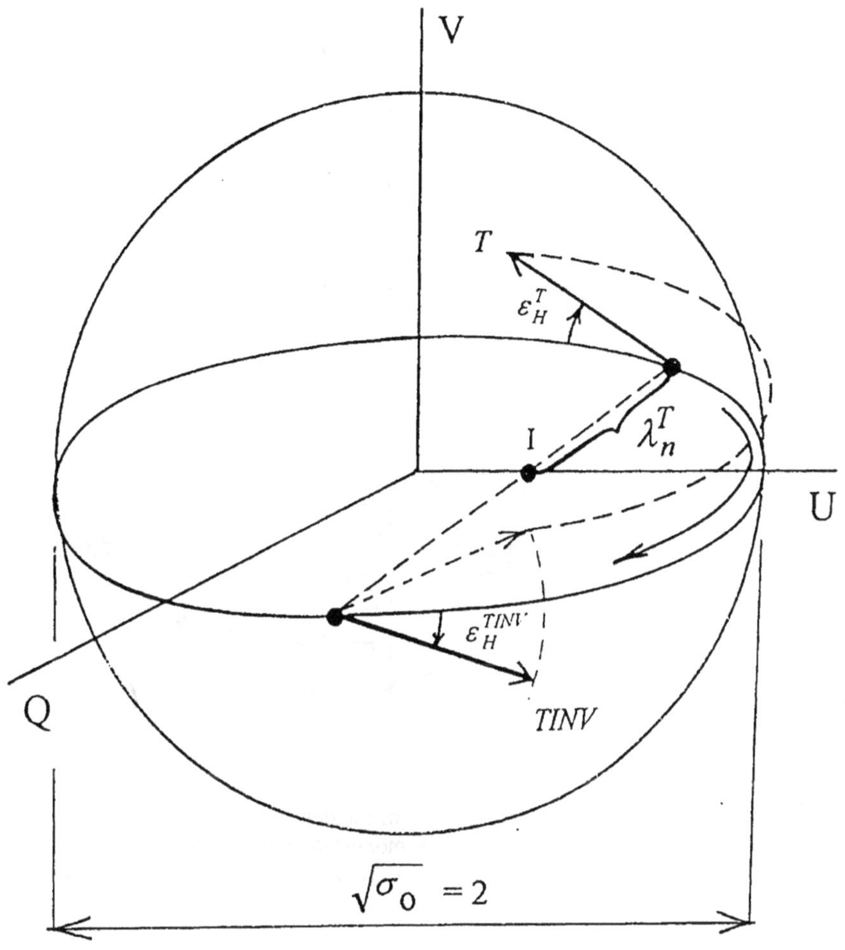

$$A_{nH}^{INV} u_H^T = \begin{bmatrix} \text{-U} & 1 \\ \text{-1} & \text{U} \end{bmatrix}_{nH}^{I} \begin{bmatrix} \cos\gamma_H^T \\ \sin\gamma_H^T \end{bmatrix} \exp\{-j\varepsilon_H^T\}$$

$$= \lambda_n^T \begin{bmatrix} \cos\gamma_H^{TINV} \\ \sin\gamma_H^{TINV} \end{bmatrix} \exp\{j\varepsilon_H^{TINV}\}$$

$$\Rightarrow \varepsilon_H^{TINV} = -\varepsilon_H^T$$

Fig. 8.1. The inversion transformation on the PP sphere of the unit radius in its equatorial plane (for the inversion point on the U_H axis, $U_H^I > 0$ and $2\delta_H^T = 0$).

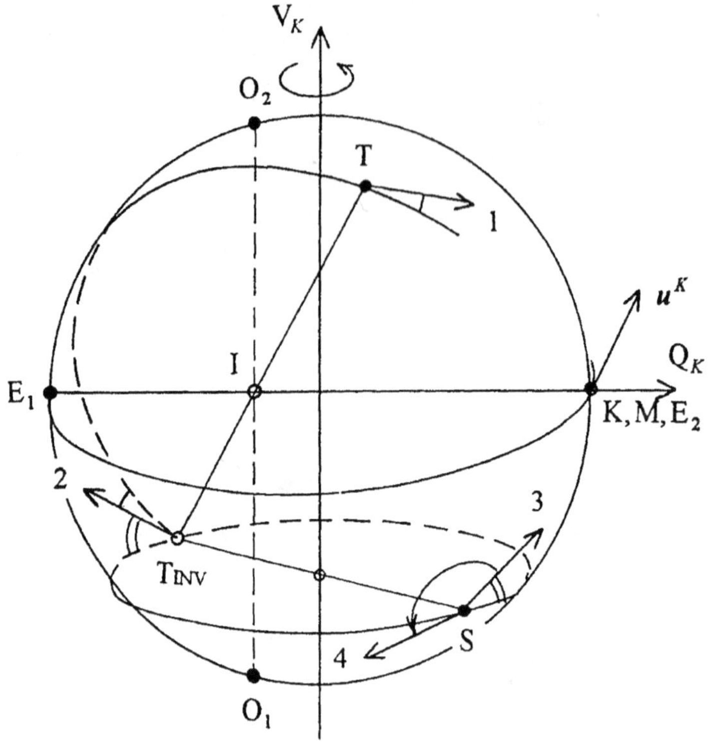

Fig. 8.2. Phasor of incident wave transformation when monostatic scattering:
1 - 2 inversion, 2 - 3 rotation, 3 - 4 change of phase.

9. Poincare Sphere Geometrical Model of the Scattering Matrix

9. 1. Special incident polarization ratios

In terms of canonical parameters (7.21b) , some special incident polarizations of the bistatic scattering matrix can be found in a simple form. The polarization ratio corresponding to the PP tangent phasor T, presented in the K basis, will be denoted by

$$\rho_K^T = \frac{b_K^T}{a_K^T} = \tan\gamma_K^T\, e^{j2\delta_K^T} \tag{9.1}$$

It can be shown (see, e.g., [14] and [45]) that neglecting the phase term ε of the PP vectors of incident waves, some special polarization ratios for those waves can be expressed by solution of a quadratic equation with three complex coefficients R_1, R_2, and R_3:

$$R_2\rho^2 - 2R_1\rho - R_3 = 0 \tag{9.2}$$

Special polarization ratios will then be presented in the form

$$\rho_K^{T_1,T_2} = \frac{R_1 \mp \sqrt{\Delta}}{R_2} \tag{9.3}$$

and the polarization ratios of the orthogonal polarizations by

$$\rho_K^{T_1x,T_2x} = \frac{R_1{}^* \mp \sqrt{\Delta}{}^*}{R_2{}^*} = -\frac{1}{\rho_K^{T_1,T_2}{}^*} \tag{9.4}$$

with

$$\Delta = R_1^2 + R_2R_3 \tag{9.5}$$

Rectangular coordinates of the corresponding polarization points can be found from the equalities (9.6):

$$\begin{aligned} q &= \frac{1-\rho\rho^*}{1+\rho\rho^*} \\ u &= \frac{\rho^*+\rho}{1+\rho\rho^*} \\ v &= j\frac{\rho^*-\rho}{1+\rho\rho^*} \end{aligned} \tag{9.6}$$

Using the above presented formulae with the coefficients given beneath, the polarization ratios (9.3) can be found for the following special incident polarizations T_1 and T_2:

- polarizations M and N, of maximum and minimum scattered power, with the coefficients

$$\begin{aligned} R_1 &= \tfrac{1}{2}\left(A_2^2 - A_1^2\right) \\ R_2 &= -B_1(A_2 - A_1) - jB_2(A_2 + A_1) = R_3{}^*; \end{aligned} \tag{9.7}$$

these polarizations are mutually orthogonal and situated at the end points of Poincare sphere diameter through the inversion point I,

- 'CO-PLO Nulls', O_1 and O_2, producing the orthogonally polarized scattered waves, with

$$R_1 = 0$$
$$R_2 = 2A_1; \quad R_3 = -2A_2 , \tag{9.8}$$

what results in an especially simple formula

$$\rho_K^{O_1,O_2} = \mp j\sqrt{\frac{A_2}{A_1}}; \tag{9.9}$$

these polarizations are represented by points on the Poincare sphere which are in the $Q_K V_K$ plane, symmetrically located against the Q_K axis in its negative part,

- 'eigenpolarizations' E_1 and E_2, with

$$R_1 = \tfrac{1}{2}\left(A_1^2 - A_2^2\right)$$
$$R_2 = B_1(A_2 + A_1) + jB_2(A_2 - A_1) = -R_3{}^* ; \tag{9.10}$$

these polarizations are represented by points symmetrically located against the Q_K =0 plane (they exist only when $\Delta \geq 0$),

- 'mutual' polarizations E_1' and E_2', with

$$R_1 = \tfrac{j}{2}(A_2^2 - A_1^2)$$
$$R_2 = B_2(A_2 - A_1) - jB_1(A_2 + A_1) = R_3{}^* ; \tag{9.11}$$

the points of these polarizations are in the Q_K=0 plane (they exist when eigenpolarizations disappear, and also when $\Delta \geq 0$),

- polarizations K and L, mutually orthogonal, corresponding to maximum and 'saddle' received power when using same polarized transmit and receive antennas, with

$$R_1 = A_2^2 - A_1^2 = \sqrt{\Delta}$$
$$R_2 = 0 = R_3{}^*, \text{ resulting in } \rho_K^{\mathrm{K}} = 0 \text{ and } \rho_K^{\mathrm{L}} = \infty; \tag{9.12}$$

the first polarization ratio, corresponding to the characteristic polarization, can be obtained after resolving the ambiguity.

Some of these, and other special incident polarizations, can be presented by even simpler formulae when using elements of the Kennaugh matrix in its canonical form (in the CCS). Also geometrical model of the scattering matrix may be better understood if presented in terms of those elements. Therefore, it is of interest to arrive at the canonical forms for both Sinclair and Kennaugh matrices.

9. 2. Transformation of the Sinclair matrix from linear to the characteristic basis

The change of basis will be done in two steps. In the first step, polarization only of the basis will be changed by the use of the following matrix (as usually, indices of the matrices are related to their elements and, in this case, to their angular parameters):

$$C_H^{K'} = \begin{bmatrix} e^{-j\delta} & 0 \\ 0 & e^{j\delta} \end{bmatrix}_H^{K'} \begin{bmatrix} \cos\gamma & -\sin\gamma \\ \sin\gamma & \cos\gamma \end{bmatrix}_H^{K'} \begin{bmatrix} e^{j\delta} & 0 \\ 0 & e^{-j\delta} \end{bmatrix}_H^{K'} \tag{9.13}$$

which shifts the basis from H to K' position parallel, preserving its phase (Fig. 9.1). After transformation, the matrix (8.1) will take the form

$$A_{K'} = \tilde{C}_H^{K'} A_H C_H^{K'} = \begin{bmatrix} A_2 & A_3 \\ A_4 & A_1 \end{bmatrix}_{K'} \equiv \begin{bmatrix} A_2' & A_3' \\ -A_3' & A_1' \end{bmatrix} \tag{9.14}$$

In the next step the phase only of the basis will be changed using the matrix

$$C_{K'}^{K}=\begin{bmatrix} e^{-j(\delta+\varepsilon)} & 0 \\ 0 & e^{j(\delta+\varepsilon)} \end{bmatrix}_{H}^{K} \qquad (9.15)$$

to arrive at the final, canonical form (7.21b) of the matrix in its characteristic K basis:

$$A_K=\widetilde{C}_{K'}^{K}A_{K'}C_{K'}^{K}=\begin{bmatrix} A_2 & B_1+jB_2 \\ -B_1-jB_2 & A_1 \end{bmatrix}_{CCS} e^{j\mu} \qquad (9.16)$$

The first step follows the transformation rule proposed in [14], adopted to obtain the result (9.14). The characteristic polarization ratio in the H basis, of the form like (9.3),

$$\rho_H^K\equiv\rho=\frac{R_1-\sqrt{R_1^2+R_2R_2{}^*}}{R_2} \qquad (9.17)$$

will be obtained with:

$$\begin{aligned} R_1 &= A_{2H}A_{2H}{}^*-A_{1H}A_{1H}{}^* \\ R_2 &= -A_{1H}(A_{3H}{}^*+A_{4H}{}^*)-A_{2H}{}^*(A_{3H}+A_{4H}) \end{aligned} \qquad (9.18)$$

The end result of the two steps for $R_2\neq 0$ is as follows:

$$\begin{aligned} A'_2 &= [A_{2H}+(A_{3H}+A_{4H})\rho+A_{1H}\rho^2]/(1+\rho\rho^*) \\ A'_3 &= [-A_{2H}\rho^*+A_{3H}-A_{4H}\rho\rho^*+A_{1H}\rho]/(1+\rho\rho^*) \\ A'_1 &= [A_{2H}\rho^{*2}-(A_{3H}+A_{4H})\rho^*+A_{1H}]/(1+\rho\rho^*) \\ \mu &= \tfrac{1}{2}(\arg A'_2+\arg A'_1) \\ 2\delta_H^K &= \arg\rho \\ 2\gamma_H^K &= 2\operatorname{arc\,tan}|\rho| \\ 2\varepsilon_H^K &= \tfrac{1}{2}(\arg A'_2-\arg A'_1)-2\delta_H^K \\ A_2 &\equiv A_{2CCS}=|A'_2|, \\ A_1 &\equiv A_{1CCS}=|A'_1|, \\ B_1+jB_2 &\equiv (B_1+jB_2)_{CCS}=A'_3\,e^{-j\mu} \end{aligned} \qquad (9.19)$$

What should be observed here is an ambiguity in determination of the canonical phase μ and the rotation angle $2\varepsilon_H^K$, because it is always possible to add 2π to the argument of A_1' or A_2', thus changing the canonical phase and the rotation angle by π. In order to omit such an ambiguity, an additional requirement will be stated:

$$B_2>0,\quad \text{or}\quad B_1\geq 0 \ \text{if}\ B_2=0,\quad \text{with}\quad A_2\geq A_1\geq 0 \qquad (9.20)$$

The above requirement should be satisfied also for $R_2=0$, in which case the whole procedure becomes simpler but direct inspection of the change of basis equation of the form

$$A_K = \tilde{C}_H^K A_H C_H^K = \begin{bmatrix} A_2 & B_1 + jB_2 \\ -B_1 - jB_2 & A_1 \end{bmatrix} e^{j\mu} \tag{9.21}$$

with the rotation matrix (5.30) is necessary.

All formulae remain valid also in the case of symmetrical matrices (monostatic scattering) with an evident simplification: $A_3' = 0$.

9. 3. Poincare sphere geometrical model of the bistatic scattering matrix

Apart from the allowed regions for the inversion point location described in Section C, main parameters of the model are: diameter of the sphere, as well as axis and angle of rotation after inversion.

The best starting point to the model analysis is the power scattering equation in the characteristic K basis

$$\mathsf{K}_K \mathsf{P}_K^T = \sigma^T \mathsf{P}_K^S \tag{9.22}$$

corresponding to the amplitude scattering equation

$$A_K u_K^T = \lambda^T u_K^S * \tag{9.23}$$

with a real coefficient

$$\sigma^T = \left(\lambda^T\right)^2 = \mathrm{a}_1 + \mathrm{b}_1 \mathrm{q}_K^T + \mathrm{b}_3 \mathrm{u}_K^T + \mathrm{b}_5 \mathrm{v}_K^T \tag{9.24}$$

presenting scattered power for the unit incident power.

From the equality (9.24) an essential dependence can be found of the scattered power σ^T on the (IT), the distance from the inversion point I to the incident polarization point T, if the square of diameter of the Poincare sphere model of the scattering matrix will be chosen as (see [45]):

$$\sigma_0 = SpanA + 2|\det A| = 2(\mathrm{a}_1 + \mathrm{a}_0) \tag{9.25}$$

where

$$\mathrm{a}_0 = |\det A| = \sqrt{\mathrm{a}_1^2 - \mathrm{b}_0^2} \tag{9.26}$$

and

$$\mathrm{b}_0^2 = \mathrm{b}_1^2 + \mathrm{b}_3^2 + \mathrm{b}_5^2 \tag{9.27}$$

At first we observe that from (9.26) - (9.27) we obtain

$$\mathrm{a}_1 = \tfrac{1}{2} SpanA = \frac{\sigma_0}{4} + \frac{\mathrm{b}_0^2}{\sigma_0} \tag{9.28}$$

Then, for the coordinates of the I point inside the sphere of the radius equal to $\sqrt{\sigma_0}/2$,

$$\begin{bmatrix} Q \\ U \\ V \end{bmatrix}_{K;\, 2r_0 = \sqrt{\sigma_0}}^{I} = \frac{-1}{\sqrt{\sigma_0}} \begin{bmatrix} \mathrm{b}_1 \\ \mathrm{b}_3 \\ \mathrm{b}_5 \end{bmatrix}_K \tag{9.29}$$

we have the square of the (IT) distance

$$
\begin{aligned}
(\mathrm{IT})^2 &= \left(\frac{\sqrt{\sigma_0}}{2}\mathrm{q}_K^{\mathrm{T}} - \mathrm{Q}_K^{\mathrm{I}}\right)^2 + \left(\frac{\sqrt{\sigma_0}}{2}\mathrm{u}_K^{\mathrm{T}} - \mathrm{U}_K^{\mathrm{I}}\right)^2 \\
&+ \left(\frac{\sqrt{\sigma_0}}{2}\mathrm{v}_K^{\mathrm{T}} - \mathrm{V}_K^{\mathrm{I}}\right)^2 \\
&= \sigma^{\mathrm{T}}
\end{aligned}
\qquad (9.30)
$$

according to (9.24).

This is exactly the result known for the monostatic scattering when the diameter of the sphere is equal to the trace $A_2 + A_1$ of the scattering matrix in its real diagonal form. However, it should be observed that such a trace is also equal to the square root of (9.25). Hence, the same formula (9.30) is applicable in both monostatic and bistatic scattering, what was also shown in [45].

9. 4. The allowed region for the inversion point inside the Poincare sphere model of the scattering matrix

There is an allowed region inside the Poincare sphere in which the inversion point I can be located (outside that region, elements of the Kennaugh matrix would become complex). The permitted coordinates of that point inside the sphere of unit radius, in the CCS, are in the ranges:

$$
\begin{gathered}
-1 \le \mathrm{Q} \le 0 \\
-\sqrt{-\mathrm{Q}-\mathrm{Q}^2} \le \mathrm{U} \le \sqrt{-\mathrm{Q}-\mathrm{Q}^2} \\
0 \le \mathrm{V} \le \begin{cases} \dfrac{1}{-\mathrm{Q}}\left[\sqrt{\left(\mathrm{Q}^2+\mathrm{U}^2\right)\left(1-\mathrm{Q}^2\right)} - |\mathrm{U}|\right] & \text{for } \mathrm{V} \ge |\mathrm{U}| \\ \sqrt{-\mathrm{Q}-\mathrm{Q}^2-\mathrm{U}^2} & \text{for } \mathrm{V} \le |\mathrm{U}| \end{cases}
\end{gathered}
\qquad (9.31)
$$

One part of that allowed region is inside an upper half (see Fig.9.2) of a 'small sphere' of radius equal to 1/2, having its diameter coinciding with the Poincare sphere radius directed along the negative part of the OQ_K axis. Moreover, there is another part of that allowed region which is above the small sphere but below a boundary surface formed by the hyperbolic curves, in Q= const planes, determined by the limiting V values of the upper part of the last equality in (9.31) with the I point coordinates as in (9.32),

$$
\begin{bmatrix} \mathrm{Q} \\ \mathrm{U} \\ \mathrm{V} \end{bmatrix}^{\mathrm{I}}_{K;\, r_0=1} = \frac{-2}{\sigma_0}\begin{bmatrix} \mathrm{b}_1 \\ \mathrm{b}_3 \\ \mathrm{b}_5 \end{bmatrix}_K = \frac{1}{\sigma_0}\begin{bmatrix} \left(A_1^2 - A_2^2\right) \le 0 \\ 2B_1\left(A_1 - A_2\right) \\ 2B_2\left(A_2 + A_1\right) \ge 0 \end{bmatrix} \qquad (9.32)
$$

These curves are tangential to the small sphere in V = |U| points.

The just mentioned boundary surface corresponds to such Sinclair matrices for which the real part of determinant of their canonical form (9.16) with $\mu = 0$ is equal to zero:

$$
R = A_2 A_1 + B_1^2 - B_2^2 = 0 \qquad (9.33)
$$

There are two branches of that boundary surface, corresponding to plus and minus U values. They are crossing along a quarter of the Poincare sphere great circle U = 0 (for Q < 0 and V > 0). For I point on the Poincare sphere surface the whole determinant of the scattering matrix is zero.

When Q = 0, the allowed region is on the positive V semiaxis only. When both U and V equal zero, i.e. the inversion point is situated on the Q_K axis (in its negative part), then the Sinclair matrix is symmetric, corresponding to monostatic scattering.

9. 5. Reconstruction of the amplitude scattering matrix for given coordinates of the inversion point in the CCS

Having coordinates Q, U, and V of the inversion point I in the allowed regions inside the Poincare sphere of unit radius, in the CCS, it is possible to reconstruct the whole Sinclair scattering matrix of strength k in the characteristic K basis. Elements of the matrix will be expressed by the use of some auxiliary parameters:

$$S = 2(a_1 + |R|) = k^2 S_n; \qquad k = \frac{\sqrt{\sigma_0}}{2}, \tag{9.34}$$

$$S_n = 2\left(1 + Q^2 + U^2 + V^2 + \sqrt{\left[1 - \left(Q^2 + U^2 + V^2\right)\right]^2 - \left(2UV/Q\right)^2}\right) \tag{9.35}$$

and

$$t = U/Q = b_3/b_1 \tag{9.36}$$

or

$$e = -Q/V = -b_1/b_5 \geq 0 \tag{9.37}$$

There are two solutions possible: I and II. Solution I is for the whole allowed region. It depends on the t parameter:

$$A_{2,1} = \frac{b_1 S \pm 2\left(b_1^2 + b_3^2\right)}{2\sqrt{S}\sqrt{b_1^2 + b_3^2}} = k\frac{S_n \mp 4Q\left(1 + t^2\right)}{2\sqrt{S_n}\sqrt{1 + t^2}} \geq 0 \tag{9.38a}$$

$$B_1 = \frac{b_3}{2}\sqrt{\frac{S}{b_1^2 + b_3^2}} = k\frac{t}{2}\sqrt{\frac{S_n}{1 + t^2}} \tag{9.38b}$$

$$B_2 = \frac{-b_5}{b_1}\sqrt{\frac{b_1^2 + b_3^2}{S}} = k2V\sqrt{\frac{1 + t^2}{S_n}} \geq 0 \tag{9.38c}$$

Solution II can be obtained only for a part of the allowed region, above the small sphere, i.e. when $V > |U|$ and $V^2 > -Q - Q^2 - U^2$. So, in that part of the region, both solutions exist. Solution II will be expressed in terms of the e parameter:

$$A_{2,1} = \frac{2\left(b_1^2 + b_5^2\right) \pm b_1 S}{2\sqrt{S}\sqrt{b_1^2 + b_5^2}} = k\frac{4V\left(e^2 + 1\right) \pm eS_n}{2\sqrt{S_n}\sqrt{e^2 + 1}} \geq 0 \tag{9.39a}$$

$$B_1 = \frac{b_3}{b_1}\sqrt{\frac{b_1^2 + b_5^2}{S}} = k(-U)\frac{2}{e}\sqrt{\frac{e^2 + 1}{S_n}} \tag{9.39b}$$

$$B_2 = \frac{-b_5}{2}\sqrt{\frac{S}{b_1^2 + b_5^2}} = k\frac{1}{2}\sqrt{\frac{S_n}{e^2 + 1}} \geq 0 \tag{9.39c}$$

In terms of canonical elements, solution I is when

$$R = A_2 A_1 + B_1^2 - B_2^2 \geq 0 \quad \text{or} \quad S = (A_2 + A_1)^2 + 4B_1^2 \geq 4k^2|Q|(1 + t^2) \tag{9.40a}$$

and solution II is when

$$R = A_2 A_1 + B_1^2 - B_2^2 \le 0 \quad \text{or} \quad S = (A_2 - A_1)^2 + 4B_2^2 \le 4k^2 \mathrm{V}(e^2+1)/e \tag{9.40b}$$

In case of the equality, common solution, I and II, exists.

9. 6. Reconstruction of the canonical Kennaugh matrix from elements of its first row

The Kennaugh matrix in canonical form depends on four real parameters only, as its Sinclair counterpart does when neglecting the phase factor. Therefore, having four elements of its first row it is possible to restore the whole matrix. Again, as in the case of reconstruction of the Sinclair matrix from coordinates of the inversion point I, two solutions are possible. This is because of the dependence of three first row elements on these coordinates:

$$\begin{bmatrix} \mathrm{b}_1 \\ \mathrm{b}_3 \\ \mathrm{b}_5 \end{bmatrix}_K = -\frac{\sigma_o}{2} \begin{bmatrix} \mathrm{Q} \\ \mathrm{U} \\ \mathrm{V} \end{bmatrix}^{\mathrm{I}}_{K;\, r_0=1} = -2k^2 \begin{bmatrix} \mathrm{Q} \\ \mathrm{U} \\ \mathrm{V} \end{bmatrix}^{\mathrm{I}}_{K;\, r_0=1} \tag{9.41}$$

Introducing magnitude of the real part of determinant of the canonical Sinclair matrix with $\mu = 0$ in the form

$$|R| = |A_2 A_1 + B_1^2 - B_2^2| = \sqrt{\mathrm{a}_1^2 - \frac{1}{\mathrm{b}_1^2}\left(\mathrm{b}_1^2 + \mathrm{b}_3^2\right)\left(\mathrm{b}_1^2 + \mathrm{b}_5^2\right)} \tag{9.42}$$

and using equalities (7.27) we obtain:

$$\mathrm{a}_2 = \frac{\mathrm{a}_1\left(\mathrm{b}_1^4 - \mathrm{b}_3^2\mathrm{b}_5^2\right) \pm \mathrm{b}_1^2\left(\mathrm{b}_5^2 - \mathrm{b}_3^2\right)|R|}{\left(\mathrm{b}_1^2 + \mathrm{b}_3^2\right)\left(\mathrm{b}_1^2 + \mathrm{b}_5^2\right)} = \frac{a_1(e^2 - t^2) \pm (1 - e^2 t^2)|R|}{(1+t^2)(e^2+1)} \tag{9.43a}$$

$$\mathrm{a}_3 = \frac{-\mathrm{a}_1\mathrm{b}_3^2 \pm \mathrm{b}_1^2|R|}{\mathrm{b}_1^2 + \mathrm{b}_3^2} = \frac{-a_1 t^2 \pm |R|}{1+t^2} \tag{9.43b}$$

$$\mathrm{a}_4 = \frac{-\mathrm{a}_1\mathrm{b}_5^2 \mp \mathrm{b}_1^2|R|}{\mathrm{b}_1^2 + \mathrm{b}_5^2} = \frac{-a_1 \mp e^2|R|}{e^2+1} \tag{9.43c}$$

$$\mathrm{b}_4 = \mathrm{b}_1\mathrm{b}_3 \frac{\mathrm{a}_1 \pm |R|}{\mathrm{b}_1^2 + \mathrm{b}_3^2} = t\frac{a_1 \pm |R|}{1+t^2} \tag{9.43d}$$

$$\mathrm{b}_6 = \mathrm{b}_1\mathrm{b}_5 \frac{\mathrm{a}_1 \mp |R|}{\mathrm{b}_1^2 + \mathrm{b}_5^2} = -e\frac{\mathrm{a}_1 \mp |R|}{e^2+1} \quad ; \quad \mathrm{b}_2 = 0 \tag{9.43e}$$

The double signs in the above equalities (9.43) correspond to solutions I and II, respectively.

9. 7. Rotation after inversion axis and angle

Having the CCS coordinates of a P phasor

$$
\begin{aligned}
q_K^P &= \cos 2\gamma_K^P = \frac{a_3 a_4 - a_0 a_2}{a_0(a_1 + a_0)} \\
u_K^P &= \sin 2\gamma_K^P \cos 2\delta_K^P = \frac{b_4(a_4 - a_0)}{a_0(a_1 + a_0)} \\
v_K^P &= \sin 2\gamma_K^P \sin 2\delta_K^P = \frac{b_6(a_3 - a_0)}{a_0(a_1 + a_0)} \geq 0
\end{aligned}
\tag{9.44}
$$

which rotated after inversion takes K position, we can find in terms of elements (96) direction of the rotation axis as a ratio of its unit vector components (n_2 along the U_K axis and n_3 along V_K , $n_2^2 + n_3^2 = 1$):

$$
\tan 2\delta_K^P = -\frac{n_2}{n_3} = -\frac{a_1 \mp |R|}{a_1 \pm |R|} \cdot \frac{2t + (1/t)(a_0 \mp |R|)}{(2/e) + e(a_0 \pm |R|)}
$$

$$
= \frac{b_6}{b_4} \cdot \frac{a_3 - a_o}{a_4 - a_o} \; ; \; n_3 \geq 0, \tag{9.45}
$$

with

$$
\left.\begin{aligned}
n_2 > 0, \quad & 0 < 2\phi \leq \pi \text{ and } 2\phi = 2\gamma_K^P && \text{for } B_1 \geq 0, \text{ i.e. } t \geq 0 \\
n_2 < 0, \quad & \pi < 2\phi < 2\pi \text{ and } 2\phi = 2\pi - 2\gamma_K^P && \text{for } B_1 < 0, \text{ i.e. } t < 0
\end{aligned}\right\} (0 \leq 2\gamma_K^P \leq \pi)
$$

and the 2ϕ angle of rotation about that axis

$$
\begin{aligned}
\cos 2\gamma_K^P &= \cos 2\phi \\
&= \frac{2a_1 t^2 - e^2(|R|^2 + a_1 a_0)(4/\sigma_o) \mp 2(1 - e^2 t^2)|R|}{2a_0(1 + t^2)(e^2 + 1)} \\
&= \frac{a_3 a_4 - a_0 a_2}{a_0(a_1 + a_0)} = \frac{2b_1^2 - a_2 \sigma_o}{a_o \sigma_o}
\end{aligned}
\tag{9.46}
$$

These formulae are very useful when considering special cases of the I point location at the boundaries of its allowed region.

9. 8. Special polarization points in terms of canonical Kennaugh matrix elements

From (9.24) it is also immediately seen that maximum and minimum scattered power correspond to the following normalized Stokes parameters of the incident wave:

$$
\begin{bmatrix} q \\ u \\ v \end{bmatrix}_K^{M,N} = \frac{\pm 1}{b_o} \begin{bmatrix} b_1 \\ b_3 \\ b_5 \end{bmatrix} \tag{9.47}
$$

Hence, values of those maximum and minimum scattered powers are

$$\sigma^{M,N} = a_1 \pm b_o \tag{9.48}$$

Equations (24)-(27) are in their form independent of the polarization basis, but in the K basis there is the following worth noticing dependence between polarization points of illuminating (M,N) and the corresponding scattered (M', N') waves

$$\begin{bmatrix} q \\ u \\ v \end{bmatrix}_K^{M',N'} = \begin{bmatrix} q \\ -u \\ -v \end{bmatrix}_K^{M,N} \tag{9.49}$$

Coordinates of the CO-POL Null points are

$$\begin{bmatrix} q \\ u \\ v \end{bmatrix}_K^{O_1,O_2} = \begin{bmatrix} -b_3/b_4 \\ 0 \\ \mp\sqrt{\dfrac{-a_4 + a_3}{a_1 + a_3}} \end{bmatrix} = \begin{bmatrix} \dfrac{A_1 - A_2}{A_1 + A_2} \\ 0 \\ \dfrac{\mp 2\sqrt{A_1 A_2}}{A_1 + A_2} \end{bmatrix} \tag{9.50}$$

Scattered wave polarizations are orthogonal, and scattered powers

$$\sigma^{O_1,O_2} = -a_4 \mp b_5 \sqrt{\frac{-a_4 + a_3}{a_1 + a_3}} = B_1^2 + \left(B_2 \pm \sqrt{A_1 A_2}\right)^2 \tag{9.51}$$

Coordinates of eigenpolarizations are

$$\begin{bmatrix} q \\ u \\ v \end{bmatrix}_K^{E_1,E_2} = \frac{1}{b_1} \begin{bmatrix} \mp\sqrt{b_1^2 - b_4^2 - b_6^2} \\ -b_4 \\ -b_6 \end{bmatrix} \tag{9.52}$$

These polarizations can exist only when $b_1^2 \geq b_4^2 + b_6^2$. The scattered powers are

$$\sigma^{E_1,E_2} = a_2 \mp \sqrt{b_1^2 - b_4^2 - b_6^2} \tag{9.53}$$

When eigenpolarizations do not exist, then mutual polarizations points appear of coordinates

$$\begin{bmatrix} q \\ u \\ v \end{bmatrix}_K^{E_1',E_2'} = \frac{1}{b_4^2 + b_6^2} \begin{bmatrix} 0 \\ -b_1 b_4 \pm b_6\sqrt{-b_1^2 + b_4^2 + b_6^2} \\ -b_1 b_6 \mp b_4\sqrt{-b_1^2 + b_4^2 + b_6^2} \end{bmatrix} = \begin{bmatrix} 0 \\ \cos 2\delta^{E_1',E_2'} \\ \sin 2\delta^{E_1',E_2'} \end{bmatrix} \tag{9.54}$$

and the corresponding scattered powers are

$$\sigma^{E_1',E_2'} = \left|A_2 + (B_1 + jB_2)\exp(j2\delta^{E_1',E_2'})\right|^2 \tag{9.55}$$

with

$$A_2 = \sqrt{\tfrac{1}{2}(a_1 + a_2) + b_1},$$
$$B_1 + jB_2 = [b_3 + b_4 - j(b_5 + b_6)]/(2A_2),$$
$$A_1 = (a_3 - a_4)/(2A_2). \tag{9.56}$$

Polarization points of the characteristic basis have components

$$\begin{bmatrix} q \\ u \\ v \end{bmatrix}_K^{K,L} = \begin{bmatrix} \pm 1 \\ 0 \\ 0 \end{bmatrix} \tag{9.57}$$

corresponding to scattered powers

$$\sigma^{K,L} = a_1 \pm b_1 . \tag{9.58}$$

Simple geometrical constructions can be presented indicating special polarization points in the CCS for the inversion points given (see [45] and [46]).

9. 9. Scattering matrix synthesis for special polarizations given

Such a synthesis can easily be performed in the characteristic coordinate system in which determination of special polarization points requires minimum parameters to be specified [55].

For example, only three parameters are necessary to determine: CO-POL nulls (one parameter), and polarization point M of maximum scattered power (two parameters). Choosing these parameters: $q^O < 0,\ u^M,\ v^M \le 0$, we arrive at the desired Sinclair and Kennaugh matrices:

$$A_K = \sqrt{\frac{-q^O b_o}{2q^M}} \begin{bmatrix} \dfrac{q^M}{-q^O}(1 - q^O) & \dfrac{u^M}{-q^O} - jv^M \\ \dfrac{u^M}{q^O} + jv^M & \dfrac{q^M}{-q^O}(1 + q^O) \end{bmatrix} e^{j\mu} \tag{9.59}$$

and K_K of the form (7.20b) with elements:

$$a_2 = -q^M b_o \left(q^O + \frac{1}{q^O} \right) - a_1$$

$$a_3 = \frac{q^M b_o}{-q^O} - a_1$$

$$a_4 = -q^M q^O b_o - a_1 \tag{9.60}$$

$$b_1 = q^M b_o$$
$$b_3 = u^M b_o$$
$$b_4 = \frac{u^M b_o}{-q^O}$$

$$b_5 = v^M b_o$$
$$b_6 = -v^M q^O b_o$$

where

$$a_1 = \frac{\sigma_o}{4} + \frac{b_o^2}{\sigma_o}$$
$$b_o = \frac{\sigma_o}{2} \cdot \frac{\sqrt{(1-r^2)^2 + 4(1+r^2)c - 8rd}}{1 + r^2 + 2(c+t)}$$
$$r = \frac{1 - q^O}{1 + q^O}$$
$$c = d_1^2 + d_2^2$$
$$d = d_1^2 - d_2^2$$
$$d_1 = u^M \frac{1+r}{2q^M}$$
$$d_2 = v^M \frac{1-r}{2q^M}$$
$$t = \sqrt{r^2 + c^2 + 2rd}$$

The same parameters can serve to compute the angle of rotation after inversion and the direction of the rotation axis:

$$\cos 2\phi = \frac{c^2 - r^2 + (2c - 1 - r^2)(t/2)}{t^2 + (2c + 1 + r^2)(t/2)}$$
$$\frac{n_2}{n_3} = \frac{v^M (q^O)^2}{-u^M} \cdot \frac{c + t - r}{c + t + r} \qquad (9.61)$$

Worth noticing is the following dependence for parameters of eigenpolarizations in the CCS if they exist:

$$u^E = u^M / (q^M q^O), \qquad v^E = v^M q^O / q^M. \qquad (9.62)$$

9. 10. Geometrical model of the Jones propagation matrix.

One of possibilities of the Jones matrix analysis is its presentation as a product of the Lorenz and rotation matrices. The Lorenz matrix can be treated as the inversion followed by the orthogonality transformation, i.e. inverse transformation to the inversion through the center of the Poincare sphere. Moreover, the Poincare sphere can be replaced by Wanielik's ellipsoid ([45], [138]) in order to present magnitudes of scattered powers proportional to the distance from one focus of the ellipsoid (coinciding with the center of the auxiliary polarization sphere) to its surface.

However, in order to obtain a geometrical model of the Jones propagation matrix in the form most convenient for presentation of special incident polarization points, the reasonable solution is to use an S1 type scattering matrix defined in the linear H basis as

$$A_H^{xo} = C_H^o\, C^x A_H = \begin{bmatrix} A_4 & A_1 \\ A_2 & A_3 \end{bmatrix}_H \qquad (9.63)$$

followed by the orthogonality transformation. Then, instead of the known expression for the Jones matrix in the H basis:

$$A_H^{\circ} = C_H^{\circ} A_H = \begin{bmatrix} -1 & 0 \\ 0 & 1 \end{bmatrix} \begin{bmatrix} A_2 & A_3 \\ A_4 & A_1 \end{bmatrix}_H \tag{9.64}$$

one obtains

$$A_H^{\circ} = C^{\times} A_H^{\times\circ} = \begin{bmatrix} 0 & -1 \\ 1 & 0 \end{bmatrix} \begin{bmatrix} A_4 & A_1 \\ A_2 & A_3 \end{bmatrix}_H \tag{9.65}$$

What has to be observed is a straightforward relationship between the Sinclair matrices in the two above equations with the rows simply interchanged.

Of course, the Poincare sphere models of these two Sinclair matrices differ. Only the sphere diameters are equal to each other. For example, the fork angles differ, what can be seen immediately when inspecting the corresponding null-polarization ratios: for the A_H matrix,

$$\rho^{O_{1,2}} = \frac{-(A_3 + A_4) \mp \sqrt{(A_3 + A_4)^2 - 4A_1A_2}}{2A_1} \tag{9.66}$$

and for the $A_H^{\times\circ}$ matrix,

$$\rho^{O_{1,2}^{\times\circ}} = \frac{-(A_1 + A_2) \mp \sqrt{(A_1 + A_2)^2 - 4A_3A_4}}{2A_3} \tag{9.67}$$

The fork angle of the first matrix becomes zero when its two null-polarizations coincide for

$$(A_3 + A_4)^2 = 4A_1A_2 \tag{9.68}$$

and then, in general,

$$(A_1 + A_2) \neq 4A_3A_4 \tag{9.69}$$

what means that the fork angle of the second matrix, with fork's prongs pointing to two different null-polarizations, is different from zero.

A simple example can show how the two models depend on each other. Consider the free space scattering matrix of the well known form:

$$A_H = C^{\circ} = \begin{bmatrix} -1 & 0 \\ 0 & 1 \end{bmatrix} \tag{9.70}$$

Then

$$A_H^{\times\circ} = \begin{bmatrix} 0 & 1 \\ -1 & 0 \end{bmatrix} = \widetilde{C}^{\times} \tag{9.71}$$

and the corresponding Jones propagation matrix of the free space becomes an identity:

$$A_H^{\circ} = C^{\times}\widetilde{C}^{\times} = \begin{bmatrix} 1 & 0 \\ 0 & 1 \end{bmatrix} \tag{9.72}$$

what ought to be expected.

Now, an analysis of special polarization points of a propagation matrix becomes very simple, as for the corresponding scattering matrix. The structure of the new scattering matrix $A^{xo}_{K^o}$ in its characteristic K^o basis, usually different from the K basis of the previous scattering matrix, is exactly the same as of the A_K matrix. The only difference is in the physical meaning of special polarization points because of an additional orthogonality transformation for the propagation matrix. So, null-polarizations of the $A^{xo}_{K^o}$ matrix become eigenpolarizations of the propagation matrix, and vice versa. Therefore, eigenpolarizations always exist in the FSA and CO-POL nulls may not appear, contrary to what is being observed in the BSA. Only incident polarizations for the maximum forward scattering (or transferred) waves remain unchanged, though they take different positions in the characteristic coordinate systems (CCSs') corresponding to the K and K^o phasors. Mutual positions of these two phasors can be found from the equality

$$u_K^{K^o} = \begin{bmatrix} a \\ b \end{bmatrix}_K^{K^o} = \begin{bmatrix} a^* & b^* \\ -b & a \end{bmatrix}_H^K \begin{bmatrix} a \\ b \end{bmatrix}_H^{K^o} = \tilde{C}_H^K * u_H^{K^o} \tag{9.73}$$

expressed in terms of the Cayley-Klein parameters:

$$\begin{aligned} a &= \cos\gamma\, e^{-j(\delta+\varepsilon)} \\ b &= \sin\gamma\, e^{j(\delta-\varepsilon)} \end{aligned} \tag{9.74}$$

for halves of the known 2γ, 2δ, 2ε Euler angles of the two phasors in the H basis.

The A^{xo} matrices represent classical examples of the bistatic scattering matrices and their geometrical models can appear especially useful when considering synthesis of forward scattering matrices for desired polarimetric properties.

9.11. Concluding Remarks

A PP vector approach to the theory of coherent bistatic radar polarimetry has been based on application of matrix calculus in the two-dimensional complex space of polarization and phase vectors. Owing to that approach it was possible to obtain simple canonical forms of bistatic scattering matrices and their Poincare sphere geometrical models. Such models, demonstrating the way of polarization and phase transformation when scattering, may become useful in various practical applications like target recognition and classification, or polarimetric analysis of microwave networks. They can also be used to synthesize scattering or propagation matrices of desired polarimetric properties.

Introduction of several new concepts appeared useful in application of matrix calculus. There were concepts of:

- moving helix model of the monochromatic EM plane wave and its spatial phase,
- conjugate PP vectors representing CAs' of waves propagating in „-z direction",
- polarization sphere of tangential phasors (representing the PP vectors) as a 2-folded Riemann surface,
- addition and multiplication of phasors,
- the ONP polarization basis defined versus local spatial coordinate system with its 'horizontal' reference plane,
- polarization phasor notation presenting the unit column PP vectors and matrices in the ONP bases,
- the 'spatial reversal' as another, after rotation, the change of basis operation ,
- the characteristic basis of a bistatic scattering matrix or its CCS,
- decomposition of a bistatic scattering matrix into product of an inversion and rotation matrices,
- 3 coordinates of the inversion point and three basis rotation angles as two kinds of parameters specifying geometrical models of scattering matrices,
- two scattering and two propagation types of amplitude matrices,
- geometrical models of propagation matrices without employing the Lorentz transformation.

It is hoped that here presented theory will serve as an extension of the pioneering fundamental work of Huynen [85] from mono- to bistatic scattering, following suggestions contained in short communication published by Kennaugh [96] whose fundamental concept of the inversion point was an inspiration for this author to further developing the Poincare sphere models of scattering matrices.

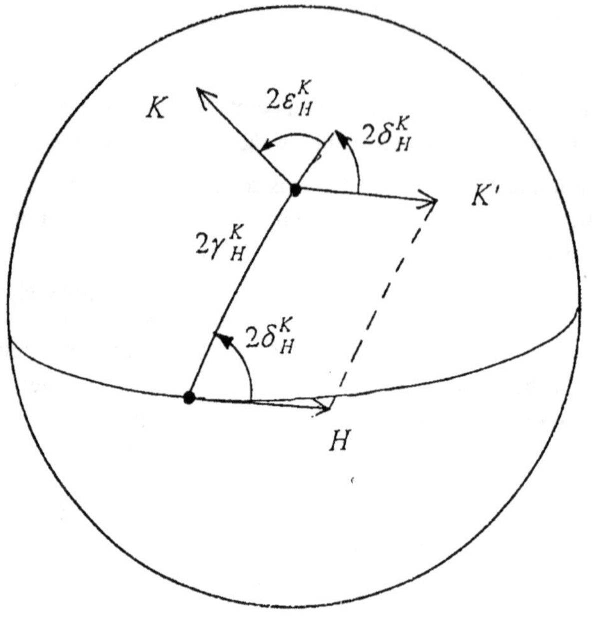

Fig. 9.1. To the two step procedure of obtaining the characteristic K basis: $H \rightarrow K'$ (change of polarization), and $K' \rightarrow K$ (change of phase).

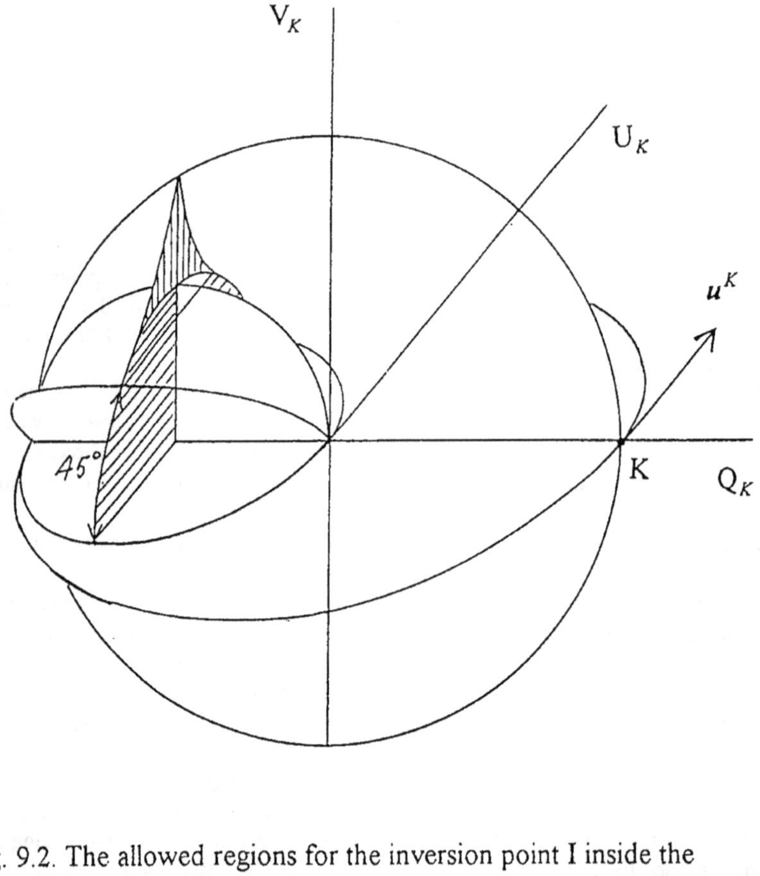

Fig. 9.2. The allowed regions for the inversion point I inside the Poincare sphere. Shadowed areas are in the crossection of those regions by a plane perpendicular to the Q_K axis.

10. Special Polarizations of the Bistatic Scattering Matrix in an Arbitrary ONP PP basis

10.1. Polarizations M and N of the maximum and minimum scattered powers

For the scattering equation

$$\mathrm{K}_H \mathrm{P}_{0H}^T = \begin{bmatrix} a_1 & b_1 & b_3 & b_5 \\ c_1 & a_2 & b_4 & b_6 \\ c_3 & c_4 & a_3 & b_2 \\ c_5 & c_6 & c_2 & a_4 \end{bmatrix}_H \begin{bmatrix} 1 \\ q^T \\ u^T \\ v^T \end{bmatrix}_H = \sigma^T \begin{bmatrix} 1 \\ q^S \\ u^S \\ v^S \end{bmatrix}_H = \sigma^T \mathrm{P}_{0H}^S . \tag{10.1}$$

the *normalized* total scattered power (corresponding to the unit incident total power) is

$$\sigma^T = a_1 + b_{1H} q_H^T + b_{3H} u_H^T + b_{5H} v_H^T \tag{10.2}$$

Consider the last three terms in (10.2) as the scalar product of two vectors:

- one with components q_H^T, u_H^T, v_H^T, of the unit magnitude, and
- another one with components b_{1H}, b_{3H}, b_{5H}, of the magnitude (compare (E.6g))

$$b_0 = (\sqrt{b_1^2 + b_3^2 + b_5^2})_H = (\sqrt{c_1^2 + c_3^2 + c_5^2})_H \tag{10.3}$$

One can see immediately that maximum and minimum scattered powers correspond to the following *full*, unit Stokes four-vectors (with the first total power component equal to unity in case of the completely polarized wave),

$$\mathrm{P}_{0H}^{\mathrm{M,N}} = \begin{bmatrix} 1 \\ q^{\mathrm{M,N}} \\ u^{\mathrm{M,N}} \\ v^{\mathrm{M,N}} \end{bmatrix}_H = \frac{\pm 1}{b_0} \begin{bmatrix} b_0 \\ b_1 \\ b_3 \\ b_5 \end{bmatrix}_H \tag{10.4}$$

The corresponding (also full, unit) four-vectors of the scattered waves can be found when introducing (10.4) to (10.1) as illumination, and making use of second equalities of each set, (E.6g), (E.6a), (E.6f), and (E.6d), for the successive components of the resulting four-vector, thus obtaining:

$$\mathrm{P}_{0H}^{\mathrm{M",N"}} = \begin{bmatrix} 1 \\ q^{\mathrm{M",N"}} \\ u^{\mathrm{M",N"}} \\ v^{\mathrm{M",N"}} \end{bmatrix}_H = \frac{\pm 1}{b_0} \begin{bmatrix} b_0 \\ c_1 \\ c_3 \\ c_5 \end{bmatrix}_H \tag{10.5}$$

Maximum and minimum normalized total scattered powers are then:

$$\begin{aligned} \sigma^{\mathrm{M}} &= a_1 + b_0 = \sigma^{\max} \\ \sigma^{\mathrm{N}} &= a_1 - b_0 = \sigma^{\min} \end{aligned} \tag{10.6}$$

The last equality undoubtedly suggest the necessary condition for physical realizability of the Kennaugh matrix in (10.1):

$$a_1 \geq b_0 . \tag{10.7}$$

Complex polarization ratios for those four-vectors are:

$$\rho_H^{M,N} = \frac{b_{3H} + jb_{5H}}{b_{1H} \pm b_0} \quad \text{and} \quad \rho_H^{M",N"} = \frac{c_{3H} + jc_{5H}}{c_{1H} \pm b_0} \tag{10.8}$$

what can be found by applying fundamental formulae (derived in Appendix O) for mutually orthogonal polarizations:

$$\rho = \frac{u + jv}{1+q} \quad \text{and} \quad \rho^X = \frac{-1}{\rho^*} = \frac{u + jv}{1-q}, \quad \text{for} \quad q^2 + u^2 + v^2 = 1. \tag{10.9}$$

Complex ratios for M and N polarizations can be expressed also in terms of the Sinclair matrix elements when applying to (10.8), for instance, some formulae from sets (E.4) and (E.5a-e). However those expressions are not such simple.

10.2. CO-POL nulls, O_1 and O_2

In this case, it is more convenient to use the Sinclair matrix elements and to solve the quadratic versus ρ transmission equation in any ONP PP basis H of the form:

$$\begin{bmatrix} 1 & \rho_H^{O_1,O_2} \end{bmatrix}_H \begin{bmatrix} A_2 & A_3 \\ A_4 & A_1 \end{bmatrix}_H \begin{bmatrix} 1 \\ \rho_H^{O_1,O_2} \end{bmatrix}_H = 0 \tag{10.10}$$

obtaining the simple end result,

$$\rho_H^{O_1,O_2} = \frac{-(A_{3H} + A_{4H}) \mp \sqrt{(A_{3H} + A_{4H})^2 - 4A_{1H}A_{2H}}}{2A_{1H}}. \tag{10.11}$$

Expressions for unit Stokes vector coordinates of those points are much more complex and will be not presented here. They can be found when applying formulae inverted versus (10.9), namely:

$$\begin{bmatrix} q \\ u \\ v \end{bmatrix} = \frac{1}{1+\rho\rho^*} \begin{bmatrix} 1-\rho\rho^* \\ \rho^* + \rho \\ j(\rho^* - \rho) \end{bmatrix}. \tag{10.12}$$

Also the normalized total scattered powers is much more convenient to present in terms of the Sinclair (or Kennaugh) matrix elements in another, characteristic ONP PP basis K, as in (9.51).

10.3. Eigenpolarizations, E_1 and E_2 and their spatial phases

Also eigenpolarizations, when expressed in an arbitrary ONP PP basis H, it is much more convenient to find in terms of the amplitude scattering matrix. Demanding to obtain identical incident and scattered unit PP vectors, we have to solve the scattering equation

$$A_H u_H^{E_1,E_2} = \lambda \left(u_H^{E_1,E_2} \right)^* \tag{10.13}$$

with λ real and nonnegative. Looking at the equation conjugate versus (10.13),

$$A_H^* \left(u_H^{E_1,E_2} \right)^* = \lambda u_H^{E_1,E_2} \tag{10.14}$$

one can see that an action of the A_H^* matrix on two sides of equation (10.13) leads to

$$A_H^* A_H u_H^{E_1,E_2} = \lambda^2 u_H^{E_1,E_2}. \tag{10.15}$$

Introducing notation

$$M_{2H} = A_{2H}A_{2H}^*$$
$$M_{1H} = A_{1H}A_{1H}^*$$

and

$$\begin{aligned}
W_2 &= M_{2H} + A_{4H}A_{3H}^* \\
W_1 &= M_{1H} + A_{3H}A_{4H}^* \\
R_1 &= W_1 - W_2 \\
R_2 &= 2(A_{1H}A_{3H}^* + A_{3H}A_{2H}^*) \\
R_3 &= 2(A_{2H}A_{4H}^* + A_{4H}A_{1H}^*) \\
\Delta &= R_1^2 + R_2R_3
\end{aligned} \tag{10.16}$$

we obtain

$$A_H A_H^* = \begin{bmatrix} W_2 & R_2/2 \\ R_3/2 & W_1 \end{bmatrix} \tag{10.17}$$

and the characteristic equation

$$\begin{aligned}
\begin{vmatrix} W_2 - \lambda^2 & R_2/2 \\ R_3/2 & W_1 - \lambda^2 \end{vmatrix} &= \lambda^4 - Tr(A_H A_H^*)\lambda^2 + \det A_H A_H^* \\
&= \lambda^4 - (W_1 + W_2)\lambda^2 + W_1W_2 - \tfrac{1}{4}R_2R_3 = 0
\end{aligned} \tag{10.18}$$

with eigenvalues

$$\lambda^2 = \tfrac{1}{2}(W_1 + W_2) \mp \sqrt{\Delta} \tag{10.19}$$

and with, applying notation as in (E.4)-(E.6),

$$\begin{aligned}
\Delta &= \left(Tr(A_H A_H^*)\right)^2 - \left(2\left|\det A_H\right|\right)^2 \\
&= (M_1 + M_2 + 2S_{34})_H^2 - 4[M_1M_2 + M_3M_4 - 2\,\mathrm{Re}(A_1A_2A_3^*A_4^*)]_H \\
&= (\mathrm{a}_1 + \mathrm{a}_{2H} + \mathrm{a}_{3H} + \mathrm{a}_{4H})^2 - 4(\mathrm{a}_1^2 - \mathrm{b}_0^2).
\end{aligned} \tag{10.20}$$

The value of Δ is real, but also should be nonnegative because λ^2 is real. Therefore, an important conclusion follows that the Sinclair matrix A_H may have no eigenpolarizations (for negative Δ values). That fact can best be observed on the Poincare sphere model of that matrix (see Appendix J). The necessary and sufficient conditions of eigenpolarizations existence are inequalities

$$Tr(A_H A_H^*) - 2\left|\det A_H\right| \geq 0 \tag{10.21}$$

or, equivalently,

$$\mathrm{a}_1 + \mathrm{a}_{2H} + \mathrm{a}_{3H} + \mathrm{a}_{4H} - 2\sqrt{\mathrm{a}_1^2 - \mathrm{b}_0^2} \geq 0. \tag{10.22}$$

From the homogeneous equation, for instanc

$$W_2 - \lambda^2 + \rho R_2/2 = 0, \tag{10.23}$$

the complex polarization ratio follows:

$$\rho_H^{\mathrm{E}_1,\mathrm{E}_2} = \frac{R_1 \mp \sqrt{\Delta}}{R_2}. \tag{10.24}$$

The normalized total scattered powers are, according to (10.19),

$$\sigma^{E_1,E_2} = \lambda^2 = \tfrac{1}{2}(W_1 + W_2) \mp \sqrt{\Delta}$$
$$= \tfrac{1}{2}(a_1 + a_{2H} + a_{3H} + a_{4H} \mp \sqrt{(a_1 + a_{2H} + a_{3H} + a_{4H})^2 - 4(a_1^2 - b_0^2)}). \tag{10.25}$$

Also spatial eigenphases, accompanying eigenpolarizations according to the scattering equation (10.13) with real λ, can be determined for known complex polarization ratios as in (10.24). For that purpose the PP vector representing eigenpolarizations with their spatial phases will be presented in the form like in, e.g., (7.7):

$$u_H^{E_1,E_2} = \begin{bmatrix} \cos\gamma e^{-j(\delta+\varepsilon)} \\ \sin\gamma e^{+j(\delta-\varepsilon)} \end{bmatrix}_H^P$$
$$= \cos\gamma_H^{E_1,E_2} \begin{bmatrix} 1 \\ \tan\gamma e^{j2\delta} \end{bmatrix}_H^{E_1,E_2} \exp[-j(\delta+\varepsilon)_H^{E_1,E_2}] \tag{10.26}$$
$$= \cos\gamma_H^{E_1,E_2} \begin{bmatrix} 1 \\ \rho \end{bmatrix}_H^{E_1,E_2} \exp(-j\nu_H^{E_1,E_2})$$

With that column vector the first component of the matrix equation (10.13) reads

$$(A_{2H} + A_{3H}\rho_H^{E_1,E_2})\exp(-j\nu_H^{E_1,E_2}) = \lambda^{E_1,E_2}\exp(+j\nu_H^{E_1,E_2})$$

where from the following values of eigenphases can be obtained:

$$\exp(+j2\nu_H^{E_1,E_2}) = \frac{1}{\lambda^{E_1,E_2}}(A_{2H} + A_{3H}\rho_H^{E_1,E_2}) \tag{10.26a}$$

and

$$2\varepsilon_H^{E_1,E_2} = 2\nu_H^{E_1,E_2} - 2\delta_H^{E_1,E_2}; \text{ with } 2\delta_H^{E_1,E_2} = \arg\rho_H^{E_1,E_2} . \tag{10.26b}$$

10.4. Polarizations X_1 and X_2, orthogonal to eigenpolarizations, and their spatial phases

Sometimes polarizations orthogonal to eigenpolarizations can be of interest. Their polarization ratios are

$$\rho_H^{X_1,X_2} = \frac{-1}{\rho_H^{E_1,E_2}*} = \frac{R_1 * \pm\sqrt{\Delta}}{R_3 *} \tag{10.27}$$

Their spatial phases can be found from the orthogonality transformation equation

$$u_H^{X_1,X_2} = C^X(u_H^{E_1,E_2})*$$

or

$$\cos\gamma_H^{X_1,X_2} \begin{bmatrix} 1 \\ \rho \end{bmatrix}_H^{X_1,X_2} \exp(-j(\delta+\varepsilon)_H^{X_1,X_2}) = \begin{bmatrix} 0 & -1 \\ 1 & 0 \end{bmatrix} \cos\gamma_H^{E_1,E_2} \begin{bmatrix} 1 \\ \rho * \end{bmatrix}_H^{E_1,E_2} \exp(+j(\delta+\varepsilon)_H^{E_1,E_2}) \tag{10.28}$$

from which, taking into account that

$$\cos\gamma_H^{X_1,X_2} = \sin\gamma_H^{E_1,E_2}, \qquad \cot\gamma_H^{E_1,E_2}\left(\rho_H^{E_1,E_2}\right)* = \exp(-j2\delta_H^{E_1,E_2}), \qquad \delta_H^{X_1,X_2} = -90^0 + \delta_H^{E_1,E_2} \tag{10.29}$$

it follows from the first component of (10.28) that

$$\exp(-j(\delta_H^{E_1,E_2} - 90^0))\exp(-j\varepsilon_H^{X_1,X_2}) = \exp(-j(\delta_H^{E_1,E_2} + \varepsilon_H^{E_1,E_2}))$$

or, finally,

$$\varepsilon_H^{X_1,X_2} = -90^0 - \varepsilon_H^{E_1,E_2}. \tag{10.30}$$

10.5. Polarizations K and L, characteristic for the scattering matrix, and their spatial phases

These characteristic for the Sinclair and Kennaugh matrix polarizations and spatial phases have been derived in Appendix G. The aim of introduction of such PP vectors, corresponding to the tangential polarization (TP) phasors and denoted as *K*, was to obtain in the ONP PP basis *K* a very simple form of the bistatic scattering matrices, Sinclair and Kennaugh (see (7.21b) and (7.20b). Obtaining of such a form of those matrices was necessary to construct their geometrical Poincare sphere models enabling one to easily observe mutual locations of all special polarization points (see Appendix J) and to discover the possibility of a special classification of bistatic scattering matrices of nondepolarizing ('point') targets (see Appendix I). Such classification is based on the inherent features of those matrices, invariant versus the ONP PP bases transformations.

See also formula (10.41) in the next Section for rectangular coordinates of K and L points, formulae (9.17) or (G.15) for ρ_H^K, and formulae (9.19) or (G.5) for the double spatial phase argument, $2\varepsilon_H^K$, uniquely determining also $2\varepsilon_H^L = -180^0 - 2\varepsilon_H^K$ (compare with (10.30)).

10.6. Recapitulation of the results obtained

The considerations just performed indicate the possibility of expressing complex polarization ratios of special polarizations through three complex in general parameters, R_1, R_2, R_3, appearing in a square equation

$$R_2\rho^2 - 2R_1\rho - R_3 = 0 \tag{10.31}$$

and being functions of the Sinclair matrix elements. With those parameters, complex polarization ratios of pairs special polarization points, $\mathrm{T}_1, \mathrm{T}_2$, and their orthogonal versions, $\mathrm{T}_1\mathrm{x}, \mathrm{T}_2\mathrm{x}$, can be presented in any ONP PP basis *H* as follows,

$$\rho_H^{\mathrm{T}_1,\mathrm{T}_2} = \left(\frac{R_1 \mp \sqrt{\Delta}}{R_2}\right)_H^{\mathrm{T}_1,\mathrm{T}_2}, \qquad \rho_H^{\mathrm{T}_1\mathrm{x},\mathrm{T}_2\mathrm{x}} = \left(\frac{R_1 * \mp \sqrt{\Delta *}}{R_3 *}\right)_H^{\mathrm{T}_1,\mathrm{T}_2} \tag{10.32a}$$

where

$$\Delta = R_1{}^2 + R_2R_3. \tag{10.32b}$$

Cartesian coordinates of those points in the Stokes parameter space can be found with the aid of formulae

$$\begin{bmatrix} \mathrm{q} \\ \mathrm{u} \\ \mathrm{v} \end{bmatrix}_H^{\mathrm{T}_1,\mathrm{T}_2} = \frac{1}{M}\begin{bmatrix} L \\ \mathrm{Re}\,N \\ \mathrm{Im}\,N \end{bmatrix} \tag{10.33}$$

with

$$\begin{aligned} L &= R_2R_2 * -KK * \\ M &= R_2R_2 * +KK * \\ N &= 2R_2 * K \end{aligned} \tag{10.34}$$

and

$$K = R_1 \mp \sqrt{\Delta}$$
$$KK^* = R_1 R_1^* + |\Delta| \mp 2\,\mathrm{Re}(R_1^* \sqrt{\Delta}) \tag{10.35}$$
$$|\Delta| = \sqrt{|R_1|^4 + |R_2 R_3|^2 + 2\,\mathrm{Re}[(R_1^*)^2 R_2 R_3]}.$$

The following parameters determine special polarization points,

- M and N=Mx:
$$R_1 = \tfrac{1}{2}(A_2 A_2^* - A_3 A_3^* + A_4 A_4^* - A_1 A_1^*) = b_1$$
$$R_2 = -(A_1 A_4^* + A_3 A_2^*) = R_3^* = -b_3 + jb_5, \tag{10.36}$$

- O_1 and O_2:
$$R_1 = -A_3 - A_4$$
$$R_2 = 2A_1, \quad R_3 = -2A_2, \tag{10.37}$$

- E_1 and E_2:
$$R_1 = A_1 A_1^* - A_2 A_2^* + A_3 A_4^* - A_4 A_3^*$$
$$R_2 = 2(A_1 A_3^* + A_3 A_2^*) \tag{10.38}$$
$$R_3 = 2(A_2 A_4^* + A_4 A_1^*), \quad \Delta \geq 0$$

- $X_1 = E_1x$ and $X_2 = E_2x$:
$$R_1 = A_2 A_2^* - A_1 A_1^* + A_3 A_4^* - A_4 A_3^*$$
$$R_2 = -2(A_1 A_4^* + A_4 A_2^*) \tag{10.39}$$
$$R_3 = -2(A_2 A_3^* + A_3 A_1^*), \quad \Delta \geq 0$$

- K and L=Kx :
$$R_1 = A_2 A_2^* - A_1 A_1^*$$
$$R_2 = -A_1(A_3^* + A_4^*) - A_2^*(A_3 + A_4) = R_3^*. \tag{10.40}$$

It should be remembered that in case of polarizations M and N, it is not advisable to use formulae (10.36). Instead, simpler formula (10.4) is more convenient.

Simple formulae determine also rectangular coordinate of points K and L in terms of parameters in (10.40):

$$\begin{bmatrix} q \\ u \\ v \end{bmatrix}_H^{K,L} = \frac{\pm 1}{\sqrt{\Delta}} \begin{bmatrix} R_1 \\ -\mathrm{Re}\,R_2 \\ \mathrm{Im}\,R_2 \end{bmatrix}_H \tag{10.41}$$

with Δ as in (10.32b).

It is worth noticing that the polarization ratios $\rho_H^{K,L}$ can also be obtained by solving the problem of eigenvector/eigenvalue for the Graves matrix [79] of the form

$$\mathsf{G}_H = \tfrac{1}{4}(A_H^* + \widetilde{A}_H^*)(A_H + \widetilde{A}_H) = \tfrac{1}{4}(A_H^* A_H + A_H^* \widetilde{A}_H + \widetilde{A}_H^* A_H + \widetilde{A}_H^* \widetilde{A}_H) \tag{10.42}$$

Solving separately eigenproblems of the four above presented products of matrices one obtains in succession the polarization ratios: $\rho_H^{E_2,E_1}$, $\rho_H^{M",N"}$, $\rho_H^{M,N}$, $\rho_H^{X_2,X_1}$.

11. Constant Received Power Curves on the Poincare Sphere

In the English language literature there is a lack of publications treating that very useful 'geometrical' presentation of scattering. Basic concepts can be found in technical reports of Kennaugh [95] (1952), in Russian book on 'Polarization of Radar Signals', chapter 9, written by D.B. Kanareykin [94] (1966), and also in the Polish paper presented by this author [41] (1970). But still attempts can be met of researchers trying to find the most convenient ways to draw on the Poincare sphere the curves of constant co-polarized received power scattered backwards from nondepolarizing targets (see for example J. Yang [142] (1999)). It is believed that formulae presented beneath will be of some help also for those intending to solve similar problems for partially polarized returns.

Application of the here presented formulae is envisaged, e. g., for estimating ranges of the allowed deviation of elliptical polarization, in monostatic radars, from its optimum values designed for efficient cancelling the rain clutter (compare Z. H. Czyz [39] (1967)).

11. 1. The CO-POL channel (the case of completely polarized scattering)

To analyze the equipower curves on the Poincare sphere it is advisable, without loss of generality, to transform (rotate) the coordinate system of three Stokes parameters to the characteristic coordinate system (CCS) of the Kennaugh matrix under consideration. The CCS corresponds to a characteristic orthonormal polarization basis K in which the Sinclair matrix has the most simple diagonal form,

$$A_K = \left\{ \begin{bmatrix} A_2 & 0 \\ 0 & A_1 \end{bmatrix}_{CCS} e^{j\mu} \right\}_K ; \qquad A_2 \geq A_1 , \tag{11.1}$$

with real positive A_2, A_1, and μ. The corresponding Kennaugh matrix is also very simple, and the resulting equation for the received power in the CO-POL channel is

$$\begin{aligned} P_c(\mathrm{q,u,v}) &= \frac{1}{2}\begin{bmatrix} 1 & \mathrm{q} & \mathrm{u} & \mathrm{v} \end{bmatrix}_K \begin{bmatrix} a_1 & b_1 & 0 & 0 \\ b_1 & a_1 & 0 & 0 \\ 0 & 0 & a_3 & 0 \\ 0 & 0 & 0 & -a_3 \end{bmatrix}_K \begin{bmatrix} 1 \\ \mathrm{q} \\ \mathrm{u} \\ \mathrm{v} \end{bmatrix}_K \\ &= \frac{1}{2}\left\{ a_1\left(1+\mathrm{q}^2\right) + 2b_1\mathrm{q} + a_3\left(\mathrm{u}^2 - \mathrm{v}^2\right) \right\}_K \end{aligned} \tag{11.2}$$

with

$$a_1 = \frac{A_2^2 + A_1^2}{2}, \qquad b_1 = \frac{A_2^2 - A_1^2}{2}, \qquad a_3 = A_2 A_1; \qquad a_1^2 - a_3^2 = b_1^2 \tag{11.3}$$

and

$$\mathrm{q}^2 + \mathrm{u}^2 + \mathrm{v}^2 = 1 . \tag{11.4}$$

The last equation determines the Poincare sphere of unit radius in the CCS, while the $P_c(\mathrm{q,u,v}) = const.$ expresses a more complex surface of rotational symmetry. Its axial crossection presents the Cassini oval with focuses at the CO-POL Null points. For $P_c = 0$ it reduces to two points, its focuses. For the double CO-POL Null and non vanishing P_c, the surface becomes just a sphere. This is in agreement with the Kennaugh's 'geometrical' formula for the CO-POL received power and can be explained as follows.

Let us introduce the geometrical model of the Kennaugh symmetrical matrix, in the CCS, which has the form of the Poincare sphere

$$\mathrm{x}^2 + \mathrm{y}^2 + \mathrm{z}^2 = r^2; \qquad \mathrm{x} = r\mathrm{q}, \;\; \mathrm{y} = r\mathrm{u}, \;\; \mathrm{z} = r\mathrm{v} \tag{11.5}$$

with the radius

$$r = \frac{A_2 + A_1}{2} , \tag{11.6}$$

and the x,y,z coordinates of the CO-POL Null points: $O_{1,2}(-e,\ 0,\ \mp d)$, where

$$d = \sqrt{A_2 A_1}, \quad e = \frac{A_2 - A_1}{2}; \quad d^2 + e^2 = r^2. \tag{11.7}$$

For transmit/receive polarization corresponding to a point P(x, y, z) on the model's sphere, the received power can be expressed by the equation

$$\begin{aligned} P_c(\mathrm{P}) = P_c(\mathrm{x,y,z}) &= \frac{1}{2r^2}\left\{a_1\left(r^2 + \mathrm{x}^2\right) + 2b_1 r\mathrm{x} + a_3\left(\mathrm{y}^2 - \mathrm{z}^2\right)\right\} \\ &= \frac{1}{2r^2}\left\{\left(r^2 + e^2\right)\left(r^2 + \mathrm{x}^2\right) + 4er^2\mathrm{x} + d^2\left(\mathrm{y}^2 - \mathrm{z}^2\right)\right\} \end{aligned} \tag{11.8a,b}$$

In such a model the x,y,z coordinates of the CO-POL Max and CO-POL Saddle points are $\mathrm{M}(1,\ 0,\ 0)$ and $\mathrm{N}(-1,\ 0,\ 0)$, respectively, the x,y,z coordinates of the X-POL Max points are $\mathrm{C}_{1,2}(0,\ 0,\ \pm r)$, and the x,y,z coordinates of the X-POL Saddle points are $\mathrm{D}_{1,2}(0,\ \mp r,\ 0)$.

There is also another very important point inside the model, the so called inversion point, I, of x,y,z coordinates $\mathrm{I}(-e,\ 0,\ 0)$. For any incident polarization represented by a P point on the model's surface the received power in the matched-pol channel is equal to the total scattered power, i.e., to the square of the (IP) distance:

$$P_c = (\mathrm{IP})^2. \tag{11,9}$$

For instance,

$$P_{c\,\max} = (\mathrm{IM})^2 = A_2^2, \tag{11.10}$$

and

$$P_{c\,\mathrm{saddle}} = (\mathrm{IN})^2 = A_1^2. \tag{11.11}$$

The I point enables one to find immediately the scattered polarization point S. That can be done by inversion of the P point, through the I point, back to the model's surface, and by rotation of the inverted point, with the sphere, by 180° about the z axis. The P_c power can then be determined when multiplying the scattered power by the square of cosine of half an angle between the transmit/receive antenna polarization point P and the scattered polarization point S (see also formula (4.15) and the Appendix C):

$$P_c(\mathrm{P}) = (\mathrm{IP})^2 \cos^2 \frac{\mathrm{SP}}{2}. \tag{11.12}$$

Kennaugh has shown (see also Appendix D) that

$$P_c(\mathrm{P}) = \frac{(\mathrm{O_1P})^2 \times (\mathrm{O_2P})^2}{(2r)^2}. \tag{11.13}$$

For any point P in the CCS, not necessarily on the model's surface, the $P_c(\mathrm{P}) = const.$ equation determines the rotational surface with an axial crossection of the form of the earlier mentioned Cassini oval, with focuses at $O_{1,2}$ points. The exact equation of that surface in the ρz cylindrical coordinate system is

$$\left.\begin{aligned} \left[\rho^2 + z^2\right]^2 + 2d^2\left[\rho^2 - z^2\right] &= c^4 - d^4 \\ \rho^2 = (x+e)^2 + y^2 & \end{aligned}\right\} \tag{11.14}$$

with a constant

$$c^4 = (O_1P)^2 \times (O_2P)^2 = 4r^2 P_c(P). \tag{11.15}$$

Crossections of those surfaces with the Poincare sphere determine the curves of constant received power. Simple formulae present projections of those curves onto the coordinate planes *xy* and *xz*. Projection onto the *yz* plane gives more complex curve and will be not presented here.

The xy plane

For $z^2 = r^2 - x^2 - y^2$, we obtain the following equation of ellipses as projections of constant CO-POL power curves onto the xy plane

$$r^2(e + x)^2 + d^2 y^2 = \frac{c^4}{4}. \tag{11.16}$$

Their semi-axes, *a* and *b*, can be found as follows:

$$\left.\begin{aligned} &\frac{4r^2(e+x)^2}{c^4} + \frac{4d^2y^2}{c^4} = 1 = \frac{(e+x)^2}{b^2} + \frac{y^2}{a^2}; \\ &a = \frac{c^2}{2d}, \quad b = \frac{c^2}{2r}; \qquad a \geq b. \end{aligned}\right\} \tag{11.17}$$

The xz plane

Similarly, taking $y^2 = r^2 - x^2 - z^2$, we obtain a set of hyperbolae as projections of the constant CO-POL power curves onto the *xz* plane:

$$\left(ex + r^2\right)^2 - d^2 z^2 = \frac{c^4}{4}. \tag{11.18}$$

Parameters *a* and *b* of those hyperbolae can be found similarly:

$$\left.\begin{aligned} &\frac{4e^2\left(x + \frac{r^2}{e}\right)^2}{c^4} - \frac{4d^2z^2}{c^4} = 1 = \frac{\left(x + \frac{r^2}{e}\right)^2}{a^2} - \frac{z^2}{b^2}; \\ &a = \frac{c^2}{2e}, \quad b = \frac{c^2}{2d}. \end{aligned}\right\} \tag{11.19}$$

The asymptotes of those hyperbolae are

$$z = \pm\frac{e}{d}(x + \frac{r^2}{e}) \tag{11.19a}$$

It is worth noticing that those asymptotes are tangent to the great circle of the polarization sphere at the points O_1 and O_2. They are crossing on the x axis at a point I' of coordinate $x = -(r^2/e)$. It means that I and I' points are 'mutual reflections' in the sphere surface.

11. 2. Computational formulae in terms of the relative power level (the completely polarized scattering case)

The parameter of constant level of the received signal versus $\left(P_c\right)_{max} = A_2^2$ can be computed from the formulae

$$L = P_c[db] = 10\log\frac{(P_c)_{\max}}{P_c} = 20\log\frac{c^2_{\max}}{c^2} \geq 0, \tag{11.20}$$

where

$$c^2 = (A_2 + A_1)\sqrt{P_c}, \qquad c^2_{\max} = 2r(r+e) = (A_2 + A_1)A_2 \tag{11.21}$$

Hyperbolae and ellipses for given L are the following functions of x :

$$z(L,\mathrm{x}) = \pm\frac{\sqrt{(e\mathrm{x}+r^2)^2 - \frac{c^4}{4}}}{d}, \qquad y(L,\mathrm{x}) = \pm\frac{\sqrt{\frac{c^4}{4} - r^2(e+\mathrm{x})^2}}{d}, \tag{11.22}$$

with

$$\frac{c^4}{4} = \left(\frac{r(r+e)}{\log^{-1}(L/20)}\right)^2. \tag{11.23}$$

Ellipses can be presented also in cylindrical coordinates by the formula

$$\rho = \frac{c^2}{2\sqrt{r^2 - e^2\sin^2\varphi}} \tag{11.24}$$

where the φ angle is being taken from the xz plane.

11. 3. The case of CO-POL returns from a partially depolarizing backscatterer

Considerations of this Section show how the partial polarization arises when incoherent superposition of scatterings from non-depolarizing targets. An example has been presented of the most simple case of the partially depolarizing scattering matrix which can be decomposed into a sum of two only nondepolarizing matrices (usually the sum of three non-depolarizing matrices is needed to present the partially polarized backscattering). Such example represents a simplified model of the rain scatterer consisting of vertically oriented spheroidal drops of different oblatenesses depending on their sizes. It may be used for estimating ('underestimating') the depolarization of a wave illuminating the rain cloud. (To represent another limiting case, of 'overestimation' of the phenomenon of depolarization, can serve the cloud of raindrops oriented stochastically.)

In the case of such a simple partially depolarizing backscatterer, its Kennaugh scattering matrix in the CCS has the form

$$\mathrm{K}_K = \begin{bmatrix} \mathrm{A} & \mathrm{B} & 0 & 0 \\ \mathrm{B} & \mathrm{A} & 0 & 0 \\ 0 & 0 & \mathrm{C} & 0 \\ 0 & 0 & 0 & -\mathrm{C} \end{bmatrix} \tag{11.25}$$

the elements of which are governed by the inequality relation

$$\mathrm{A}^2 \geq \mathrm{B}^2 + \mathrm{C}^2, \tag{11.26}$$

indicating that the condition of retaining complete polarization may not be fulfilled (only the equality corresponds to the previously analyzed nondepolarizing scattering). Geometrical model for such a scattering matrix has been presented in Fig. 11.1, showing its crossection by the plane $y = 0$. It can be considered as an 'incoherent sum' of two models in the same CCS and of the same diameter,

$$2r = \sqrt{\mathrm{A} + \mathrm{C}}, \tag{11.27}$$

but of two different inversion points, $\mathrm{I}^{(1)}$ and $\mathrm{I}^{(2)}$, and the corresponding two pairs of the CO-POL Null points, $\mathrm{O}_{1,2}^{(1)}$ and $\mathrm{O}_{1,2}^{(2)}$. These models correspond to two nondepolarizing (Sinclair) matrices in the CCS,

$$A_{CCS}^{(i)} = \begin{bmatrix} A_2^{(i)} & 0 \\ 0 & A_1^{(i)} \end{bmatrix}_{CCS}, \quad i = 1,\, 2 \tag{11.28}$$

creating the above defined Kennaugh matrix:

$$\mathrm{K}_K = \tilde{\mathrm{U}}(A_{CCS}^{(1)} \otimes A_{CCS}^{(1)}{}^*)\mathrm{U} + \tilde{\mathrm{U}}(A_{CCS}^{(2)} \otimes A_{CCS}^{(2)}{}^*)\mathrm{U} \tag{11.29}$$

Elements of those Sinclair matrices in terms of elements of that Kennaugh matrix can be found to be

$$\begin{aligned}
A_1^{(1)} &= \frac{1}{2\sqrt{\mathrm{A}+\mathrm{C}}}(\mathrm{A}-\mathrm{B}+\mathrm{C}-\sqrt{\mathrm{A}^2-(\mathrm{B}^2+\mathrm{C}^2)}) \\
A_1^{(2)} &= \frac{1}{2\sqrt{\mathrm{A}+\mathrm{C}}}(\mathrm{A}-\mathrm{B}+\mathrm{C}+\sqrt{\mathrm{A}^2-(\mathrm{B}^2+\mathrm{C}^2)}) \\
A_2^{(1)} &= \frac{1}{2\sqrt{\mathrm{A}+\mathrm{C}}}(\mathrm{A}+\mathrm{B}+\mathrm{C}+\sqrt{\mathrm{A}^2-(\mathrm{B}^2+\mathrm{C}^2)}) \\
A_2^{(2)} &= \frac{1}{2\sqrt{\mathrm{A}+\mathrm{C}}}(\mathrm{A}+\mathrm{B}+\mathrm{C}-\sqrt{\mathrm{A}^2-(\mathrm{B}^2+\mathrm{C}^2)})
\end{aligned} \tag{11.30}$$

Introducing parameters:

$$\begin{aligned}
c_1^4 &= (\mathrm{O}_1^{(1)}\mathrm{P})^2 \times (\mathrm{O}_2^{(1)}\mathrm{P})^2 = (2r)^2 P_c^{(1)} = (\mathrm{A}+\mathrm{C})P_c^{(1)} \\
c_2^4 &= (\mathrm{O}_1^{(2)}\mathrm{P})^2 \times (\mathrm{O}_2^{(2)}\mathrm{P})^2 = (2r)^2 P_c^{(2)} = (\mathrm{A}+\mathrm{C})P_c^{(2)}
\end{aligned} \tag{11.31}$$

and

$$c^4 = \tfrac{1}{2}(c_1^4 + c_2^4) \tag{11.32}$$

we obtain the following formula for the total co-polarization received power:

$$\begin{aligned}
P_c &= \tfrac{1}{2}\begin{bmatrix} 1 & \mathrm{q} & \mathrm{u} & \mathrm{v} \end{bmatrix}_K^T \mathrm{K}_K \begin{bmatrix} 1 \\ \mathrm{q} \\ \mathrm{u} \\ \mathrm{v} \end{bmatrix}_K^T \\
&= P_c^{(1)} + P_c^{(2)} = \frac{2c^4}{(2r)^2} = \frac{2c^4}{\mathrm{A}+\mathrm{C}}
\end{aligned} \tag{11.33}$$

Having constant $P_c^{(1)}$ and $P_c^{(2)}$ power curves as crossections, with the sphere $\mathrm{x}^2 + \mathrm{y}^2 + \mathrm{z}^2 = r^2$, of elliptical cylinders

$$r^2(e_1 + \mathrm{x})^2 + d_1^2\mathrm{y}^2 = \frac{c_1^4}{4} \tag{11.34}$$

and

$$r^2(e_2 + \mathrm{x})^2 + d_2^2\mathrm{y}^2 = \frac{c_2^4}{4} \tag{11.35}$$

or hyperbolical cylinders

$$\left(e_1 x + r^2\right)^2 - d_1^2 z^2 = \frac{c_1^4}{4} \tag{11.36}$$

and

$$\left(e_2 x + r^2\right)^2 - d_2^2 z^2 = \frac{c_2^4}{4} \tag{11.37}$$

we can find the corresponding curves for the constant total power, P_c. Denoting (see Fig. 11.1):

$$\begin{aligned} d^2 &= \tfrac{1}{2}(d_1^2 + d_2^2) = \frac{\mathrm{C}}{2} \\ f^2 &= \tfrac{1}{2}(e_1^2 + e_2^2) = \frac{\mathrm{A} - \mathrm{C}}{4} \\ e &= \tfrac{1}{2}(e_1 + e_2) = \frac{\mathrm{B}}{2\sqrt{\mathrm{A} + \mathrm{C}}} \\ w &= e_1 - e_2 = \sqrt{\frac{\mathrm{A}^2 - (\mathrm{B}^2 + \mathrm{C}^2)}{\mathrm{A} + \mathrm{C}}} \end{aligned} \tag{11.38}$$

and performing addition of expressions for the corresponding cylinders, we arrive at similar cuves for the total received power. They are also ellipses:

$$r^2 (x + e)^2 + d^2 y^2 = \frac{c^4 - r^2 w^2}{4} \geq 0 \tag{11.39}$$

or hyperbolae

$$\left(f x + \frac{e}{f} r^2\right)^2 - d^2 z^2 = \frac{c^4}{4} - \frac{r^4 w^2}{4 f^2} \geq 0 \tag{11.40}$$

though slightly modified. In terms of the Kennaugh matrix elements they are - ellipses:

$$(\mathrm{A} + \mathrm{C})(x + \frac{\mathrm{B}}{2\sqrt{\mathrm{A} + \mathrm{C}}})^2 + 2\mathrm{C}y^2 = c^4 - \frac{\mathrm{A}^2 - (\mathrm{B}^2 + \mathrm{C}^2)}{4} \tag{11.41}$$

and hyperbolae

$$(\mathrm{A} - \mathrm{C})(x + \frac{\mathrm{B}\sqrt{\mathrm{A} + \mathrm{C}}}{2(\mathrm{A} - \mathrm{C})})^2 - 2\mathrm{C}z^2 = c^4 - \frac{\mathrm{A}^2 - (\mathrm{B}^2 + \mathrm{C}^2)}{4} \times \frac{\mathrm{A} + \mathrm{C}}{\mathrm{A} - \mathrm{C}}. \tag{11.42}$$

The above equations yield

$$\begin{aligned} c_{\max}^4 &= r^2 [w^2 + 4(r + e)^2] = \tfrac{1}{2}(\mathrm{A} + \mathrm{B})(\mathrm{A} + \mathrm{C}), \\ c_{\min}^4 &= r^2 w^2 = \tfrac{1}{4}[\mathrm{A}^2 - (\mathrm{B}^2 + \mathrm{C}^2)] \end{aligned} \tag{11.43}$$

We observe that the minimum received power is non vanishing now, corresponding to

$$P_{c\min} = \frac{2c_{\min}^4}{(2r)^2} = \frac{w^2}{2} = \frac{\mathrm{A}^2 - (\mathrm{B}^2 + \mathrm{C}^2)}{2(\mathrm{A} + \mathrm{C})}; \quad \text{for } x = -e, \ y = 0 \ \text{ and } \ z^2 = r^2 - e^2, \tag{11.44}$$

and the maximum received power is

$$P_{c\max} = \frac{2c_{\max}^4}{(2r)^2} = \frac{w^2}{2} + (e + r)^2 = \mathrm{A} + \mathrm{B}; \quad \text{for } x = r \ \text{ and } \ y = z = 0. \tag{11.45}$$

Expressing the equation for projection of the curve of constant received power onto the xy plane of the CCS in the form of an ellipse

$$\frac{(x+e)^2}{b^2}+\frac{y^2}{a^2}=1 \tag{11.46}$$

we arrive at its great and small semiaxes:

$$a=\sqrt{\frac{c^4-r^2w^2}{4d^2}}=\sqrt{\frac{4c^4-A^2+(B^2+C^2)}{8C}},$$
$$b=\sqrt{\frac{c^4-r^2w^2}{4r^2}}=\sqrt{\frac{4c^4-A^2+(B^2+C^2)}{4(A+C)}} \tag{11.47}$$

and their ratio

$$\frac{a}{b}=\frac{r}{d}=\sqrt{\frac{(A+C)}{2C}}. \tag{11.48}$$

Similar dependencies for hyperbolae are:

$$\frac{(x+e\frac{r^2}{f^2})^2}{b^2}-\frac{z^2}{a^2}=1, \tag{11.49}$$

$$a^2=\mp\frac{c^4f^2-r^4w^2}{4f^2d^2},\quad b^2=\mp\frac{c^4f^2-r^4w^2}{4f^4},\quad \frac{a}{b}=\frac{f}{d}. \tag{11.50}$$

Here the upper signs correspond to the hyperbolae above the asymptotes

$$z=\pm\frac{f}{d}(x+e\frac{r^2}{f^2}) \tag{11.51}$$

(for their positive z values) which always cross the $x^2+z^2=r^2$ great circle. They become tangent to it for w = 0 what means the case of nondepolarizing scattering considered earlier. Those asymptotes correspond to the parameter

$$c^4_{asympt}=\frac{r^4w^2}{f^2}=\frac{A^2-(B^2+C^2)}{4}\times\frac{A+C}{A-C}. \tag{11.52}$$

It is interesting to observe that

$$\frac{c^4_{\max}}{c^4_{\min}}=1+4\left(\frac{r+e}{w}\right)^2=2\frac{(A+B)(A+C)}{A^2-(B^2+C^2)}$$
$$=\frac{c^4_{\max}}{c^4_{asympt}}\times\frac{r^2}{f^2}. \tag{11.53}$$

11. 4. Computational formulae in terms of the relative power level for partially polarized backscattered returns

Curves of the constant received power can be computed also in terms of the power level using formulae similar to (11.22):

$$y(L,x)=\pm\frac{\sqrt{\frac{c^4}{4}-[r^2(e+x)^2+\frac{r^2w^2}{4}]}}{d}\quad \text{(ellipses)} \tag{11.22'}$$

$$z(L,x) = \pm \frac{\sqrt{(fx + \frac{e}{f}r^2)^2 + \frac{r^4 w^2}{4f^2} - \frac{c^4}{4}}}{d} \quad \text{(hyperbolae)},$$

with

$$\frac{c^4}{4} = \frac{r^2[w^2 + 4(r+e)^2]}{4\log^{-1}(L/10)}; \quad L = P_c[db] = 10\log\frac{(P_c)_{\max}}{P_c} = 10\log\frac{c^4_{\max}}{c^4} > 0. \qquad (11.23')$$

See Fig. 11.1 for a numerical example of the asymptotes of hyperbolae of the constant received power level curves projected onto the zx CCS coordinate plane.

11. 5. The received power in the X-POL channel (the completely polarized backscattering case)

The $P_x = C$, or constant X-POL power curves on the Poincare sphere model of the symmetrical Sinclair matrix, can be found when applying a similar procedure.

Starting from the expression for the X-POL received power:

$$P_x(\mathrm{q,u,v}) = \frac{1}{2}\begin{bmatrix}1 & \mathrm{-q} & \mathrm{-u} & \mathrm{-v}\end{bmatrix}\begin{bmatrix} a_1 & b_1 & 0 & 0 \\ b_1 & a_1 & 0 & 0 \\ 0 & 0 & a_3 & 0 \\ 0 & 0 & 0 & -a_3 \end{bmatrix}\begin{bmatrix}1 \\ \mathrm{q} \\ \mathrm{u} \\ \mathrm{v}\end{bmatrix} = \frac{1}{2}\left\{a_1\left(1-\mathrm{q}^2\right) - a_3\left(\mathrm{u}^2 - \mathrm{v}^2\right)\right\}$$

$$= \frac{1}{2}\left\{\frac{A_2^2 + A_1^2}{2}\left(1-\mathrm{q}^2\right) - A_2 A_1\left(\mathrm{u}^2 - \mathrm{v}^2\right)\right\} = \frac{1}{2}\left\{\left(r^2 + e^2\right)\left(1-\mathrm{q}^2\right) - d^2\left(\mathrm{u}^2 - \mathrm{v}^2\right)\right\}$$

(11.54)

or

$$P_x(\mathrm{x,y,z}) = \frac{1}{2r^2}\left\{-\left(r^2 + e^2\right)\mathrm{x}^2 - d^2\left(\mathrm{y}^2 - \mathrm{z}^2\right) + r^2\left(r^2 + e^2\right)\right\}; \quad d^2 = r^2 - e^2, \qquad (11.55)$$

one obtains the equation of hyperboloids, with parameter P_x changing from 0 to r^2 (for $\mathrm{x}^2 + \mathrm{y}^2 + \mathrm{z}^2 \le r^2$), i.e.,

one-sheeted hyperboloids of the form:

$$\frac{\mathrm{x}^2}{a^2} + \frac{\mathrm{y}^2}{b^2} - \frac{\mathrm{z}^2}{b^2} = 1 \qquad \text{for} \qquad 2P_x < r^2 + e^2 \qquad (11.56a)$$

or two-sheeted hyperboloids of the form:

$$-\frac{\mathrm{x}^2}{a^2} - \frac{\mathrm{y}^2}{b^2} + \frac{\mathrm{z}^2}{b^2} = 1 \qquad \text{for} \qquad 2P_x > r^2 + e^2 \qquad (11.56b)$$

with parameters

$$a = r\sqrt{\frac{|r^2 + e^2 - 2P_x|}{r^2 + e^2}} \quad \text{and} \quad b = r\sqrt{\frac{|r^2 + e^2 - 2P_x|}{r^2 - e^2}} \qquad (11.57)$$

where $a/b < 1$, and with Oz being always the axis of symmetry.

The following projections of lines of crossections of those hyperboloids with the Poincare sphere onto the coordinate planes may be of interest.

The xz plane. Substituting $y^2 = r^2 - x^2 - z^2$ in the above formulae we obtain hyperbolae in the *xz* plane,

$$e^2x^2 - d^2z^2 = r^2\left(e^2 - P_x\right). \quad (11.58)$$

The xy plane. Substituting $z^2 = r^2 - x^2 - y^2$, we obtain ellipses in the *xy* plane,

$$r^2x^2 + d^2y^2 = r^2\left(r^2 - P_x\right). \quad (11.59)$$

The yz plane. Similarly, substituting $x^2 = r^2 - y^2 - z^2$, we obtain other ellipses, in the *yz* plane,

$$e^2y^2 + r^2z^2 = r^2P_x . \quad (11.60)$$

11. 6. The curves of equal CO-POL and X-POL received powers

Starting from the above derived formulae:

$$P_c(x,y,z) = \frac{1}{2r^2}\left\{\left(r^2 + e^2\right)x^2 + 4er^2x + d^2\left(y^2 - z^2\right) + r^2(r^2 + e^2)\right\} \quad (11.8b)$$

$$P_x(x,y,z) = \frac{1}{2r^2}\left\{-\left(r^2 + e^2\right)x^2 - d^2\left(y^2 - z^2\right) + r^2(r^2 + e^2)\right\} \quad (11.55)$$

and comparing both expressions

$$P_c(x,y,z) = P_x(x,y,z) \Rightarrow$$

$$\Rightarrow \left(r^2 + e^2\right)x^2 + 4er^2x + d^2\left(y^2 - z^2\right) + r^2\left(r^2 + e^2\right) = -\left(r^2 + e^2\right)x^2 - d^2\left(y^2 - z^2\right) + r^2(r^2 + e^2)$$

we obtain the desired equation for those curves,

$$\left(r^2 + e^2\right)x^2 + 2er^2x + d^2\left(y^2 - z^2\right) = 0 \quad (11.61)$$

which depends on two parameters: r and e ($d^2 = r^2 - e^2$), and is independent of the received power level. It may be interesting to find projections of those curves, for $r^2 = x^2 + y^2 + z^2$, onto the coordinate planes xy, yz and xz, as presented beneath.

The yz plane. For $x^2 = r^2 - y^2 - z^2$, we obtain

$$\left(r^2 + e^2\right)\left(r^2 - y^2 - z^2\right) + 2er^2\sqrt{r^2 - y^2 - z^2} + d^2\left(y^2 - z^2\right) = 0 \quad (11.62)$$

or

$$r^2 - y^2 - z^2 =$$

$$= \frac{1}{\left(2er^2\right)^2}\left\{\left\{\left(r^2 + e^2\right)\left(r^2 - (y^2 + z^2)\right)\right\}^2 + \left\{d^2\left(y^2 - z^2\right)\right\}^2 + 2d^2\left(r^2 + e^2\right)\left(r^2 - (y^2 + z^2)\right)\left(y^2 - z^2\right)\right\}$$

$$= \frac{1}{\left(2er^2\right)^2}\Big\{\left(r^2+e^2\right)^2\left(r^4+y^4+z^4+2y^2z^2-2r^2y^2-2r^2z^2\right)+d^4y^4+d^4z^4-2d^4y^2z^2+$$

$$+2r^2d^2\left(r^2+e^2\right)y^2-2r^2d^2\left(r^2+e^2\right)z^2-2d^2\left(r^2+e^2\right)y^4+2d^2\left(r^2+e^2\right)z^4\Big\},$$

and finally,

$$\frac{1}{\left(2er^2\right)^2}\Big\{\left[\left(r^2+e^2\right)^2-2d^2\left(r^2+e^2\right)+d^4\right]y^4+\left[\left(r^2+e^2\right)^2+2d^2\left(r^2+e^2\right)+d^4\right]z^4+$$

$$+2\left[\left(r^2+e^2\right)^2-d^4\right]y^2z^2+$$

$$+2r^2\left\{\left(r^2+e^2\right)\left[d^2-\left(r^2+e^2\right)\right]+2e^2r^2\right\}y^2-2r^2\left\{\left(r^2+e^2\right)\left[d^2+\left(r^2+e^2\right)\right]-2e^2r^2\right\}z^2+r^4d^4\Big\}=0$$

or

$$4\left(e^2y^2+r^2z^2\right)^2-4r^2\left(e^4y^2+r^4z^2\right)+r^4d^4=0. \qquad (11.63)$$

The xy plane. For $z^2=r^2-x^2-y^2$, we obtain the equation of an ellipse

$$\frac{\left(x-x_0\right)^2}{a^2}+\frac{y^2}{b^2}=1 \qquad (11.64)$$

with

$$x_0=-\frac{e}{2}, \quad a=\frac{\sqrt{r^2+d^2}}{2}, \quad b=\frac{r\sqrt{r^2+d^2}}{2d}; \quad r^2=d^2+e^2, \quad \frac{b}{a}=\frac{r}{d}\geq 1. \qquad (11.65)$$

It is worth noticing that for $e=0$ $(d=r)$, and $x=0$, we obtain $x_0=0$, and $a=b=\frac{r}{\sqrt{2}}$.

For $d=0$ $(e=r)$, and $x=0$, we have $a=-x_0=\frac{r}{2}$ (compare same results for the xz plane).

The xz plane. For $y^2=r^2-x^2-z^2$, we obtain hyperbolae:

$$\mp\frac{\left(x-x_0\right)^2}{a^2}\pm\frac{z^2}{c^2}=1 \qquad (11.66)$$

with upper signs for $d>e$, lower signs for $d<e$, and with

$$x_0=-\frac{r^2}{2e}, \quad a=\frac{r\sqrt{\left|d^2-e^2\right|}}{2e}, \quad c=\frac{r\sqrt{\left|d^2-e^2\right|}}{2d}; \quad r^2=d^2+e^2, \quad \frac{c}{a}=\frac{e}{d}. \qquad (11.67)$$

It is worth noticing that for $e \to 0$, $d \to r$, $\frac{x_0}{a} \to -1$ and $c \to \frac{r}{2}$ the result is $z \to \pm\frac{r}{\sqrt{2}}$, for any $|x| \le r$ denoting the crossection of the sphere with two parallel (asymptotic) plates.

For $d = 0$ $(e = r)$, and x = 0, we have $a = -x_0 = \frac{r}{2}$ and $z = 0$; in that case the resulting curve is the great circle in the x = 0 plane.

The above presented results can be explained most simply when analyzing the geometrical model of the scattering matrix and the 'geometrical' formulae for the received power:

$$P_c(P) = (IP)^2 \cos^2 \frac{SP}{2} \quad \text{and} \quad P_x(P) = (IP)^2 \cos^2 \frac{SP^x}{2} \tag{11.68}$$

Here P^x means the polarization point of the receiving antenna, antipodal to P. One can immediately see that the obtained curves of equal CO-POL and X-POL received power correspond to equal anglular distances, $SP = SP^x$, for scattered polarization points S, obtained for incident polarization point P after inversion and rotation of the polarization sphere.

11. 7. The case of the bistatic scattering

The problem has been solved for symmetrical Sinclair and Kennaugh matrices. In cases of bistatic scattering we deal with nonsymmetrical matrices. In that case the Sinclair matrix in its characteristic ONP PP basis can be considered as a sum of the diagonal real matrix and the matrix of the orthogonality transformation, both multiplied by a phase factor. Therefore, when using the same transmit/receive antenna, the problem remains exactly the same like for the symmetrical matrices. That is evident because the antenna will not receive the orthogonally polarized component of the scattered wave. So, what should be done, it is to take as the Sinclair scattering matrix its symmetrical part only. However, when considering the bistatically scattered wave received in the cross-polarized channel, one should apply a modified procedure. Again, only the symmetrical part can be taken into account but of a different Kennaugh matrix, after the orthogonality transformation. According to the known rule, the received power in the cross-polarized channel can be expressed as beneath:

$$P_r = \tilde{P}^x KP = \tilde{P}^x D^x D^x KP = \tilde{P} K^x P \tag{11.69}$$

That leads to another characteristic ONP PP basis, for the K^x matrix, and to consideration of the 'co-polarized' reception. In that case, however, formulae obtained for the curves of equal CO-POL and X-POL received powers cannot be applied directly because of dealing with two different scattering matrices.

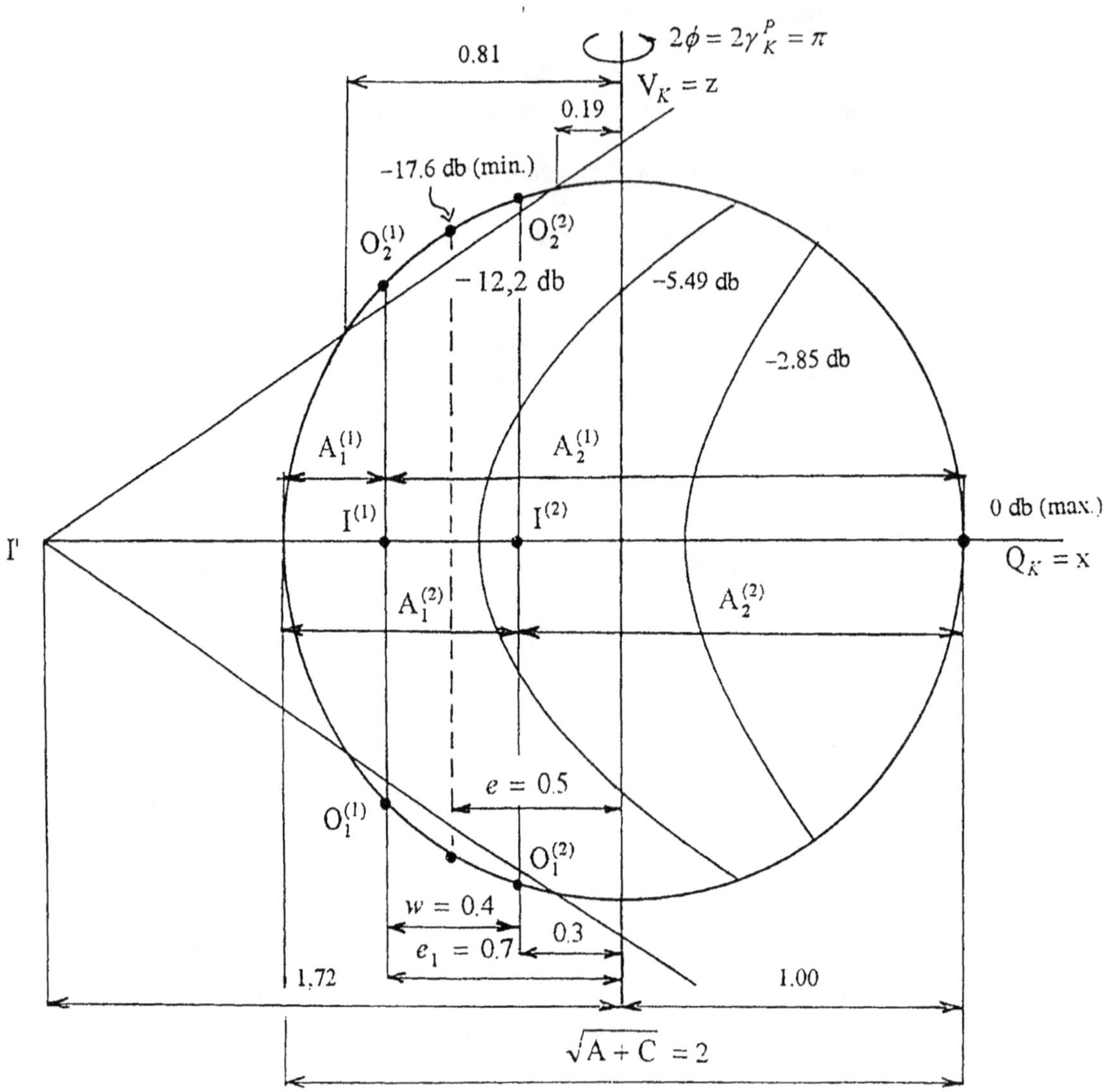

Fig. 11.1. Geometrical model of the scattering matrix with the curves of constant received power levels for a partially depolarizig target determined by three parameters. A numerical example for the input data:

$$r = 1, \quad e_1 = 0.7, \quad e_2 = 0.3 .$$

The secondary parameters: $e = \frac{1}{2}(e_1 + e_2) = 0.5$, $w = e_1 - e_2 = 0.4$, $f^2 = \frac{1}{2}(e_1^2 + e_2^2) = 0.29$,

$$d_1 = \sqrt{r^2 - e_1^2} = \sqrt{0.51}, \quad d_2 = \sqrt{r^2 - e_2^2} = \sqrt{0.91}, \quad d^2 = \tfrac{1}{2}(d_1^2 + d_2^2) = 0.71; \quad f^2 + d^2 = r^2 .$$

A_{ccs} and K_K matrix elements: $A_1^{(1)} = r - e_1 = 0.3$, $A_2^{(1)} = r + e_1 = 1.7$, $A_1^{(2)} = r - e_2 = 0.7$,

$A_2^{(2)} = r + e_2 = 1.3$; $\mathrm{A} = (2r)^2 - 2d^2 = 2.58$, $\mathrm{B} = 4re = 2$, $\mathrm{C} = 2d^2 = 1.42$; $\sqrt{\mathrm{A} + \mathrm{C}} = 2r$.

$$P_{c\min} = \frac{\mathrm{A}^2 - (\mathrm{B}^2 + \mathrm{C}^2)}{2(\mathrm{A} + \mathrm{C})} = 0.08 \ (L = 17.58 \text{ db, for } \mathrm{x} = -e = -0.5),$$

$$P_{c\max} = \mathrm{A} + \mathrm{B} = 4.58 \ (L = 0 \text{ db, for } \mathrm{x} = 1); \quad \tan 2\alpha_{opt} = \frac{\sqrt{r^2 - e^2}}{e} = \sqrt{\frac{\mathrm{A} + \mathrm{C} - \mathrm{B}}{\mathrm{A} + \mathrm{C} + \mathrm{B}}} = \tan 60^0 .$$

$$\mathrm{z}_{asympt} = \pm\frac{f}{d}(\mathrm{x} + e\frac{r^2}{f^2}) = \pm 0.639(\mathrm{x} + 1.72), \quad L_{asympt} = \frac{P_{c\max}}{P_{c\min}} \times \frac{f^2}{r^2} = 12{,}20 \text{ db} .$$

Exemplary hyperbolae for: $L = 5.493$ db ($z = 0$ for $\mathrm{x} = -0.4$ and $\mathrm{z} = 0.706$ for $\mathrm{x} = 0$),

$L = 2.845$ db ($z = 0$ for $\mathrm{x} = +0.18$ and $\mathrm{z} = 0.852$ for $\mathrm{x} = 0.6$).

12. The Basis-Invariant Decompositions of the Sinclair Matrix

12. 1. Preliminary considerations on 'elementary' models of the Sinclair matrices

Before approaching the ONP PP basis-rotation-invariant decomposition, or simply 'basis-invariant' decomposition of the Sinclair bistatic scattering matrix, it is instructive to present at first the Kennaugh and Sinclair matrices and their geometrical models corresponding to separate Huynen parameters of the symmetric Kennaugh matrix of the form

$$K_H = \begin{bmatrix} a_1 & b_1 & b_3 & b_5 \\ b_1 & a_2 & b_4 & b_6 \\ b_3 & b_4 & a_3 & b_2 \\ b_5 & b_6 & b_2 & a_4 \end{bmatrix}_H = \begin{bmatrix} A_0+B_0 & C & H & F \\ C & A_0+B & E & G \\ H & E & A_0-B & D \\ F & G & D & -A_0+B_0 \end{bmatrix}_H . \tag{12.1}$$

The dependence of the Sinclair matrix elements on those of the Kennaugh matrix for monostatic scattering (compare with (E.8) for the bistatic scattering case) is:

$$\begin{aligned} 2A_{2H} &= \sqrt{2(a_1+a_2+2b_1)_H} && = \sqrt{2(2A_0+B_0+B+2C)} \\ A_{3H} &= A_{4H} = [b_3+b_4-j(b_5+b_6)]_H/(2A_{2H}) && -[H+E-j(F+G)]/(2A_{2H}) \\ A_{1H} &= [a_3-a_4-j2b_2]_H/(2A_{2H}) && = (2A_0-B_0-B-j2D)/(2A_{2H}) \end{aligned} \tag{12.2}$$

For further purposes it is convenient to assume the phase of the first element equal to $\phi+\phi_0$ with

$$\phi_0 = \arg(2A_0+C+jD). \tag{12.3}$$

The phase ϕ may be chosen arbitrarily. The ϕ_0 argument indicates the phase difference between the two, (12.4) and (12.5), beneath shown amplitude representations of the Kennaugh matrix (12.1).

So, the Sinclair matrix just obtained is

$$A_H = \frac{e^{j(\phi+\phi_0)}}{\sqrt{2(2A_0+B_0+B+2C)}} \begin{bmatrix} 2A_0+B_0+B+2C & H+E-j(F+G) \\ H+E-j(F+G) & 2A_0-B_0-B-j2D \end{bmatrix}_H . \tag{12.4}$$

Its equivalent is

$$A_H = \frac{e^{j\phi}}{2\sqrt{A_0}} \begin{bmatrix} 2A_0+C+jD & H-jG \\ H-jG & 2A_0-C-jD \end{bmatrix}_H \tag{12.5}$$

The last form can be obtained when applying to (12.4) the following selected conditions, (12.a), (12.g), and (12.h) for preservation of the complete polarization expressed in terms of the Huynen parameters and selected from their complete set (see also[85] or (E.14a) and (14.d)):

$$C^2+D^2 = 2A_0(B_0+B) \tag{12.6a}$$

$$H^2+G^2 = 2A_0(B_0-B) \tag{12.6b}$$

$$F^2+E^2 = B_0^2-B^2 \tag{12.6c}$$

$$C(B_0-B) = EH+FG \tag{12.6d}$$

$$D(B_0-B) = -EG+FH \tag{12.6e}$$

$$C(B_0 - B) = EH + FG \tag{12.6d}$$

$$D(B_0 - B) = -EG + FH \tag{12.6e}$$

$$H(B_0 + B) = CE + DF \tag{12.6f}$$

$$2A_0E = CH - DG \tag{12.6g}$$

$$2A_0F = CG + DH \tag{12.6h}$$

$$G(B_0 + B) = CF - DE \tag{12.6i}$$

The Sinclair matrices in the right-circular basis will be also considered. The Euler angles chosen for that basis:

$$2\delta_H^R = 2\gamma_H^R = -2\varepsilon_H^R = \pi/2 \tag{12.7}$$

yield the column PP vector

$$u_H^R = \begin{bmatrix} \cos\gamma_H^R \, e^{-j(\delta_H^R + \varepsilon_H^R)} \\ \sin\gamma_H^R \, e^{j(\delta_H^R - \varepsilon_H^R)} \end{bmatrix} = \frac{1}{\sqrt{2}} \begin{bmatrix} 1 \\ j \end{bmatrix}_H \tag{12.8}$$

and the change-of-basis matrix

$$C_H^R = \begin{bmatrix} u_H^R & u_H^{R\times} \end{bmatrix} = \frac{1}{\sqrt{2}} \begin{bmatrix} 1 & j \\ j & 1 \end{bmatrix}_H . \tag{12.9}$$

The resulting transformation of (12.5) to the R basis gives:

$$A_R = \widetilde{C}_H^R A_H C_H^R = \frac{e^{j\phi}}{2\sqrt{A_0}} \begin{bmatrix} G + C + j(H + D) & j\sqrt{A_0} \\ j\sqrt{A_0} & G - C + j(H - D) \end{bmatrix}_R \tag{12.10}$$

or, the use of another Huynen's condition, (12.6b), yields

$$A_R = e^{j\phi} \begin{bmatrix} \sqrt{B_0 + F} \, e^{j\arg[G + C + j(H + D)]} & j\sqrt{A_0} \\ j\sqrt{A_0} & \sqrt{B_0 - F} \, e^{j\arg[G - C + j(H - D)]} \end{bmatrix}_R . \tag{12.11}$$

Similar transformation of the (12.4) matrix gives:

$$A_R = \frac{e^{j(\phi + \phi_0)}}{\sqrt{2(2A_0 + B_0 + B + 2C)}} \times \\ \times \begin{bmatrix} F + G + B_0 + B + C + j(H + E + D) & -D + j(2A_0 + C) \\ D + j(2A_0 + C) & F + G - B_0 - B - C + j(H + E - D) \end{bmatrix}_R \tag{12.12}$$

The submitted Table 12.1. presents Sinclair and Kennaugh matrices in the horizontal linear H basis and the Sinclair matrices in the right-circular R basis (always with $\phi = 0$). The individual targets correspond to separate Huynen's parameters (named in the first column) and values of the remaining non-zero parameters are listed (second column). The Sinclair matrices have been obtained by direct use of forms (12.5) or (12.4) and of (12.13) or (12.10). Poincare sphere models of matrices are presented in the H basis. They show the CO-POL Null and inversion points, always for the sphere of unit radius. Coordinates of the inversion point can be found from the equality

$$\begin{bmatrix} Q \\ U \\ V \end{bmatrix}_H^I = \frac{-1}{A_0 + B_0 + \sqrt{(A_0 + B_0)^2 - C^2 - H^2 - F^2}} \begin{bmatrix} C \\ H \\ F \end{bmatrix} = \frac{-1}{2} \begin{bmatrix} C \\ H \\ F \end{bmatrix} . \tag{12.13}$$

They coincide with the double CO-POL Nulls if the I point occurs on the surface of the sphere. In another possible case, for the I point in the center of the sphere, the CO-POL Nulls are situated at the antipodal points, on the axis of rotation (by 180^0) after inversion, and their spherical coordinates versus the H phasor are determined by the formula

$$\rho \equiv \rho_H^{O_{1,2}} = \tan \gamma_H^{O_{1,2}} \exp\{2\delta_H^{O_{1,2}}\} = \frac{-A_{3H} \mp \sqrt{A_{3H}^2 - A_{1H}A_{2H}}}{A_{1H}} . \qquad (12.14)$$

Sinclair matrix elements can be found from the equalities (12.2) which are rather simple in cases under consideration. Having those complex polarization ratios ρ, the corresponding Stokes coordinates can be found also from the well known relations:

$$\begin{bmatrix} Q \\ U \\ V \end{bmatrix}_H^{O_{1,2}} = \frac{1}{1+\rho\rho^*} \begin{bmatrix} 1-\rho\rho^* \\ \rho^*+\rho \\ j(\rho^*-\rho) \end{bmatrix} . \qquad (12.15)$$

The list of elementary symmetrical amplitude matrices of Table 12.1 should be supplemented with one non-symmetrical, but of special significance, the orthogonality amplitude matrix and its power counterpart, both of the form

$$A_H = A_R = C^{\times} = \begin{bmatrix} 0 & -1 \\ 1 & 0 \end{bmatrix}; \qquad \mathrm{K}_H = \mathrm{K}_R = \mathrm{D}^{\times} = \begin{bmatrix} 1 & 0 & 0 & 0 \\ 0 & -1 & 0 & 0 \\ 0 & 0 & -1 & 0 \\ 0 & 0 & 0 & -1 \end{bmatrix} \qquad (12.16)$$

Its model has the inversion point in the center of the sphere, but no rotation after inversion axis exists in that case. As a result the whole surface of the sphere presents the CO-POL Null points.

Target defining element	Other non-zero elements	K_H	A_H	A_R	Poincaré sphere model	Target
$A_0 = 1$	-	$\begin{bmatrix}1&&&\\&1&&\\&&1&\\&&&-1\end{bmatrix}$	$\begin{bmatrix}1&0\\0&1\end{bmatrix}$	$\begin{bmatrix}0&j\\j&0\end{bmatrix}$	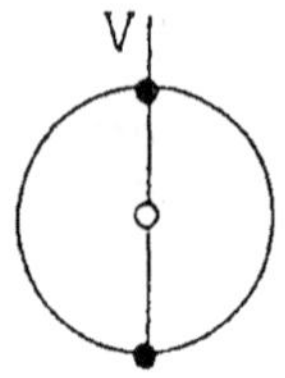	sphere
$B = 1$	$B_0 = 1$	$\begin{bmatrix}1&&&\\&1&&\\&&-1&\\&&&1\end{bmatrix}$	$\begin{bmatrix}1&0\\0&-1\end{bmatrix}$	$\begin{bmatrix}1&0\\0&-1\end{bmatrix}$	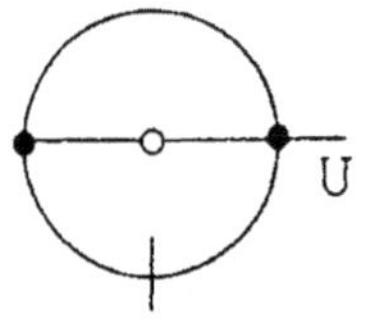	diplane 0°
$B = -1$	$B_0 = 1$	$\begin{bmatrix}1&&&\\&-1&&\\&&1&\\&&&1\end{bmatrix}$	$\begin{bmatrix}0&1\\1&0\end{bmatrix}$	$\begin{bmatrix}j&0\\0&j\end{bmatrix}$	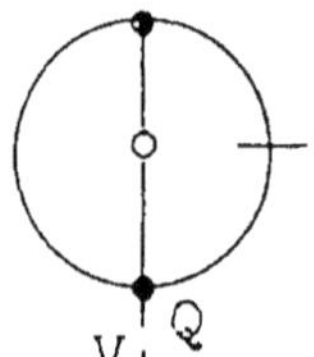	diplane 45°
$D = 1$	$A_0 = B_0 = B = \frac{1}{2}$	$\begin{bmatrix}1&&&\\&1&&\\&&0&+1\\&&1&0\end{bmatrix}$	$\frac{1}{\sqrt{2}}\begin{bmatrix}1+j&0\\0&1-j\end{bmatrix}$	$\frac{1}{\sqrt{2}}\begin{bmatrix}j&j\\j&-j\end{bmatrix}$	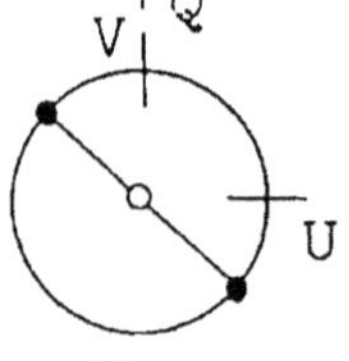	sphere and +j diplane 0°
$D = -1$	$A_0 = B_0 = B = \frac{1}{2}$	$\begin{bmatrix}1&&&\\&1&&\\&&0&-1\\&&-1&0\end{bmatrix}$	$\frac{1}{\sqrt{2}}\begin{bmatrix}1-j&0\\0&1+j\end{bmatrix}$	$\frac{1}{\sqrt{2}}\begin{bmatrix}-j&j\\j&j\end{bmatrix}$	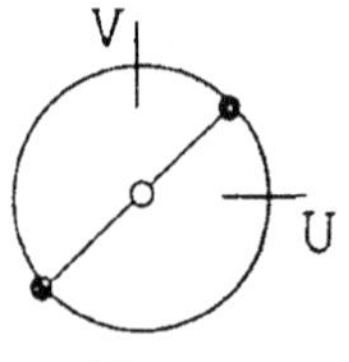	sphere and -j diplane 0°
$G = 1$	$A_0 = B_0 = -B = \frac{1}{2}$	$\begin{bmatrix}1&&&\\&0&&1\\&&1&\\&1&&0\end{bmatrix}$	$\frac{1}{\sqrt{2}}\begin{bmatrix}1&-j\\-j&1\end{bmatrix}$	$\frac{1}{\sqrt{2}}\begin{bmatrix}1&j\\j&1\end{bmatrix}$	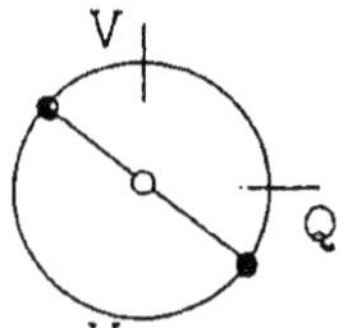	sphere and -j diplane 45°
$G = -1$	$A_0 = B_0 = -B = \frac{1}{2}$	$\begin{bmatrix}1&&&\\&0&&-1\\&&1&\\&-1&&0\end{bmatrix}$	$\frac{1}{\sqrt{2}}\begin{bmatrix}1&j\\j&1\end{bmatrix}$	$\frac{1}{\sqrt{2}}\begin{bmatrix}-1&j\\j&-1\end{bmatrix}$	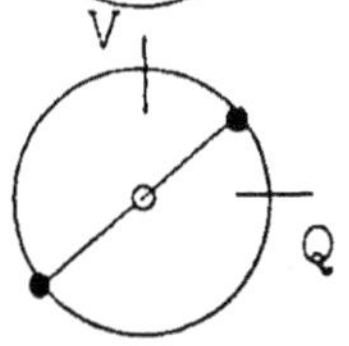	sphere and +j diplane 45°
$E = 1$	$B_0 = 1$	$\begin{bmatrix}1&&&\\&0&1&\\&1&0&\\&&&1\end{bmatrix}$	$\frac{1}{\sqrt{2}}\begin{bmatrix}1&1\\1&-1\end{bmatrix}$	$\frac{1}{\sqrt{2}}\begin{bmatrix}1+j&0\\0&-1+j\end{bmatrix}$	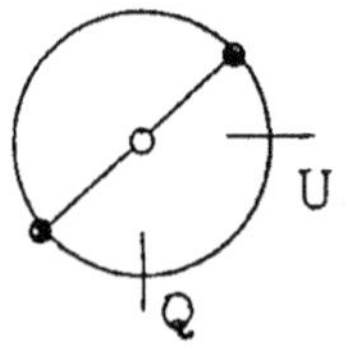	diplane $+22.5^{\circ}$

Table 12. 1. Scattering properties of elementary targets /to be continued/

Target defining element	Other non-zero elements	K_H	A_H	A_R	Poincare sphere model	Target
E =-1	B_0= 1	$\begin{bmatrix} 1 & & & \\ & 0 & -1 & \\ & -1 & 0 & \\ & & & 1 \end{bmatrix}$	$\frac{1}{\sqrt{2}}\begin{bmatrix} 1 & -1 \\ -1 & -1 \end{bmatrix}$	$\frac{1}{\sqrt{2}}\begin{bmatrix} 1-j & 0 \\ 0 & -1-j \end{bmatrix}$	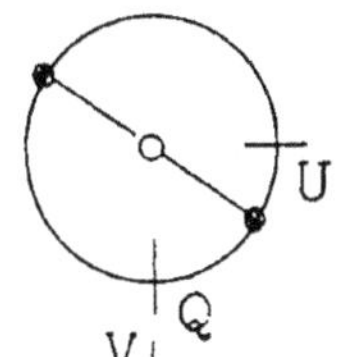	diplane -22.5°
F = 2	B_0= 2	$\begin{bmatrix} 2 & & & 2 \\ & 0 & & \\ & & 0 & \\ 2 & & & 2 \end{bmatrix}$	$\begin{bmatrix} 1 & -j \\ -j & -1 \end{bmatrix}$	$\begin{bmatrix} 2 & 0 \\ 0 & 0 \end{bmatrix}$	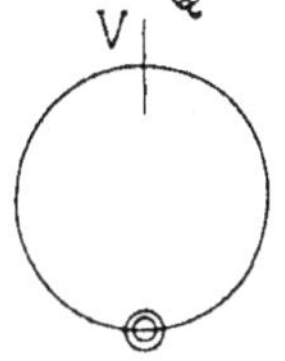	right helix
F =-2	B_0= 2	$\begin{bmatrix} 2 & & & -2 \\ & 0 & & \\ & & 0 & \\ -2 & & & 2 \end{bmatrix}$	$\begin{bmatrix} 1 & j \\ j & -1 \end{bmatrix}$	$\begin{bmatrix} 0 & 0 \\ 0 & -2 \end{bmatrix}$	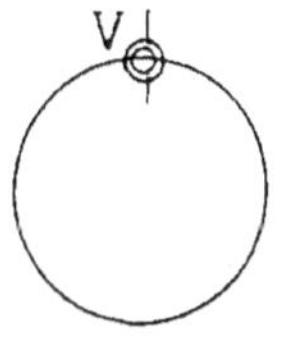	left helix
C = 2	$A_0=B_0=B^0=1$	$\begin{bmatrix} 2 & 2 & & \\ 2 & 2 & & \\ & & 0 & \\ & & & 0 \end{bmatrix}$	$\begin{bmatrix} 2 & 0 \\ 0 & 0 \end{bmatrix}$	$\begin{bmatrix} 1 & j \\ j & -1 \end{bmatrix}$	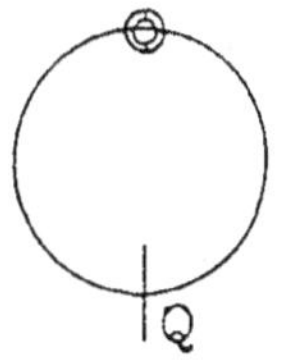	hori-zontal dipole
C =-2	$A_0=B_0$ =B =1	$\begin{bmatrix} 2 & -2 & & \\ -2 & 2 & & \\ & & 0 & \\ & & & 0 \end{bmatrix}$	$\begin{bmatrix} 0 & 0 \\ 0 & 2 \end{bmatrix}$	$\begin{bmatrix} -1 & j \\ j & 1 \end{bmatrix}$	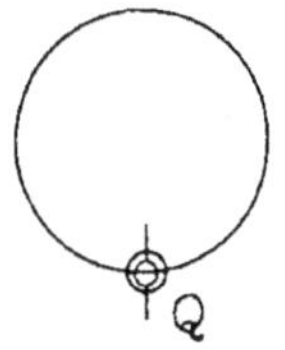	verti-cal dipole
H = 2	$A_0=B_0=-B=1$	$\begin{bmatrix} 2 & & 2 & \\ & 0 & & \\ 2 & & 2 & \\ & & & 0 \end{bmatrix}$	$\begin{bmatrix} 1 & 1 \\ 1 & 1 \end{bmatrix}$	$\begin{bmatrix} j & j \\ j & j \end{bmatrix}$	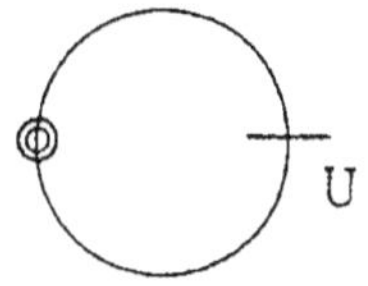	tilted dipole +45°
H =-2	$A_0=B_0=-B=1$	$\begin{bmatrix} 2 & & -2 & \\ & 0 & & \\ -2 & & 2 & \\ & & & 0 \end{bmatrix}$	$\begin{bmatrix} 1 & -1 \\ -1 & 1 \end{bmatrix}$	$\begin{bmatrix} -j & j \\ j & -j \end{bmatrix}$	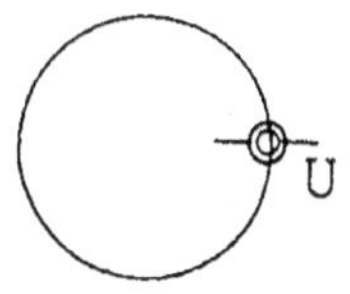	tilted dipole -45°

Table 12.1: Scattering properties of elementary targets /continued/

o - the inversion point, • - the null-polarization point

◎ - the inversion point and the double null-polarization point /for dipoles and helices/

12. 2. The Krogager's decomposition of the Sinclair matrix into matrices of the sphere, diplane, and helix

After Krogager ([102], [103]) the linear combination can be proposed of three amplitude matrices, of the sphere, diplane, and helix, representing any stable scatterer. It will be shown that such decomposition is entirely roll-invariant.

Matrices A_R of Table 1 will be taken into consideration because in the $\boldsymbol{u}^R$ basis there is a very simple form of the roll transformation by the 2θ angle about the V_H axis:

$$C_R^{ROLL} = \widetilde{C}_H^R * \begin{bmatrix} \cos\theta & \sin\theta \\ -\sin\theta & \cos\theta \end{bmatrix}_H C_H^R = \begin{bmatrix} e^{j\theta} & 0 \\ 0 & e^{-j\theta} \end{bmatrix}_R \quad (12.17)$$

Under the roll transformation,

$$A_{R\,ROLL} = \widetilde{C}_R^{R\,ROLL} A_R C_R^{R\,ROLL} \quad (12.18)$$

matrices are changing as follows:

$$A_{R\,\text{sphere}}: \quad \begin{bmatrix} 0 & j \\ j & 0 \end{bmatrix} \to \begin{bmatrix} 0 & j \\ j & 0 \end{bmatrix} \quad \text{- no change} \quad (12.18a)$$

$$A_{R\,\text{diplane}\,0^0}: \quad \begin{bmatrix} 1 & 0 \\ 0 & -1 \end{bmatrix} \to \begin{bmatrix} e^{j2\theta} & 0 \\ 0 & -e^{-j2\theta} \end{bmatrix} \quad (12.18b)$$

$$A_{R\,\text{right helix}}: \quad \begin{bmatrix} 2 & 0 \\ 0 & 0 \end{bmatrix} \to \begin{bmatrix} 2 & 0 \\ 0 & 0 \end{bmatrix} e^{j2\theta} \quad (12.18c)$$

$$A_{R\,\text{right helix}}: \quad \begin{bmatrix} 0 & 0 \\ 0 & -2 \end{bmatrix} \to \begin{bmatrix} 0 & 0 \\ 0 & -2 \end{bmatrix} e^{-j2\theta} \quad (12.18d)$$

All matrices in Table 1 are of strength $k = 1$. Altogether six parameters are at our disposal to construct the sum matrix. So, using the right helix, we can write

$$\begin{aligned} A_R^+ &= e^{j\varphi}\left\{k_s e^{j\varphi_s}\begin{bmatrix} 0 & j \\ j & 0 \end{bmatrix} + k_d^+\begin{bmatrix} e^{j2\theta} & 0 \\ 0 & -e^{-j2\theta} \end{bmatrix} + k_h e^{j2\theta}\begin{bmatrix} 2 & 0 \\ 0 & 0 \end{bmatrix}\right\} \\ &= e^{j\varphi}\begin{bmatrix} (k_d^+ + 2k_h)e^{j2\theta} & jk_s e^{j\varphi_s} \\ jk_s e^{j\varphi_s} & -k_d^+ e^{-j2\theta} \end{bmatrix} \end{aligned} \quad (12.19)$$

Similarly, with the left helix, we obtain

$$\begin{aligned} A_R^- &= e^{j\varphi}\left\{k_s e^{j\varphi_s}\begin{bmatrix} 0 & j \\ j & 0 \end{bmatrix} + k_d^-\begin{bmatrix} e^{j2\theta} & 0 \\ 0 & -e^{-j2\theta} \end{bmatrix} + k_h e^{-j2\theta}\begin{bmatrix} 0 & 0 \\ 0 & -2 \end{bmatrix}\right\} \\ &= e^{j\varphi}\begin{bmatrix} k_d^- e^{j2\theta} & jk_s e^{j\varphi_s} \\ jk_s e^{j\varphi_s} & -(k_d^- + 2k_h)\, e^{-j2\theta} \end{bmatrix} \end{aligned} \quad (12.19)$$

Using the (12.12) form of the Sinclair matrix in the R basis with

$$\phi = \varphi + \varphi_s \quad (12.20)$$

we obtain

$$\begin{aligned} &k_s = |A_{3R}| = \sqrt{A_0}, \\ &k_d = \sqrt{B_0 - |F|}, \text{ i.e. } k_d^+ = |A_{1R}| = \sqrt{B_0 - F} \text{ and } k_d^- = |A_{2R}| = \sqrt{B_0 + F} \\ &k_h = \tfrac{1}{2}\left||A_{2R}| - |A_{1R}|\right| = \sqrt{(B_0 - \sqrt{B_0^2 - F^2})/2} \end{aligned} \quad (12.21a)$$

$$\varphi = \tfrac{1}{2}(\arg A_{2R} + \arg A_{1R} - \pi)$$
$$2\theta = \tfrac{1}{2}(\arg A_{2R} - \arg A_{1R} + \pi) \qquad (12.21b)$$
$$\varphi_s = \arg A_{3R} - \tfrac{1}{2}(\arg A_{2R} + \arg A_{1R})$$

The chirality of the scatterer is:

$$\text{right, if } |A_{2R}| > |A_{1R}|, \qquad (12.22)$$
$$\text{left, if } |A_{1R}| > |A_{2R}|.$$

The k_s, k_d and k_s parameters are obviously roll-invariant because they depend on the roll-invariant Huynen parameters A_0, B_0 and F only.

Worth noticing are relations between elements of amplitude matrices in the two bases, *R* and *H*:

$$A_{2R} = jA_{3H} + \tfrac{1}{2}(A_{2H} - A_{1H}),$$
$$A_{1R} = jA_{3H} - \tfrac{1}{2}(A_{2H} - A_{1H}), \qquad (12.23)$$
$$A_{3R} = \tfrac{j}{2}(A_{2H} + A_{1H}).$$

These relations are useful when the matrix is known in the *H* basis and one wants to compute parameters of the decomposition from equalities (12.21) and (12.22).

Analytically, a sufficient condition for the roll-invariance is conservation of identical relations between strengths of matrices of the component targets under the roll transformation, independently of the ONP PP basis.

That roll-invariance can be explained also geometrically. When looking at the Poincare sphere models of the component matrices we observe their axial symmetry about the V_H axis. No doubt the model of the matrix of a sphere exhibits such a symmetry having the inversion point in the center of the model and with the axis of rotation after inversion coinciding with the V_H axis coming through the two null polarization points., Models of matrices for the right and left helices, with their double polarization points coinciding with the inversion points and located at the poles of the model on the V_H axis, show similar symmetry.

Though helices have axes of rotation after inversion perpendicular to the V_H axis and of direction dependent on their phases, nevertheless these directions have no influence on the magnitude of the scattered wave amplitude depending only on (proportional to) the distance between the transmit/receive polarization point and the inversion point.

The simplest roll-invariant decomposition is into sum of a sphere and two helices, right and left-handed, of strength $\sqrt{B_0 + F/2}$ and $\sqrt{B_0 - F/2}$, as seen from (12.11) and (12.18c and d). However, if it is desired to have only one helix component determining chirality of the scatterer, then one should observe that the sum of right and left helices of equal strength forms a diplane, rolled by an angle dependent on the phase difference between helices, according to (12.18b-d). The model of the diplane is characterized by the inversion point in its center and rotation after inversion axes are perpendicular to the V_H axis similarly like for helices. That way one can arrive geometrically at the roll-invariant Krogager decomposition because the roll-symmetry of all the component models makes the received power from all component targets independent of the targets roll-angle.

Owing to the roll-symmetry of the component models, they strengths can be discerned by applying in measurements only circularly polarized, right (R) and left (L), transmit and receive antennas. Considering scattering mechanisms of inversion and rotation it is immediately seen that: RR and RL transmit-receive combinations exclude reception from the left helix, LL and RL from the right helix, RR and LL from the sphere, and RL from the diplane. In all these cases no return from the target can be explained or by coincidence of transmit and/or receive polarization with the inversion point (helices - antipodal inversion points), or by the orthogonality of the scattered wave versus the receive polarization (sphere and diplane - rotation after inversion axes mutually perpendicular).

It should be observed that in the *H* basis amplitude matrices of the sphere, diplane 0^0, diplane 45^0, and orthogonalizer (the last multiplied by *j*) are the succeeding Pauli matrices, algebraically corresponding to the unit quaternions or orthogonal column vectors of the unitary U 4x4 transformation matrix. That orthogonal set of matrices can be exchanged for another set, also orthogonal, consisting of the same first and fourth matrix but instead of the second and third one including their sum and difference. These two new matrices represent the right and left helices. So, instead of nonothogonal set of matrices forming the Krogager's decomposition it is possible to apply the alternative decomposition employing the just proposed new orthogonal set. Both decompositions can also include the orthogonality matrix which will enable one to deal with decomposition of the nonsymmetrical Sinclair matrices.

13. Decomposition of the Partially Depolarizing Kennaugh Matrix into Four Non-Depolarizing Components

There are many decompositions possible but most interesting are those which are the ONP-PP-basis-invariant. From the mathematical point of view, the unique such decomposition, into the sum of four orthogonal non-depolarizing matrices of bistatic scattering, has been proposed by Cloude [33]. However, Cloude's component matrices are all non-symmetric, each one dependent on 16 parameters, and mutual relations between them have not yet been presented. Mainly two cumulative parameters of such a set of matrices, the 'entropy' and the so-called α parameter representing polarization properties of a monostatic or bistatic scattering object, have been used in practice for classification of 'distributed' targets [35]. Therefore it seems desirable to find alternative decompositions, though for matrices non-orthogonal, however exhibiting other important features. Some attempts of such trials have been described in a review paper by Cloude and Pottier [34]. Here, an alternative roll-invariant decomposition for the bistatic scattering will be presented in which the component matrices depend on 7, 5, 3 and 1 mutually independent parameters, the 16 parameters of the whole set.

13. 1. The decomposition into matrices depending on 7, 5, 3, and 1 parameters

Three steps of that decomposition will be considered. The first one divides the Kennaugh matrix of a bistatically scattering distributed target (BDT) into two parts: the non-symmetrical matrix, corresponding to nondepolarizing bistatically scattering 'point' target (BPT), preserving the complete polarization, and the symmetrical matrix, like of a 'monostatically scattering' distributed target (MDT), partially depolarizing the illuminating wave. Such decomposition is fully basis invariant because symmetrical matrices retain that form under any rotation of the polarization basis, similarly as nonsymmetrical matrices do.

The second step strictly follows the well known Huynen's decomposition [85] of the MDT matrix into the monostatically scattering point target (MPT) and the distributed 'polarimetric noise' target (DNT) matrices. That decomposition is also basis invariant because the feature of retaining the complete polarization does not depend on the change-of-basis procedure.

The last step, of decomposing the DNT matrix into two point noise target (PNT) matrices is only roll-invariant and can be manifold, as will be seen soon.

To analyze that decomposition the original matrix of 16 independent real parameters (elements) will be considered using the Cloude's notation [33, 34] being an extension of Huynen's notation from the monostatic to the bistatic scattering case. After the two first steps the decomposition of the BDT matrix into three matrices, of the BPT, MPT, and DNT, can be presented in any ONP PP basis , for instance the H basis, in the form:

$$\mathrm{K}_H = \mathrm{K}_H^B + \mathrm{K}_H^M + \mathrm{K}_H^N \tag{13.1}$$

with the original BDT matrix

$$\mathrm{K}_H = \begin{bmatrix} a_1 & b_1 & b_3 & b_5 \\ c_1 & a_2 & b_4 & b_6 \\ c_3 & c_4 & a_3 & b_2 \\ c_5 & c_6 & c_2 & a_4 \end{bmatrix}_H = \begin{bmatrix} A_0+B_0 & C+N & H+L & F+I \\ C-N & A+B & E+J & G+K \\ H-L & E-J & A-B & D+M \\ F-I & G-K & D-M & -A_0+B_0 \end{bmatrix}_H , \tag{13.2}$$

and the component matrices, of BPT,

$$\mathrm{K}_H^B = \begin{bmatrix} A_0^B+B_0^B & C^B+N & H^B+L & F^B+I \\ C^B-N & A^B+B^B & E^B+J & G^B+K \\ H^B-L & E^B-J & A^B-B^B & D^B+M \\ F^B-I & G^B-K & D^B-M & -A_0^B+B_0^B \end{bmatrix}_H , \tag{13.3}$$

MPT,

$$\mathbf{K}_H^M = \begin{bmatrix} A_0^M + B_0^M & C^M & H^M & F^M \\ C^M & A_0^M + B^M & E^M & G^M \\ H^M & E^M & A_0^M - B^M & D^M \\ F^M & G^M & D^M & -A_0^M + B_0^M \end{bmatrix}_H, \tag{13.4}$$

and DNT,

$$\mathbf{K}_H^N = \begin{bmatrix} B_0^N & 0 & 0 & F^N \\ 0 & B^N & E^N & 0 \\ 0 & E^N & -B^N & 0 \\ F^N & 0 & 0 & B_0^N \end{bmatrix}_H. \tag{13.5}$$

The way of construction of the first component matrix is straightforward. That matrix depends on 7 only parameters of the BDT matrix: 6 parameters of asymmetry, namely I, J, K, L, M, N, and the difference $A_0 - A$ disappearing for the monostatic scattering. All remaining parameters can be found when using the right side of equalities (E.6) representing conditions for preservation of complete polarization. So, one obtains:

$$\begin{aligned} A_0^B &= \frac{I^2 + J^2 + (A_0 - A)^2}{2(A_0 - A)} \geq 0, & C^B &= (IK + JL)/(A_0 - A), \\ A^B &= \frac{I^2 + J^2 - (A_0 - A)^2}{2(A_0 - A)}, & D^B &= (-IL + JK)/(A_0 - A), \\ & & E^B &= (KM - LN)/(A_0 - A), \\ B_0^B &= \frac{K^2 + L^2 + M^2 + N^2}{2(A_0 - A)} \geq 0, & F^B &= -(KN + LM)/(A_0 - A), \\ & & G^B &= -(IN + JM)/(A_0 - A), \\ B^B &= \frac{K^2 + L^2 - M^2 - N^2}{2(A_0 - A)}, & H^B &= (IM - JN)/(A_0 - A). \end{aligned} \tag{13.6}$$

Elements of the $\mathbf{K}_H^M$ matrix can be found by taking the difference between five elements of the $\mathbf{K}_H$ and $\mathbf{K}_H^B$ matrices. So, altogether five parameters define the $\mathbf{K}_H^M$ matrix and the remaining ones can be obtained by applying the equalities (E.6) again. That yields the following result:

$$\begin{aligned} A_0^M &= A_0 - A_0^B \geq 0, & B_0^M &= [(C^M)^2 + (D^M)^2 + (G^M)^2 + (H^M)^2]/(4A_0^M), \\ C^M &= C - C^B, & B^M &= [(C^M)^2 + (D^M)^2 - (G^M)^2 - (H^M)^2]/(4A_0^M), \\ D^M &= D - D^B, & & \\ G^M &= G - G^B, & E^M &= (C^M H^M - D^M G^M)/(2A_0^M), \\ H^M &= H - H^B, & F^M &= (C^M G^M + D^M H^M)/(2A_0^M). \end{aligned} \tag{13.7}$$

Elements of the remaining DNT matrix can be obtained by subtraction of the BPT and MPT matrices from the BDT matrix:

$$\mathrm{K}_H^N = \mathrm{K}_H - \mathrm{K}_H^B - \mathrm{K}_H^M \tag{13.8}$$

what yields

$$\begin{aligned} B_0^N &= B_0 - B_0^B - B_0^M, \\ B^N &= B - B^B - B^M, \\ E^N &= E - E^B - E^M, \\ F^N &= F - F^B - F^M. \end{aligned} \tag{13.9}$$

Decomposition of the DNT matrix into two PNT matrices will be presented in the following four alternative forms:

$$\mathrm{K}_H^N = \mathrm{K}_H^3 + \mathrm{K}_H^4 = \begin{bmatrix} B_0^3 & 0 & 0 & F^3 \\ 0 & B^N & E^N & 0 \\ 0 & E^N & -B^N & 0 \\ F^3 & 0 & 0 & B_0^3 \end{bmatrix}_H + \begin{bmatrix} B_0^4 & 0 & 0 & B_0{}^4 \\ 0 & 0 & 0 & 0 \\ 0 & 0 & 0 & 0 \\ B_0{}^4 & 0 & 0 & B_0^4 \end{bmatrix}_H \tag{13.10}$$

$$\mathrm{K}_H^N = \mathrm{K}_H^5 + \mathrm{K}_H^6 = \begin{bmatrix} B_0^5 & 0 & 0 & F^5 \\ 0 & B^N & E^N & 0 \\ 0 & E^N & -B^N & 0 \\ F^5 & 0 & 0 & B_0^5 \end{bmatrix}_H + \begin{bmatrix} B_0^6 & 0 & 0 & -B_0{}^6 \\ 0 & 0 & 0 & 0 \\ 0 & 0 & 0 & 0 \\ -B_0{}^6 & 0 & 0 & B_0^6 \end{bmatrix}_H \tag{13.11}$$

$$\mathrm{K}_H^N = \mathrm{K}_H^7 + \mathrm{K}_H^8 = \begin{bmatrix} B_0^7 & 0 & 0 & F^N \\ 0 & B^N & E^7 & 0 \\ 0 & E^7 & -B^N & 0 \\ F^N & 0 & 0 & B_0^7 \end{bmatrix}_H + \begin{bmatrix} B_0^8 & 0 & 0 & 0 \\ 0 & 0 & B_0^8 & 0 \\ 0 & B_0^8 & 0 & 0 \\ 0 & 0 & 0 & B_0^8 \end{bmatrix}_H \tag{13.12}$$

$$\mathrm{K}_H^N = \mathrm{K}_H^9 + \mathrm{K}_H^{10} = \begin{bmatrix} B_0^9 & 0 & 0 & F^N \\ 0 & B^N & E^9 & 0 \\ 0 & E^9 & -B^N & 0 \\ F^N & 0 & 0 & B_0^9 \end{bmatrix}_H + \begin{bmatrix} B_0^{10} & 0 & 0 & 0 \\ 0 & 0 & -B_0^{10} & 0 \\ 0 & -B_0^{10} & 0 & 0 \\ 0 & 0 & 0 & B_0^{10} \end{bmatrix}_H \tag{13.13}$$

with parameters (elements):

$$B_0^3 = \frac{(B_0^N - F^N)^2 + (B^N)^2 + (E^N)^2}{2(B_0^N - F^N)} \geq 0, \tag{13.14a}$$

$$B_0^4 = \frac{(B_0^N)^2 - (F^N)^2 - (B^N)^2 - (E^N)^2}{2(B_0^N - F^N)} \geq 0, \tag{13.14b}$$

$$B_0^5 = \frac{(B_0^N + F^N)^2 + (B^N)^2 + (E^N)^2}{2(B_0^N + F^N)} \geq 0, \quad (13.14c)$$

$$B_0^6 = \frac{(B_0^N)^2 - (F^N)^2 - (B^N)^2 - (E^N)^2}{2(B_0^N + F^N)} \geq 0, \quad (13.14d)$$

$$B_0^7 = \frac{(B_0^N - E^N)^2 + (B^N)^2 + (F^N)^2}{2(B_0^N - E^N)} \geq 0, \quad (13.14e)$$

$$B_0^8 = \frac{(B_0^N)^2 - (E^N)^2 - (B^N)^2 - (F^N)^2}{2(B_0^N - E^N)} \geq 0, \quad (13.14f)$$

$$B_0^9 = \frac{(B_0^N + E^N)^2 + (B^N)^2 + (F^N)^2}{2(B_0^N + E^N)} \geq 0, \quad (13.14g)$$

$$B_0^{10} = \frac{(B_0^N)^2 - (E^N)^2 - (B^N)^2 - (F^N)^2}{2(B_0^N + E^N)} \geq 0, \quad (13.14h)$$

$$F^3 = -\frac{(B_0^N - F^N)^2 - (B^N)^2 - (E^N)^2}{2(B_0^N - F^N)}, \quad (13.14i)$$

$$F^5 = \frac{(B_0^N + F^N)^2 - (B^N)^2 - (E^N)^2}{2(B_0^N + F^N)}, \quad (13.14j)$$

$$E^7 = -\frac{(B_0^N - E^N)^2 - (B^N)^2 - (F^N)^2}{2(B_0^N - E^N)}, \quad (13.14k)$$

$$E^9 = \frac{(B_0^N + E^N)^2 - (B^N)^2 - (F^N)^2}{2(B_0^N + E^N)}. \quad (13.14l)$$

Though each of the matrices K_H^3, K_H^5, K_H^7, and K_H^9 has 4 different elements, each one depends on 3 only independent parameters because of the conditions for preservation of complete polarization (see the upper equality of the right side of (E.6j)):

$$(B_0^3)^2 - (F^3)^2 = (B_0^5)^2 - (F^5)^2 = (B^N)^2 + (E^N)^2 \quad (13.15)$$

and

$$(B_0^7)^2 - (E^7)^2 = (B_0^9)^2 - (E^9)^2 = (B^N)^2 + (F^N)^2. \quad (13.16)$$

The above four decompositions of the K_H matrix of the BDT, each depending successively on 7, 5, 3, and 1 parameters (altogether 16 mutually independent parameters), were for the first time explicitly presented by this author in [54] and [59]. They can be derived, for example, when using the procedure proposed by Barnes [4] who discovered that there are always three roll-invariant decompositions possible (see also Holm and Barnes [81]). His second and third decomposition was here applied to the DNT obtained after the first decomposition of the MDT matrix. However, the second or third Barnes' decomposition can be also applied to the MDT matrix. Then, other obtained that way DNT matrices can undergo the third and first or the first and second decomposition, accordingly. Altogether three groups of four different roll-invariant decompositions are possible for the MDT matrix *in any ONP PP basis* , each group depending on four sets of 5+3+1 = 9 parameters. Of course the decomposition of the BDT matrix into matrices of MDT and BPT is unique.

13. 2. Physical realizability of the BDT matrix

Physical realizability of the original BDT matrix can be checked easily by the use of the above presented decomposition. For instance, it is *sufficient* to ascertain that all four component matrices have determined non-negative values of their first elements of general form of $a_1 = A_0 + B_0$. That is evident because all other elements fulfill conditions of preservation of the complete polarization what ensures their realizability. So, the following inequalities should hold for elements of the original matrix and matrices of the presented decomposition:

$$\begin{aligned} & A_0 + B_0 > 0 \\ & A_0 - A \geq 0, \\ & A_0 - A_0^B \geq 0 \text{ or, equivalently, } A_0^2 \geq A^2 + I^2 + J^2 + 2(A_0 - A)^2, \\ & B_0^N \geq 0, \\ & (B_0^N)^2 \geq (B^N)^2 + (E^N)^2 + (F^N)^2. \end{aligned} \tag{13.17}$$

13. 3. An example of decomposition of the completely depolarizing matrix

The simplest example of the proposed procedure is the decomposition of the matrix completely depolarizing each incoming wave. It may be presented, e. g., as follows:

$$\mathrm{K}_H = \begin{bmatrix} 4 & 0 & 0 & 0 \\ 0 & 0 & 0 & 0 \\ 0 & 0 & 0 & 0 \\ 0 & 0 & 0 & 0 \end{bmatrix}_H = \mathrm{K}_H^B + \mathrm{K}_H^M + \mathrm{K}_H^5 + \mathrm{K}_H^6 = \mathrm{K}_H^B + \mathrm{K}_H^M + \mathrm{K}_H^4 + \mathrm{K}_H^3$$

$$= \begin{bmatrix} 1 & 0 & 0 & 0 \\ 0 & -1 & 0 & 0 \\ 0 & 0 & -1 & 0 \\ 0 & 0 & 0 & -1 \end{bmatrix}_H + \begin{bmatrix} 1 & 0 & 0 & 0 \\ 0 & 1 & 0 & 0 \\ 0 & 0 & 1 & 0 \\ 0 & 0 & 0 & -1 \end{bmatrix}_H + \begin{bmatrix} 1 & 0 & 0 & 1 \\ 0 & 0 & 0 & 0 \\ 0 & 0 & 0 & 0 \\ 1 & 0 & 0 & 1 \end{bmatrix}_H + \begin{bmatrix} 1 & 0 & 0 & -1 \\ 0 & 0 & 0 & 0 \\ 0 & 0 & 0 & 0 \\ -1 & 0 & 0 & 1 \end{bmatrix}_H . \tag{13.17}$$

The component matrices in the ONP PP *H* basis of the reversed order represent successively an orthogonalizer, a sphere, and two helices: right- and left-handed.

13. 4. Exemplary presentation of two Poincare sphere models of matrices depending on 1 and 3 parameters

Exemplary presentations of matrices models will better explain the decomposition and, hopefully, will confirm usefulness and efficiency of previously elaborated methods of the Poincare sphere transformations.

Models of the K_H^3 and K_H^4 matrices will be presented. For simplicity reasons the upper indices of matrix elements, here *N*, 3, and 4, will be omitted.

<u>The K_H^4 matrix model.</u>

The K_H^4 matrix represents the right-handed helix (contrary to K_H^6 representing the left-handed helix). The inversion point is in the lower pole of its Poincare sphere model and the 'RR' received power, corresponding to the right circular polarizations of the transmit/receive antenna and the unit incident power, is equal to the square of the sphere diameter:

$$\mathrm{RR} = \tfrac{1}{2}\begin{bmatrix}1 & 0 & 0 & 1\end{bmatrix}_H \begin{bmatrix} B_0 & 0 & 0 & B_0 \\ 0 & 0 & 0 & 0 \\ 0 & 0 & 0 & 0 \\ B_0 & 0 & 0 & B_0 \end{bmatrix}_H \begin{bmatrix}1\\0\\0\\1\end{bmatrix}_H = \tfrac{1}{2}\begin{bmatrix}1 & 0 & 0 & 1\end{bmatrix}_H 2B_0 \begin{bmatrix}1\\0\\0\\1\end{bmatrix}_H = 2B_0 . \quad (13.18)$$

It is also equal to the span of the corresponding Sinclair matrix of the form:

$$A_H = \sqrt{\frac{B_0}{2}}\begin{bmatrix} 1 & -j \\ -j & 1 \end{bmatrix}_H ; \quad SpanA_H = 2a_{1H} = 2B_0 . \quad (13.19)$$

<u>The K_H^3 matrix model.</u>

The model of the K_H^3 matrix (identical considerations refer also to K_H^5, K_H^7, and K_H^9 matrices of the same structure), depends on four algebraic parameters, $B_0,\ B,\ E,\ F$, joined with one relation:

$$F^2 + B^2 = B_0^2 - E^2 . \quad (13.20)$$

It corresponds to the more complex Sinclair matrix (see (12.4))

$$A_H = \frac{e^{j(\phi+\phi_0)}}{\sqrt{2(B_0+B)}}\begin{bmatrix} B_0 + B & E - jF \\ E - jF & -(B_0+B) \end{bmatrix}_H \quad (13.21)$$

One of its geometrical parameters is the square of the Poincare sphere diameter:

$$\sigma_0 = 2(a_1 + a_0) = 2(B_0 + \sqrt{B_0^2 - F^2}) \quad (13.22)$$

The inversion point remains on the V_H axis and its one only non-zero coordinate inside the sphere of unit radius (see Fig. 13. 1. for $F < 0$),

$$\mathrm{V}_H^{\mathrm{I}} = -\frac{2}{\sigma_0}F = \sin 2\alpha_H^{\mathrm{O}_2} = \frac{\sqrt{B_0^2 - F^2} - B_0}{F} , \quad (13.23)$$

presents the second geometrical parameter. The cord joining the CO-POL Nulls, coming through the inversion point and perpendicular to the V_H axis, is inclined versus the Q_H axis at an angle $2\beta_H^{\mathrm{O}_2}$ which is the third geometrical parameter of that matrix (see for example Fig. 13. 1).

The angle $2\beta_H^{\mathrm{O}_2}$ can be found when searching for the characteristic *K* basis of the Sinclair matrix (13.21). Formulae (9.18) and (9.17) of the standard procedure lead to the characteristic polarization ratios

$$\rho_H^K = -j \ \text{ for } \ F < 0, \ \text{ and } \ \rho_H^K = j \ \text{ for } \ F > 0. \quad (13.24,25)$$

Then, realizing the first step of the transformation one obtains, after (9.19), the following matrix in the K' basis

$$A_{K'} = \begin{bmatrix} A_2' & 0 \\ 0 & A_1' \end{bmatrix}_{K'} = \frac{1}{\sqrt{2(B_0+B)}} \begin{bmatrix} B_0 + B \mp F \mp jE & 0 \\ 0 & -B_0 - B \mp F \mp jE \end{bmatrix} \tag{13.26}$$

and the Euler angles

$$2\delta_H^K = \mp 90^0, \quad 2\gamma_H^K = 90^0, \tag{13.27}$$

and

$$2\varepsilon_H^K = \tfrac{1}{2}[\arg A_2' - \arg A_1'] \pm 90^0 = \pm 90^0 - 2\beta_H^{O_2} = 2\beta_H^K, \tag{13.28}$$

with upper signs for the negative and lower for positive F values. There is an ambiguity of 180^0 in the last expression. It can be omitted by taking limited value of general phase of the matrix

$$|\mu| = \tfrac{1}{2}|\arg A_2' + \arg A_1'| < 90^0. \tag{13.29}$$

Owing to such limitation, the sense of the rotation after inversion axis will be precisely determined.

To clearly present derivation of final formulae, the intermediate parameter will be introduced:

$$t = \tan(-2\beta_H^{O_2}). \tag{13.30}$$

In turn one can write

$$\arg A_2' - \arg A_1' = \tan(-4\beta_H^{O_2}) = \frac{2t}{1+t^2}. \tag{13.31}$$

Now, for known values of arguments

$$\begin{aligned} \arg A_2' &= \arg(B_0 + B \mp F \mp jE) \\ \arg A_1' &= \arg(-B_0 - B \mp F \mp jE) \end{aligned} \tag{13.32}$$

one can find

$$\tan(-4\beta_H^{O_2}) = -\frac{E}{B} \tag{13.33}$$

independently of sign of the F element. That enables one to find simple solution of equation (13.31) in the form

$$t = \frac{B \pm \sqrt{B^2 + E^2}}{E} = \frac{B \pm \sqrt{B_0^2 + F^2}}{E} \tag{13.34}$$

That result can be rewritten in a more useful form of two equalities

$$t = \frac{\sqrt{B_0^2 + F^2} + B}{E} \tag{13.35A}$$

$$t = \frac{\sqrt{B_0^2 + F^2} - B}{-E} \tag{13.35B}$$

It is immediately seen that numerators of the two expressions (13.35), in view of (13.34), are positive. Therefore t and E are of the same sign when using (13.35A), and of opposite sign in (13.35B). Also, both expressions lead to the same relation between B and E:

$$B = -En, \tag{13.36}$$

with

$$n = \frac{1-t^2}{2t}, \tag{13.37}$$

what agrees with (13.31) and (13.33). The elements B and E are of the same sign when $t > 0$ and $t^2 > 1$, or when $t < 0$ and $t^2 < 1$, and negative in the other cases. The sign of t is opposite to the sign of $2\beta_H^{O_2}$. Combining signs of t, B, and E with expressions (13.35) the following final formulae for the $2\beta_H^{O_2}$ angle become evident:

$$2\beta_H^{O_2} = \arctan\frac{\sqrt{B_0^2 - F^2} - B}{E}; \quad \text{with} \quad \begin{cases} \mathrm{E} < 0, B > 0,\ 2\beta_H^{O_2} < 0,\ |2\beta_H^{O_2}| < 45^0 \\ \mathrm{E} > 0, \mathrm{B} < 0,\ 2\beta_H^{O_2} > 0,\ |2\beta_H^{O_2}| > 45^0 \end{cases} \tag{13.38a}$$

or

$$2\beta_H^{O_2} = \arctan\frac{\sqrt{B_0^2 - F^2} + B}{-E}; \quad \text{with} \quad \begin{cases} \mathrm{E} < 0, B < 0,\ 2\beta_H^{O_2} > 0,\ |2\beta_H^{O_2}| < 45^0 \\ \mathrm{E} > 0, \mathrm{B} > 0,\ 2\beta_H^{O_2} < 0,\ |2\beta_H^{O_2}| > 45^0 \end{cases} \tag{13.38b}$$

Beneath, also the inverse formulae will be presented enabling computation of the K_H^3 matrix elements, B_0, F, E, and B, for given geometrical parameters,

$$\begin{aligned} &r_0 > 0, \\ &2\alpha = 2\alpha_H^{O_2} = 90^0 - 2\alpha_K^{O_2} \\ &2\beta = 2\beta_H^{O_2} = -(2\beta_H^K - 90^0). \end{aligned} \tag{13.39}$$

Knowing the auxiliary parameters:

$$\begin{aligned} m &= \frac{2\sin 2\alpha}{1 + \sin^2 2\alpha}, \\ t &= -\tan 2\beta > 0, \\ n &= \frac{1 - t^2}{2t}, \end{aligned} \tag{13.40}$$

one obtains

$$\begin{aligned} B_0 &= \frac{4r_0^4}{4r_0^2 + m} > 0, \\ F &= -B_0 m. \end{aligned} \tag{13.41}$$

Therefore from (13.24), and for the definition of t in (13,30), we obtain the two remaining elements expressed as follows:
Moreover, combination of (13.36):

$$B = -En$$

and (13.20) yields

$$|E| = \sqrt{\frac{B_0^2 - F^2}{1 + n^2}}. \tag{13.42}$$

The sign of E can be found from (13.38) as a function of $2\beta_H^{O_2}$ and $|2\beta_H^{O_2}|$, similarly as the sign of B.

Fig. 13.1. presents results of computations and structure of the Poincare sphere model of the K_H^3 matrix corresponding to chosen geometrical parameters.

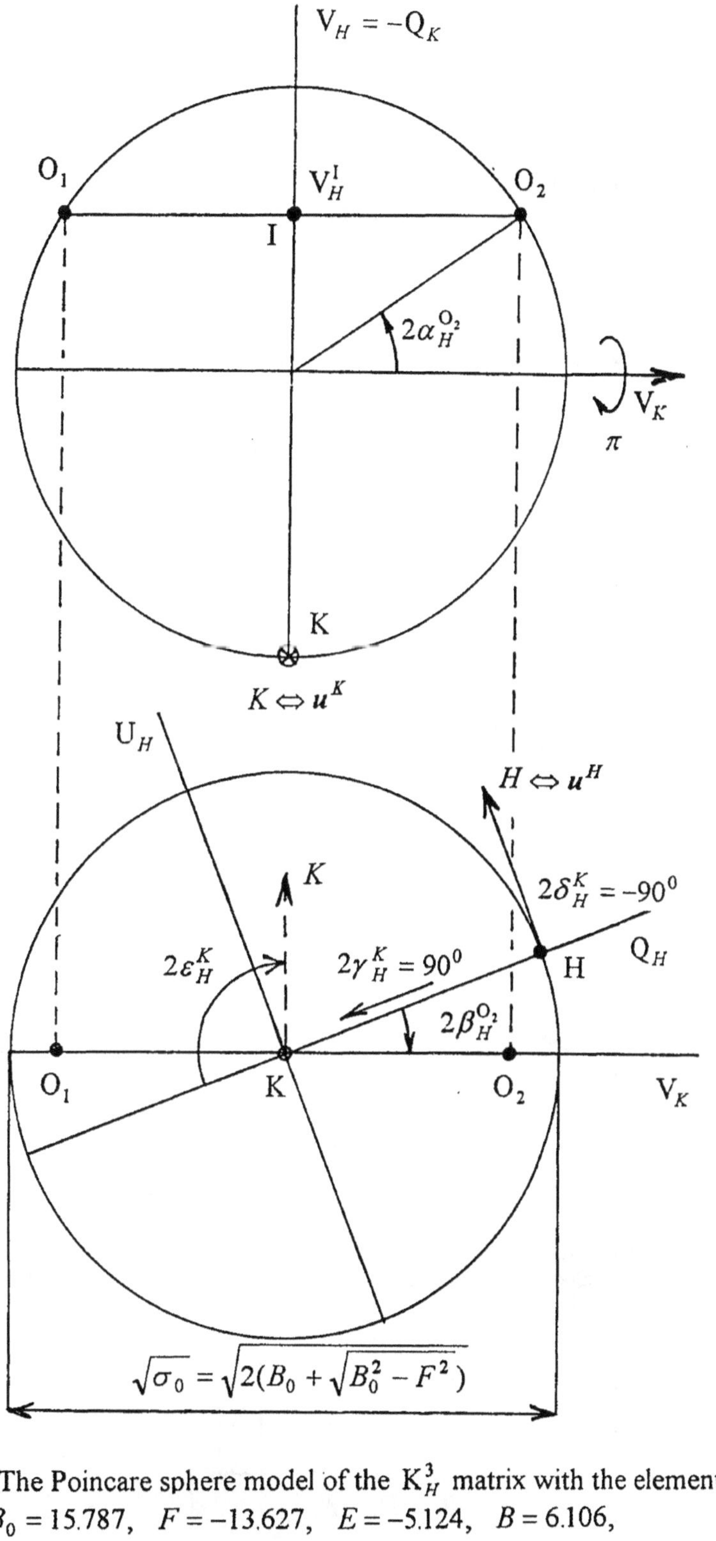

Fig. 13.1. The Poincare sphere model of the K_H^3 matrix with the elements:
$B_0 = 15.787, \quad F = -13.627, \quad E = -5.124, \quad B = 6.106,$
computed for the geometrical parameters: $r_0 = 4, \quad 2\alpha = 35^0, \quad 2\beta = -20^0$.

14. The Polarimetric Two-Ports

An important example of application of the Poincare sphere transformations is the theory of the polarimetric two-ports. Network theory of microwave four-ports can be extended to the domain of completely polarized fields of plane waves and used to polarimetric analysis of the electromagnetic two-ports. For that purpose, incoming and outgoing waves in two ports on each side of the four-port have been combined to create the polarization and phase vectors. Their transformation by two reflectance and two transmittance Sinclair matrices will be considered. Then, directions of the propagation z-axes at both ports will be reversed by rotation of spatial coordinate systems to form also two transmittance Jones matrices, thus allowing for cascading connection of the polarimetric two-ports.

Considerations will be limited to the losses, reciprocal systems and mutual relations between elements of the Sinclair and Jones matrices will be presented. A geometrical model of the scattering matrix of the whole two-port has been build up. It is of the form of four polarization spheres of tangential phasors. Each sphere represents one of four Sinclair matrices. Its diameter, inversion point, and rotation after inversion axis and angle are determined, shape and orientation of the polarization fork presented, and some special incident polarizations specified.

14. 1. Fundamental equations of the polarimetric two-ports

Scattering equation of the two-port in the coordinate systems as in Fig. 14.1, and in the linear ONP PP basis H , will be presented in the following form

$$\begin{bmatrix} S_H & \widetilde{T}_H \\ T_H & R_H \end{bmatrix}\begin{bmatrix} u_H^A \\ u_H^B \end{bmatrix}=\begin{bmatrix} \lambda_{AA}u_H^{SAA}* + \lambda_{AB}u_H^{SAB}* \\ \lambda_{BA}u_H^{SBA}* + \lambda_{BB}u_H^{SBB}* \end{bmatrix}=\begin{bmatrix} \lambda_A u_H^{SA}* \\ \lambda_B u_H^{SB}* \end{bmatrix}. \tag{14.1}$$

with the real positive coefficients λ, Sinclair back-scattering (symmetrical) reflectance matrices

$$S_H=\begin{bmatrix} S_2 & S_3 \\ S_3 & S_1 \end{bmatrix}_H, \qquad R_H=\begin{bmatrix} R_2 & R_3 \\ R_3 & R_1 \end{bmatrix}_H, \tag{14.1a}$$

and the Sinclair forward scattering, and therefore nonsymmetrical, transmittance matrices

$$T_H=\begin{bmatrix} T_2 & T_3 \\ T_4 & T_1 \end{bmatrix}_H, \qquad \widetilde{T}_H=\begin{bmatrix} T_2 & T_4 \\ T_3 & T_1 \end{bmatrix}_H. \tag{14.1b}$$

The same scattering equation, with the Sinclair scattering matrices expressed by their elements and with components of the electric field amplitude PP vectors defined as in (2.1), introduced instead of column PP unit u vectors, is of the form

$$\begin{bmatrix} S_2 & S_3 & T_2 & T_4 \\ S_3 & S_1 & T_3 & T_1 \\ T_2 & T_3 & R_2 & R_3 \\ T_4 & T_1 & R_3 & R_1 \end{bmatrix}_H \begin{bmatrix} E_{0y}^A \\ E_{0x}^A \\ E_{0y}^B \\ E_{0x}^B \end{bmatrix} = \begin{bmatrix} E_{0y}^{SA}* \\ E_{0x}^{SA}* \\ E_{0y}^{SB}* \\ E_{0x}^{SB}* \end{bmatrix}. \tag{14.2}$$

The complex and symmetrical scattering matrix of the whole two-port, presented in the above equation, depends on 20 real parameters. That number can be reduced to 10 (one half) by taking no losses case, apart from the previously assumed reciprocity. The resulting matrix has to be unitary and therefore its component

Sinclair matrices should fulfill equations (the transposition has been shown for the nonsymmetrical matrices only):

$$S_H * S_H + \widetilde{T}_H * T_H = T_H * \widetilde{T}_H + R_H * R_H = \begin{bmatrix} 1 & 0 \\ 0 & 1 \end{bmatrix} \tag{14.2a,b}$$

and

$$S_H * \widetilde{T}_H + \widetilde{T}_H * R_H = \begin{bmatrix} 0 & 0 \\ 0 & 0 \end{bmatrix} \tag{14.2c}$$

They result in the following fundamental equations for the two-port:

$$SpanS = SpanR = 2 - SpanT \tag{14.3a}$$

or

$$SpanS + SpanT = SpanR + SpanT = 2\,,$$

$$|\det S|^2 = |\det R|^2 = 1 - SpanT + |\det T|^2 \tag{14.3b}$$

or

$$|\det T|^2 = 1 - SpanS + |\det S|^2 = 1 - SpanR + |\det R|^2$$

and

$$\arg\det S + \arg\det R = 2\arg\det T\,. \tag{14.3c}$$

Here, indices of matrices have been omitted because span and determinant are independent of the ONP PP basis choice.

Two fundamental equations, (14.3a) and (14.3b), govern diameters of all four Sinclair matrices, because squares of those diameters are

$$\begin{aligned} \sigma_{oS} &= \sigma_{oR} = SpanS + 2|\det S| \\ \sigma_{oT} &= \sigma_{o\widetilde{T}} = SpanT + 2|\det T| \end{aligned} \tag{14.4}$$

The last fundamental equation, (14.3c), bounds phases of the component Sinclair matrices of the reciprocal lossless two-port depending on 10 real parameters.

14.2. A physical interpretation of the phase relations between the component Sinclair matrices of the two-port

There may be an interesting physical interpretation of equation (14.3c). If we call the overall „electrical length" of the two-port as equal to the phase of its transmittance matrix,

$$\xi = \tfrac{1}{2}\arg\det T = \xi_0 + \mu\,, \tag{14.5a}$$

being a sum of its „canonical phase" of the matrix in its *characteristic coordinate system* corresponding to the characteristic ONP PP basis K,

$$\xi_0 = \tfrac{1}{2}\arg\det T_{CCS}; \qquad T_{CCS} = \begin{bmatrix} A_1 & B_1 + jB_2 \\ -B_1 - jB_2 & A_2 \end{bmatrix}, \qquad T_K = T_{CCS}e^{j\mu}\,, \tag{14.5b}$$

and an „additional phase" μ, then we may consider backscatterings by the two-port as scatterings from a plane inside the two-port, at an electrical distance α from the port A,

$$\alpha = \tfrac{1}{4}\arg\det S = \alpha_0 + \sigma\,, \tag{14.6a}$$

and from the same plane, on the other side of the two-port, at a distance β from the port B ,

$$\beta = \tfrac{1}{4}\arg\det R = \beta_0 + \rho \tag{14.6b}$$

All the above mentioned phases and the corresponding distances are connected with the equalities

$$\xi_0 = \alpha_0 + \beta_0, \quad \xi = \alpha + \beta, \quad \text{and} \quad \mu = \sigma + \rho. \tag{14.7}$$

The canonical phase of reflection at the port A, α_0, and the additional phase σ, will be defined in the next section.

14.3. The five-parameter model of the two-port

Further reduction of the number of independent real parameters is necessary to obtain the geometrical model of the two-port. That can be reached by changing the ONP H basis for the characteristic basis, KT, of the T matrix. Of course, the KT basis will be neither the characteristic basis, KS, of the S matrix, nor the characteristic basis, KR, of the R matrix. The resulting canonical form of the transmittance scattering matrix can be presented as

$$T_{KT} = \widetilde{C}_H^{KT} T_H C_H^{KT} = \begin{bmatrix} A_2 & B_1 + jB_2 \\ -B_1 - jB_2 & A_1 \end{bmatrix} e^{j\mu} \tag{14.8}$$

while the form of the reflectance Sinclair matrices in the new KT basis will not undergo any simplifications and will read:

$$S_{KT} = \widetilde{C}_H^{KT} S_H C_H^{KT} = \begin{bmatrix} S_2 & S_3 \\ S_3 & S_1 \end{bmatrix}_{KT} \tag{14.9a}$$

and

$$R_{KT} = \widetilde{C}_H^{KT} R_H C_H^{KT} = \begin{bmatrix} R_2 & R_3 \\ R_3 & R_1 \end{bmatrix}_{KT} \tag{14.9b}$$

Now, the number of parameters has been reduced to 7. That is 3 parameters less, because of 3 Euler angles of the basis rotation for the transmittance matrix. What can be done more is neglecting two phases: one, corresponding to the „additional electrical length" of the two-port, it is μ, and another one, responsible for the „additional electrical distance" of the scattering plane inside the two-port from the port A., it is σ. By the following equation we shall include σ in the definition of the S matrix with S_3 element, in the KT basis, defined as real:

$$S_{KT} = \left\{ \begin{bmatrix} S_2 & S_3 \\ S_3 & S_1 \end{bmatrix} e^{j2\sigma} \right\}_{KT} ; \qquad S_3 = S_3{}^{*}. \tag{14.10a}$$

Now, the canonical phase for the port A can also be defined as:

$$\alpha_0 = \tfrac{1}{4}\arg\det S_{KT} - \sigma = \tfrac{1}{4}\arg(S_2 S_1 - S_3^2). \tag{14.10b}$$

When neglecting additional phases, μ and σ (and ρ, in virtue of (14.7)), not essential for further considerations, the resultant scattering matrix of the two-port and its four-sphere geometrical model can be considered as depending on five only real parameters: A_2, A_1, B_1, B_2, and S_3.

14.4. The allowed range for the S_3 parameter and the *S* matrix dependence on *T* matrix elements

The S_3 value of equation (14.10) cannot be chosen arbitrarily. It should be contained in the range determined as follows:

$$SpanS - 2|\det S| \le \left(\frac{2S_3}{\sin\theta}\right)^2 \le SpanS + 2|\det S| = \sigma_{0S} \tag{14.11}$$

where θ is an angle between the Q_{KT} axis of the CCS of the *T* matrix and the (OI) vector, from the center O of the Poincare sphere model of that matrix to its inversion point I. The range (14.11) is a direct consequence of the unitarity of the two-port scattering matrix. In order to show that, the two-port scattering matrix (14.12) will be considered, corresponding to the matrix (14.2) transformed to the *KT* basis and with the additional phases μ and σ neglected,

$$\begin{bmatrix} S_2 & S_3 & A_2 & -B_1 - jB_2 \\ S_3 & S_1 & B_1 + jB_2 & A_1 \\ A_2 & B_1 + jB_2 & R_2 & R_3 \\ -B_1 - jB_2 & A_1 & R_3 & R_1 \end{bmatrix}. \tag{14.12}$$

By inspection of a Hermitian product of the two first columns of that unitary matrix with S_3 real:

$$0 = S_2 {}^* S_3 + S_3 S_1 + B_1(A_2 - A_1) + jB_2(A_2 + A_1) \tag{14.13}$$

and considering the C^1 plane with complex numbers expressed by vectors: $S_2 {}^* S_3$, $S_3 S_1$, and

$$\begin{aligned} W &= B_1(A_2 - A_1) + jB_2(A_2 + A_1) \\ &= -C\, e^{j\gamma}, \end{aligned} \tag{14.14}$$

we arrive at a vector diagram on that plane, as in Fig. 2. It shows that two solutions of the equation (14.13) for an S_3 given are possible. They exist when only the length *H* of a straight line segment, as shown in Fig. 2, is real.

In order to express two elements, S_2 and S_1, and the length *H*, as a function of the five real parameters of the two-port model, we will introduce the following auxiliary real parameters:

$$\begin{aligned} P &= 1 - B_1^2 - B_2^2 - A_2^2 \ge 0, \\ Q &= 1 - B_1^2 - B_2^2 - A_1^2 \ge P, \end{aligned} \tag{14.15a}$$

$$\begin{aligned} C^2 &= WW^*, \\ D^2 &= PQ - C^2, \\ E &= \frac{Q - P}{2C}. \end{aligned} \tag{14.15b}$$

In terms of those parameters we obtain:

$$|S_2|^2 = P - S_3^2,$$
$$|S_1|^2 = Q - S_3^2 \geq |S_2|^2, \tag{14.16}$$

$$H = \tfrac{1}{2}\sqrt{2(P+Q)S_3^2 - 4(1+E^2)S_3^4 - C^2}\,. \tag{14.17}$$

and, introducing additional parameters,

$$A = \frac{C}{2} - ES_3^2,$$
$$B = C - A, \tag{14.18a}$$

$$\cos\gamma = -\operatorname{Re} W / C,$$
$$\sin\gamma = -\operatorname{Im} W / C. \tag{14.18b}$$

the following two solutions for the arguments of S_2 and S_1 elements can be found:

$$\arg S_2 = \arg\{A\cos\gamma \pm H\sin\gamma - j(A\sin\gamma \mp H\cos\gamma)\},$$
$$\arg S_1 = \arg\{B\cos\gamma \mp H\sin\gamma + j(B\sin\gamma \pm H\cos\gamma)\}. \tag{14.19}$$

The length of H, as expressed by (14.17), is real for the S_3 in the range

$$\frac{1}{2}\sqrt{\frac{Q+P-2D}{1+E^2}} \leq S_3 \leq \frac{1}{2}\sqrt{\frac{Q+P+2D}{1+E^2}} = \frac{\sqrt{\sigma_{oS}}}{2}\sin\theta \tag{14.20}$$

what corresponds to (14.11) because

$$Q + P = Span S\,, \tag{14.21}$$

$$D = |\det S|\,, \tag{14.22}$$

and

$$E = \cot\theta\,. \tag{14.23}$$

14.5. The *R* matrix dependence on the *S* and *T* matrices

Using again the matrix equations (14.2a,b,c) and after transformation of their Sinclair matrices to the *TK* basis, as in (14.8) and (14.9), we arrive at the following expressions for the *R* matrix elements:

$$R_2 = \frac{1}{\det T^*}[-S_2{}^* A_1 A_2 - S_3 B_1 (A_2 + A_1) + jS_3 B_2 (A_2 - A_1) - S_1{}^*(B_1^2 + B_2^2)],$$
$$R_1 = \frac{1}{\det T^*}[-S_2{}^*(B_1^2 + B_2^2) + S_3 B_1 (A_2 + A_1) + jS_3 B_2 (A_2 - A_1) - S_1{}^* A_1 A_2],$$
$$R_3 = \frac{1}{\det T^*}[S_2{}^* A_2 (B_1 - jB_2) - S_3 (A_2^2 - B_1^2 - B_2^2) - S_1{}^* A_2 (B_1 + jB_2)]. \tag{14.24}$$

The R matrix, corresponding to the T and S matrices as in (14.8) and (14.10), is of the form:

$$R_{KT} = \left\{ \begin{bmatrix} R_2 & R_3 \\ R_3 & R_1 \end{bmatrix} e^{j2(\mu-\sigma)} \right\}_{KT} . \tag{14.25}$$

A numerical example verifying the above presented formulae can be found in [61, 68, 69].

14.6. Geometrical parameters of the two-port Sinclair matrices' models

A few of useful formulae can be helpful in computing geometrical parameters of the models. Some of them are in terms of additional parameters presented in equations (14.15).

Diameters of the polarization spheres are:

$$\begin{aligned} \sqrt{\sigma_{oS}} &= \sqrt{\sigma_{oR}} = \sqrt{Q+P+2D}, \\ \sqrt{\sigma_{oT}} &= \sqrt{\sigma_{o\tilde{T}}} = \sqrt{2-(P+Q)+2\sqrt{1-(P+Q)+D^2}}\,. \end{aligned} \tag{14.26}$$

CO-POL Null polarization ratios in the KT basis:

$$\begin{aligned} \rho_{KT}^{OS} &= (-S_3 \mp \sqrt{-\det S})/S_1, \\ \rho_{KT}^{OR} &= (-R_3 \mp \sqrt{-\det R})/R_1, \\ \rho_{KT}^{OT} &= \rho_{KT}^{O\tilde{T}} = \mp j\sqrt{A_2/A_1}\,. \end{aligned} \tag{14.27}$$

They correspond to the fork angles (subtended between prongs pointing to the CO-POL Null polarizations):

$$\begin{aligned} \left(4\gamma^F\right)_S &= \left(4\gamma^F\right)_R = 2\arcsin\sqrt{\frac{4D}{Q+P+2D}}, \\ \left(4\gamma^F\right)_T &= \left(4\gamma^F\right)_{\tilde{T}} = 2\arcsin\frac{2\sqrt{A_1A_2}}{A_2+A_1}\,. \end{aligned} \tag{14.28}$$

Polarization ratios in the KT basis for maximum and minimum transferred powers (corresponding to minimum and maximum powers scattered at the ports) are:

$$\rho_{KT}^{MT,NT} = \left[E \mp \sqrt{1+E^2}\right] e^{j[\arg(-W^* - 90^\circ \pm 90^\circ)} \tag{14.29}$$

The corresponding powers related to the incoming power at each port equal to one are:

- transferred powers:

$$P_{transf.} = [1-(P+Q)/2] \pm C\sqrt{1+E^2}\,, \tag{14.30}$$

-powers reflected at the ports:

$$P_{refl.} = [(P+Q)/2] \mp C\sqrt{1+E^2}\,. \tag{14.31}$$

14.7. Mutual orientations of the Sinclair matrices' models of the two-port

Mutual orientations of these models can be found by computation of two pairs of the three Euler angles by which the KT basis should be rotated to the characteristic bases, KS or KR, of the two reflectance matrices.

The whole procedure of obtaining the S and R matrices in their characteristic bases is rather simple because of known values of the polarization ratios

$$\rho_{KT}^{KS} = -\rho_{KT}^{KR} = \rho_{KT}^{NT} \tag{14.32}$$

according to (14.29). What has to be done is just to apply formulae (9.17) - (9.19) remembering that for monostatic matrices there is always $A_3' = 0$, and that the indices of the Euler angles should be changed appropriately. The reflectance matrices will take forms

$$S_{KS} = \left\{ \begin{bmatrix} S_2 & 0 \\ 0 & S_1 \end{bmatrix} e^{j\mu_S} \right\}_{KS} \tag{14.33a}$$

and

$$R_{KR} = \left\{ \begin{bmatrix} R_2 & 0 \\ 0 & R_1 \end{bmatrix} e^{j\mu_R} \right\}_{KR} \tag{1433b}$$

with real nonnegative parameters

$$\{S_2\}_{KS} = \{R_2\}_{KR}\,, \quad \{S_1\}_{KS} = \{R_1\}_{KR} \tag{14.34}$$

and with the phase angles, according to (14.3c), (14.10) and (14.25), satisfying the equalities

$$\mu_S + \mu_R = 2\xi_o; \qquad \mu_S = 2\alpha_o, \quad \mu_R = 2\beta_o = \tfrac{1}{2}\arg\det R_{KT} - 2(\mu - \sigma). \tag{14.35}$$

It should be observed that the models of mutually transposed transmittance matrices differ only by 180° rotation about the Q_{KT} axis, and by the same rotation differ also the reflectance matrices. Fig. 14.3 illustrates mutual orientations of such models corresponding to the numerical example presented in [61, 68, 69].

14.8. Four types of scattering and cascading matrices of two-ports

When analyzing the cascade of two-ports, it is advisable to use rather the complex amplitude (CA) column vectors, representing directional Jones vectors [92, 93, 107], instead of the PP vectors, in order to apply traditional notation of the network theory.

So, the scattering equation for the two-port (in the local spatial coordinate systems with z-axes directed to the two-port on its both sides) will be used in the following form of that new polarimetric notation. Instead of (14.1) we will write

$$\begin{bmatrix} S & \tilde{T} \\ T & R \end{bmatrix} \begin{bmatrix} a_1 \\ a_2 \end{bmatrix} = \begin{bmatrix} b_1 \\ b_2 \end{bmatrix} \tag{14.36}$$

where the lower indices indicating the ONP PP basis, still applicable, have been omitted for simplicity reasons. Here, complex amplitude column vectors of the incoming and outgoing waves are designated by a_i and b_i, respectively, with the lower index 'i' denoting the port number (1 or 2). One has only to remember about two slightly different rules of transformation which those CA vectors undergo under change of the ONP PP basis or under reversal (by rotation) of the spatial coordinate system. The b values (contrary to the a values) used in the scattering equation (14.36) are conjugate values of the corresponding PP vector components as representing waves propagating in the -z direction of the local coordinate system. (However, the later used values a^0 and b^0, expressed for the spatially reversed z-axes of the local coordinate systems, will behave oppositely. Namely, the a^0's will be conjugate values of the corresponding PP vector components.)

By rearranging the matrix equation (14.36), two different *cascading* matrices and the corresponding equations can be obtained. They depend on at which port, 1 or 2, the incoming and outgoing wave has to be transformed by the cascading matrix to the other port. Such matrices can take the following forms (for comparison with microwave two-port equations see [82]):

$$\begin{bmatrix} \tilde{T}-ST^{-1}R & ST^{-1} \\ -T^{-1}R & T^{-1} \end{bmatrix} \begin{bmatrix} a_2 \\ b_2 \end{bmatrix} = \begin{bmatrix} b_1 \\ a_1 \end{bmatrix} \quad \text{or} \quad \begin{bmatrix} T-R\tilde{T}^{-1}S & R\tilde{T}^{-1} \\ -\tilde{T}^{-1}S & \tilde{T}^{-1} \end{bmatrix} \begin{bmatrix} a_1 \\ b_1 \end{bmatrix} = \begin{bmatrix} b_2 \\ a_2 \end{bmatrix} \qquad (14.37a,b)$$

The first one of these two forms will be chosen for further applications.

Other types of scattering and cascading matrices of the two-ports are also possible (the names of these types have been introduced as for the types of the 2X2 complex matrices, according to similar dependence on the z-axis reversal at the 'output' and 'input'). Table 14.1 presents four types of those matrices in their transformation equations. These types differ by reversal (by rotation) of the spatial coordinate system at the ports: 1, 2, and 1 and 2, successively.

14.9. Transformation rules for reflectance and transmittance matrices

Together with the z-axis reversal, also some Sinclair scattering matrices and some CA column vectors are properly transformed. The S1 type Sinclair matrices are transformed according to formulae under z-axis reversal, and using formulae under basis rotation. The complete set of transformations for the 2X2 complex amplitude matrices contained in the scattering matrices of the two-port is given below,

$$\begin{aligned} S_H &= \tilde{C}_K^H \, S_K \, C_K^H \\ R_H &= \tilde{C}_K^H \, R_K \, C_K^H \\ T_H &= \tilde{C}_K^H \, T_K \, C_K^H \\ \tilde{T}_H &= \tilde{C}_K^H \, \tilde{T}_K \, C_K^H \end{aligned} \qquad (14.38a)$$

$$\begin{aligned} T_H^{o} &= C_H^K \, T_K^{o} \, C_K^H \, ; & T_K^{o} &= C_K^{o} \, {*}T_K \\ \tilde{T}_H^{o} &= C_H^K {*} \, \tilde{T}_K^{o} \, C_K^H {*} \, ; & \tilde{T}_K^{o} &= \tilde{T}_K \, C_K^{o} \, {*} \\ {}^{o}T_H &= C_H^K {*} \, {}^{o}T_K \, C_K^H {*} \, ; & {}^{o}T_K &= T_K \, C_K^{o} \, {*} \\ {}^{o}\tilde{T}_H &= C_H^K \, {}^{o}\tilde{T}_K \, C_K^H \, ; & {}^{o}\tilde{T}_K &= C_K^{o} {*} \, \tilde{T}_K \end{aligned} \qquad (14.38b)$$

$$\begin{aligned} {}^{o}S_H^{o} &= C_H^K \, {}^{o}S_K^{o} \, \tilde{C}_H^K \, ; & {}^{o}S_K^{o} &= C_K^{o} {*} \, S_K \, C_K^{o} {*} \\ {}^{o}R_H^{o} &= C_H^K \, {}^{o}R_K^{o} \, \tilde{C}_H^K \, ; & {}^{o}R_K^{o} &= C_K^{o} {*} \, R_K \, C_K^{o} \end{aligned} \qquad (14.38c)$$

where

$$\tilde{C}_K^H = C_H^K {*} \quad \text{and} \quad C_K^{o} = \tilde{C}_H^K \begin{bmatrix} -1 & 0 \\ 0 & 1 \end{bmatrix} C_H^K \qquad (14.38d)$$

14.10. Transformation rules for complex amplitudes

As mentioned in Section 14.8, the difference in comparison with previously used formulae has to be taken into account when dealing with the complex amplitudes instead of PP vectors.

According to definitions of the Table 14.1, the incoming CA vectors a represent waves propagating in the $+z$ directions, so their transformations are the same as those of the PP vectors, independently of the port numbers. On the other hand, the b vectors correspond to waves propagating in $-z$ direction and, being CAs, they require different kind of transformation (that is one of reasons why the use of PP vectors is preferred in general theory of polarimetry). Summarizing, the following rules of transformations under change of the PP basis and under z-axis reversal apply:

$$\begin{aligned} a_{iH} &= C_H^K \, a_{iK}, \\ b_{iH} &= C_H^K * \, b_{iK}, \\ a_{iH}^{o} &= C_H^K * \, a_{iK}^{o}; & a_{iK}^{o} &= C_K^{o} \, a_{iK}, \\ b_{iH}^{o} &= C_H^K \, b_{iK}^{o}; & b_{iK}^{o} &= C_K^{o} * \, b_{iK}. \end{aligned} \tag{14.39}$$

14.11. An example of a cascade of three two-ports

A cascade of three two-ports will be considered with z-axes at the input and output ports of the cascade directed to the cascade and between the two-ports directed to the central two-port (see Table 14.II). Scattering matrices of the successive two-ports, No.1, 2, and 3, are of the type P1, S1, and P2, respectively. Wave amplitudes between the two-ports are then:

$$\begin{aligned} a_{2(1)}^{o} &= b_{1(2)} & a_{2(2)} &= b_{1(3)}^{o} \\ b_{2(1)}^{o} &= a_{1(2)} & b_{2(2)} &= a_{1(3)}^{o} \end{aligned} \tag{14.40}$$

The amplitude transformation equation for the cascade has been also shown in Table 14.2.

14.12. Concluding remarks

The here presented polarization sphere approach to the theory of polarimetric two-ports, developed by Czyz and Boerner [16, 69], has been based on matrix calculus in the two-dimensional complex space of the polarization and phase vectors and/or complex amplitude vectors. Owing to that approach it was possible to obtain simple canonical forms of bistatic (forward) scattering transmittance matrices and their polarization sphere geometrical models, together with models of two reflectance matrices.

These models depend on no more than five real parameters, the fifth of which, S_3, governs rotation only of the S and R matrix models about their characteristic O-KS and O-KR axes, what was presented in a most simple way using the characteristic coordinate system of the transmittance T matrix.

The presented example of the polarization sphere approach, which follows suggestions contained in a short communication by Kennaugh [96], indicates physical reasons for the extension of the pioneering work of Huynen [85] from the mono- to bistatic scattering.

The theory of cascading connection of polarimetric two-ports can also be treated as an application of the theory of cascading connection of microwave two-ports presented by Horton and Wenzel [82] and continuation of an early work on similar subject published by this author in 1955 [38].

Table 14.1

Four types of scattering matrices of the reciprocal lossless two-port and the corresponding cascading matrices in their transformation equations

THE S1 TYPE MATRIX

$$\begin{bmatrix} S & \tilde{T} \\ T & R \end{bmatrix}\begin{bmatrix} a_1 \\ a_2 \end{bmatrix} = \begin{bmatrix} b_1 \\ b_2 \end{bmatrix}$$

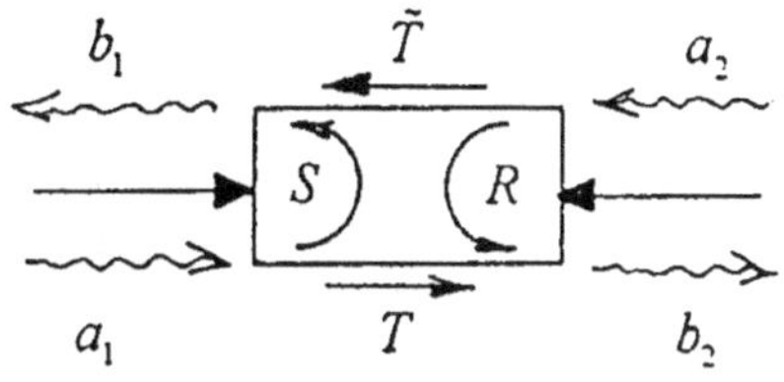

$$\begin{bmatrix} \tilde{T} - S\,T^{-1}R & S\,T^{-1} \\ -T^{-1}R & T^{-1} \end{bmatrix}^{(N)}\begin{bmatrix} a_2 \\ b_2 \end{bmatrix}^{(N)} = \begin{bmatrix} b_1 \\ a_1 \end{bmatrix}^{(N)} = \begin{bmatrix} a_2^\circ \\ b_2^\circ \end{bmatrix}^{(N-1)}$$

THE P1 TYPE MATRIX

$$\begin{bmatrix} S & \tilde{T}^\circ \\ T^\circ & {}^\circ R^\circ \end{bmatrix}\begin{bmatrix} a_1 \\ a_2^\circ \end{bmatrix} = \begin{bmatrix} b_1 \\ b_2^\circ \end{bmatrix}$$

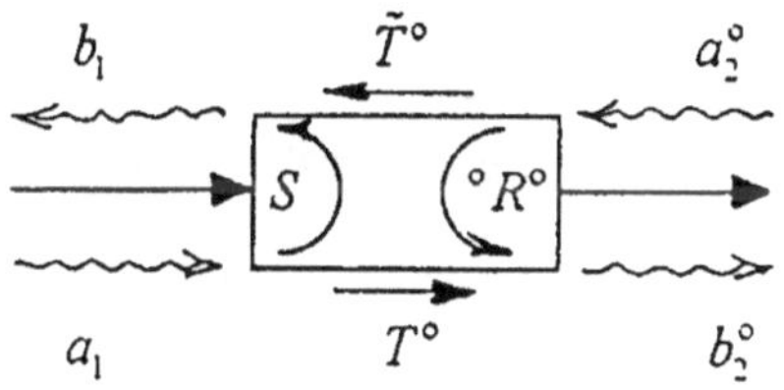

$$\begin{bmatrix} \tilde{T}^\circ - ST^{\circ-1}\,{}^\circ R^\circ & ST^{\circ-1} \\ -T^{\circ-1}\,{}^\circ R^\circ & T^{\circ-1} \end{bmatrix}^{(N)}\begin{bmatrix} a_2^\circ \\ b_2^\circ \end{bmatrix}^{(N)} = \begin{bmatrix} b_1 \\ a_1 \end{bmatrix}^{(N)} = \begin{bmatrix} a_2^\circ \\ b_2^\circ \end{bmatrix}^{(N-1)}$$

THE P2 TYPE MATRIX

$$\begin{bmatrix} {}^\circ S^\circ & {}^\circ\tilde{T} \\ {}^\circ T & R \end{bmatrix}\begin{bmatrix} a_1^\circ \\ a_2 \end{bmatrix} = \begin{bmatrix} b_1^\circ \\ b_2 \end{bmatrix}$$

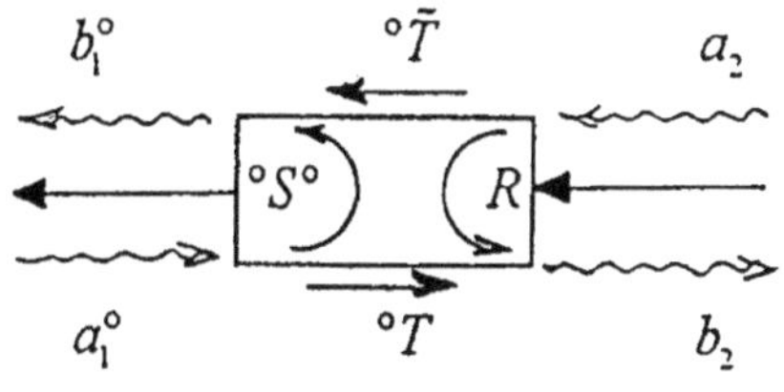

$$\begin{bmatrix} {}^\circ\tilde{T} - S\,{}^\circ T^{-1}R & {}^\circ S^\circ\,{}^\circ T^{-1} \\ -{}^\circ T^{-1}R & {}^\circ T^{-1} \end{bmatrix}^{(N)}\begin{bmatrix} a_2 \\ b_2 \end{bmatrix}^{(N)} = \begin{bmatrix} b_1^\circ \\ a_1^\circ \end{bmatrix}^{(N)} = \begin{bmatrix} a_2 \\ b_2 \end{bmatrix}^{(N-1)}$$

THE S2 TYPE MATRIX

$$\begin{bmatrix} {}^\circ S^\circ & {}^\circ\tilde{T}^\circ \\ {}^\circ T^\circ & {}^\circ R^\circ \end{bmatrix}\begin{bmatrix} a_1^\circ \\ a_2^\circ \end{bmatrix} = \begin{bmatrix} b_1^\circ \\ b_2^\circ \end{bmatrix}$$

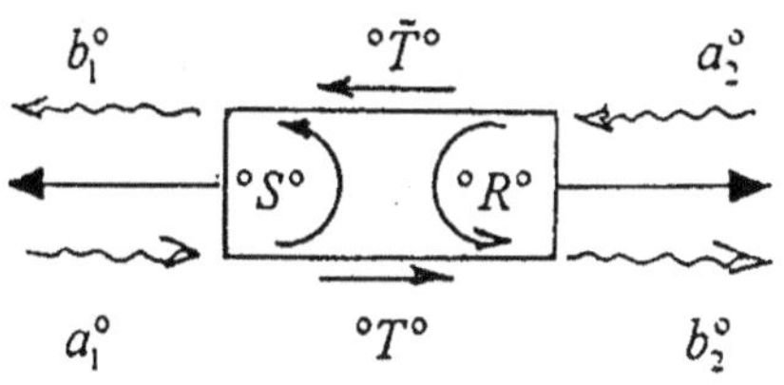

$$\begin{bmatrix} {}^\circ\tilde{T}^\circ - {}^\circ S^\circ\,{}^\circ T^{\circ-1}\,{}^\circ R^\circ & {}^\circ S^\circ\,{}^\circ T^{\circ-1} \\ -{}^\circ T^{\circ-1}\,{}^\circ R^\circ & {}^\circ T^{\circ-1} \end{bmatrix}^{(N)}\begin{bmatrix} a_2^\circ \\ b_2^\circ \end{bmatrix}^{(N)} = \begin{bmatrix} b_1^\circ \\ a_1^\circ \end{bmatrix}^{(N)} = \begin{bmatrix} a_2 \\ b_2 \end{bmatrix}^{(N-1)}$$

Table 14.2

The cascade of three two-ports

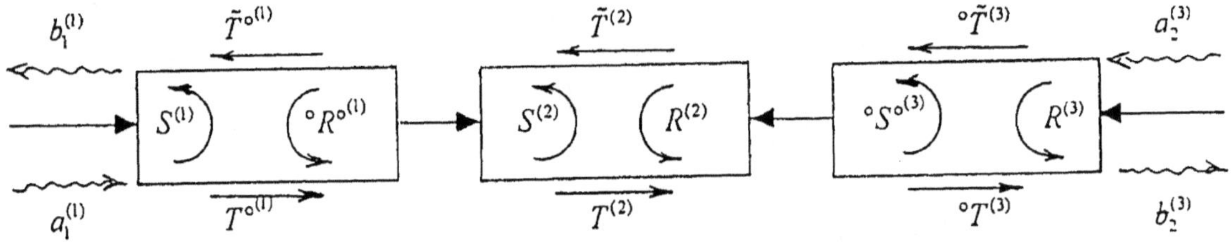

$$\begin{bmatrix} \tilde{T}^{\circ} - S T^{\circ -1}\,{}^{\circ}R^{\circ} & S T^{\circ -1} \\ -T^{\circ -1}\,{}^{\circ}R^{\circ} & T^{\circ -1} \end{bmatrix}^{(1)} \begin{bmatrix} \tilde{T} - S\,T^{-1}R & S\,T^{-1} \\ -T^{-1}R & T^{-1} \end{bmatrix}^{(2)} \begin{bmatrix} {}^{\circ}\tilde{T} - S\,{}^{\circ}T^{-1}R & {}^{\circ}S^{\circ}\,{}^{\circ}T^{-1} \\ -{}^{\circ}T^{-1}R & {}^{\circ}T^{-1} \end{bmatrix}^{(3)} \begin{bmatrix} a_2 \\ b_2 \end{bmatrix}^{(3)} = \begin{bmatrix} b_1 \\ a_1 \end{bmatrix}^{(1)}$$

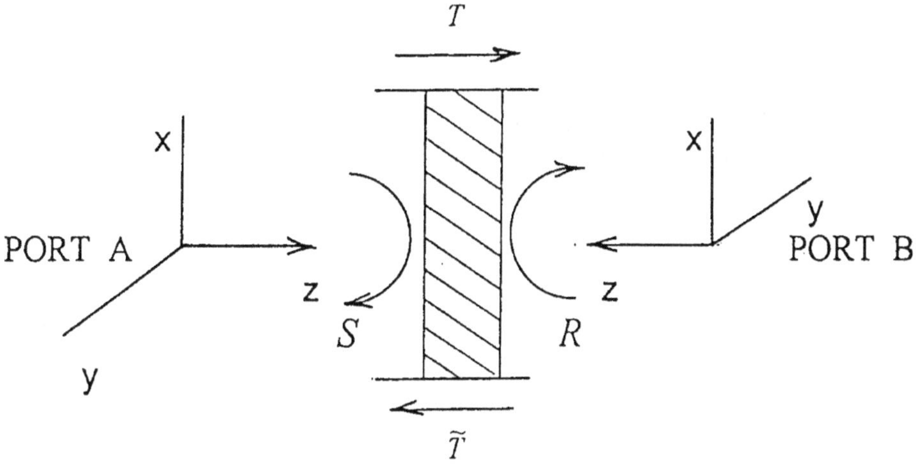

Fig. 14.1. The polarimetric two-port with the T and $\tilde{T}$ transmittance and S and R reflectance matrices, and the two local xyz coordinate systems at its A and B ports.

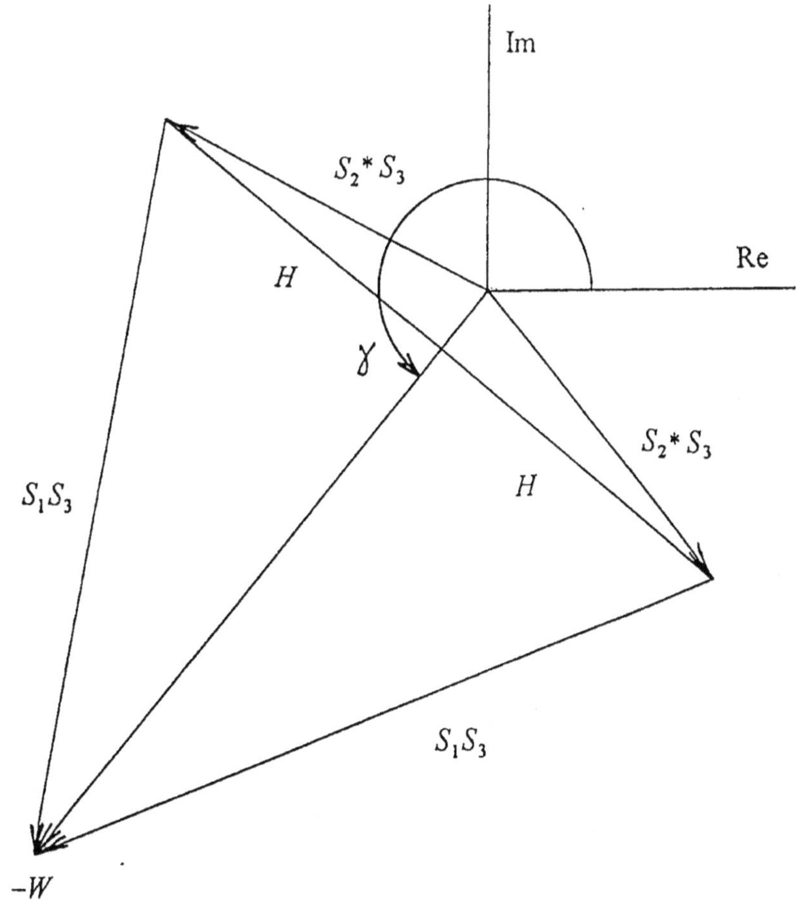

Fig. 14.2. Vector diagram for explanation of the allowed range for the S_3 real parameter

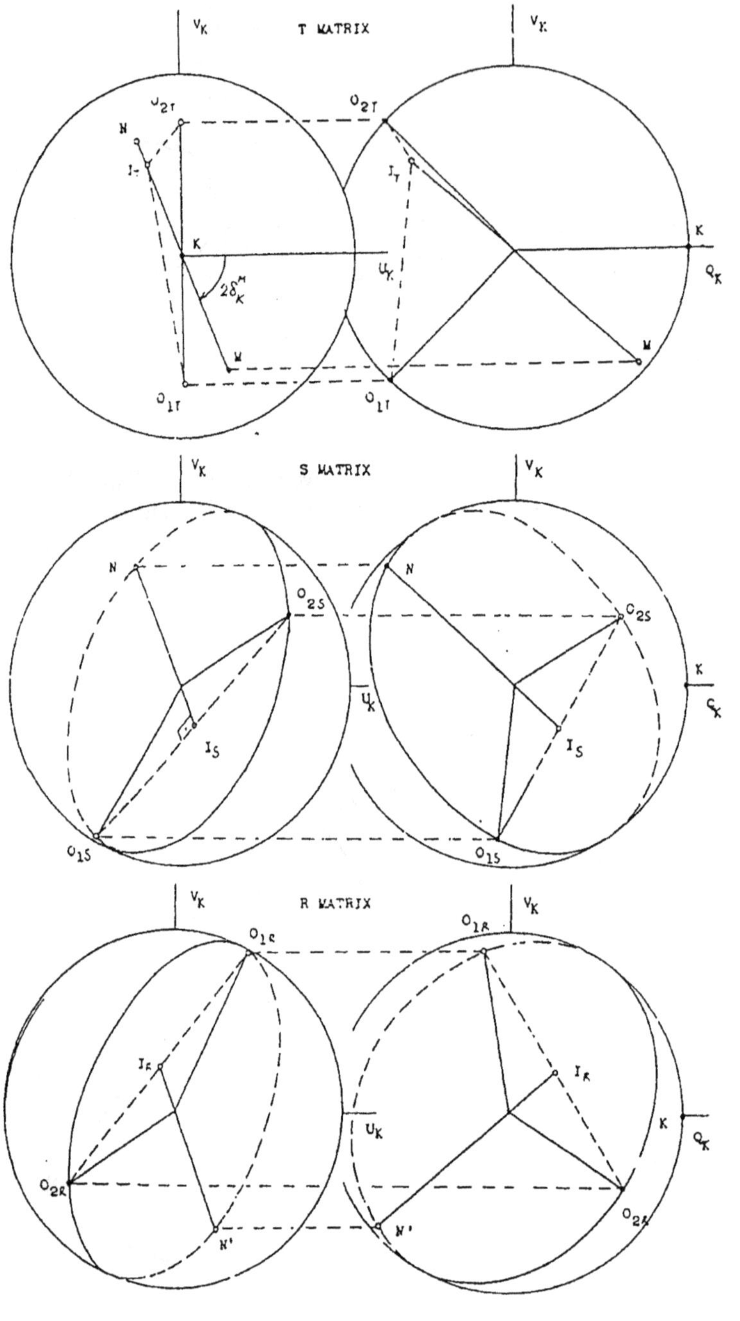

Fig. 14.3. Mutual orientations of the polarization sphere geometrical models of the transmittance and reflectance matrices (from numerical examole in [61] and [68]).

15. The Four-Sphere of Partial Polarizations and Its Applications

15.1. The Stokes' four-vector and power equations of scattering and transmission

The unit Stokes' four-vector of a partially polarized scattered wave will be presented in the form first announced by Landau and Lifschitz [104] but slightly modified, compatible with the here proposed notation for complete polarizations and independent of the direction of propagation along the z-axis of the spatial xyz right-handed coordinate system:

$$\mathbf{P}_H^S = \begin{bmatrix} \cos 2\theta \\ \sin 2\theta \cos 2\gamma \\ \sin 2\theta \sin 2\gamma \cos 2\delta \\ \sin 2\theta \sin 2\gamma \sin 2\delta \end{bmatrix}_H^S ; \quad |\mathbf{P}_H^S| = 1 . \tag{15.1}$$

Here, as usually,

- the lower index H denotes any ONP PP basis determining the coordinate system of three Stokes' parameters: Q_H, U_H, and V_H,
- the upper index S refers to the polarization point on the Poincare sphere of the completely polarized part of the wave corresponding to any S phasor (tangent to the sphere at that point), or any $\boldsymbol{u}^S$ PP vector of an arbitrary phase (the information about the phase is lost in case of the Stokes vector representation of waves, also partially polarized), and
- 2γ and 2δ are known angular coordinates of that polarization point in that basis,

The only new element is the 2θ angle, independent of the ONP PP basis of the wave's polarized part. The $\tan 2\theta$ function denotes the wave's degree of polarization. Contrary to the usual practice the double θ angle has been introduced in accordance with the notation for all other polarization sphere angles, including the 2δ angle, also usually being expressed by the single δ.

Such a unit Stokes' four-vector can be used to present the *full* Stokes four-vector:

$$\mathbf{I}_{0H}^S = \begin{bmatrix} I \\ Q \\ U \\ V \end{bmatrix}_H^S = I_0^S \mathbf{P}_H^S ; \quad I_0^S = \sqrt{(I^2 + Q^2 + U^2 + V^2)_H^S} . \tag{15.2}$$

With the 'degree of polarization',

$$p = \tan 2\theta = \frac{\sqrt{Q^2 + U^2 + V^2}}{I} , \tag{15.3}$$

the full Stokes four-vector of a partially polarized wave can be expressed also in another useful form,

$$\mathbf{I}_{0H}^S = I_0^S \cos 2\theta^S \begin{bmatrix} 1 \\ p\cos 2\gamma \\ p\sin 2\gamma \cos 2\delta \\ p \sin 2\gamma \sin 2\delta \end{bmatrix}_H^S = I^S \begin{bmatrix} 1 \\ pq \\ pu \\ pv \end{bmatrix}_H^S ; \quad q^2 + u^2 + v^2 = 1 , \tag{15.4}$$

with the total power

$$I^S = I_0^S \cos 2\theta^S . \tag{15.5}$$

For the *complete polarizations*: $2\theta = 45^0$, $p = 1$, and we arrive at the known forms of the unit, full, and efficient Stokes four-vector of the wave or receiving antenna in the H basis:

$$\mathbf{P}_H^R = \frac{1}{\sqrt{2}} \begin{bmatrix} 1 \\ \cos 2\gamma \\ \sin 2\gamma \cos 2\delta \\ \sin 2\gamma \sin 2\delta \end{bmatrix}_H^R = \frac{1}{\sqrt{2}} \begin{bmatrix} 1 \\ q \\ u \\ v \end{bmatrix}_H^R , \tag{15.6}$$

$$\mathbf{I}_{0H}^R = \mathrm{I}_0^R \mathbf{P}_H^R = \frac{\mathrm{I}_0^R}{\sqrt{2}} \begin{bmatrix} 1 \\ \cos 2\gamma \\ \sin 2\gamma \cos 2\delta \\ \sin 2\gamma \sin 2\delta \end{bmatrix}_H^R = \mathrm{I}^A \begin{bmatrix} 1 \\ q \\ u \\ v \end{bmatrix}_H^R , \tag{15.7}$$

15.2. The integrated Stokes four-vector

Generally, Stokes vectors should be spectral functions, similarly as elements of the partially depolarizing Kennaugh matrix. However, for many applications integrated (over the frequency band) components of the full Stokes vector of the quasimonochromatic wave, known as Stokes parameters, may appear sufficient to describe the partial polarization state of the wave ([28, 118, 120, 123, 132, 30, 11, 94]).

The integrated full Stokes four-vector can be expressed in terms of the temporary electric field PP vectors,

$$\boldsymbol{E}_0(t,z) = E_0(t)\, \boldsymbol{u}(t)\, e^{j(\omega_0 - kz)}; \qquad |\boldsymbol{u}(t)| = 1, \tag{15.8}$$

which, for example, can be represented by a complex quasimonochromatic transversal wave PP column vector of the amplitude stochastic (ergodic) process, assumed to be zero mean and complex normal distributed [139, 70], of the form in the $H =$ (y,x) basis:

$$\begin{aligned} [E_0(t,z)]_{(y,z)} &= \begin{bmatrix} E_{0y}(t) \\ E_{0x}(t) \end{bmatrix} e^{j(\omega_0 - kz)} \\ &= E_0(t)\, u_{(y,x)}(t)\, e^{j(\omega_0 - kz)}. \end{aligned} \tag{15.9}$$

The procedure of obtaining the required integrated full Stokes four-vector can be as follows

$$\mathbf{I}_{0(y,x)} = \sqrt{2}\tilde{\mathrm{U}}^* \left\langle E_0^2(t) (u(t) \otimes u^*(t))_{(y,x)} \right\rangle =$$

$$= \begin{bmatrix} \langle E_{0y}(t)E_{0y}(t)^* + E_{0x}(t)E_{0x}(t)^* \rangle \\ \langle E_{0y}(t)E_{0y}(t)^* - E_{0x}(t)E_{0x}(t)^* \rangle \\ \langle E_{0y}(t)E_{0x}(t)^* + E_{0x}(t)E_{0y}(t)^* \rangle \\ j\langle E_{0y}(t)E_{0x}(t)^* - E_{0x}(t)E_{0y}(t)^* \rangle \end{bmatrix} = \begin{bmatrix} \mathrm{I} \\ \mathrm{Q} \\ \mathrm{U} \\ \mathrm{V} \end{bmatrix}_{(y,x)} \tag{15.10}$$

with

$$U = \frac{1}{\sqrt{2}}\begin{bmatrix} 1 & 1 & 0 & 0 \\ 0 & 0 & 1 & -j \\ 0 & 0 & 1 & j \\ 1 & -1 & 0 & 0 \end{bmatrix}, \quad I = \langle E_0^2(t)\rangle, \quad \text{and} \quad \langle\ldots\rangle = \lim_{T\to\infty}\int_0^T \ldots\, dt \tag{15.11}$$

15.3. Power received from the partially polarized wave. The efficient Stokes four-vector

It should be observed that the reception of a partially polarized wave by an antenna is being expressed not by the product of their full Stokes four-vectors but rather with the use of smaller, the so-called efficient Stokes four-vectors. That becomes evident when considering the averaged square of Hermitian product of two PP vectors, one of which, that of scattered wave, being partially polarized (the ONP PP basis indicated here as H can be of course arbitrary):

$$\begin{aligned} P_r &= \langle |V_r|^2\rangle = \langle |E_0^S(t)h_0^R(\tilde{u}^R u^S * (t))_H|^2\rangle \\ &= \langle E_0^S(t)h_0^R(\tilde{u}^R u^S * (t))_H \otimes (\tilde{u}^R * u^S(t))_H\rangle \\ &= \tfrac{1}{2}[\sqrt{2}\, h_0^R(\tilde{u}^R \otimes \tilde{u}^R *)_H \mathrm{U} * \sqrt{2}\,\tilde{\mathrm{U}}\langle E_0^S(t)(u^S * (t) \otimes u^S(t))_H\rangle] \\ &= \tfrac{1}{2}\tilde{\mathbf{I}}_{0H}^{\mathbf{R}}\mathbf{I}_{0H}^{\mathbf{S}} \\ &= \tilde{\mathbf{I}}_{\mathrm{eff}H}^{\mathbf{R}}\mathbf{I}_{\mathrm{eff}H}^{\mathbf{S}}, \end{aligned} \tag{15.12}$$

where the effective Stokes vectors and effective powers have been defined in terms of the full Stokes vectors and its magnitude or its total power and the 2θ angle as follows,

$$\mathbf{I}_{\mathrm{eff}H}^{\mathrm{S}} = \frac{1}{\sqrt{2}}\mathbf{I}_{0H}^{\mathrm{S}} = \frac{1}{\sqrt{2}}\begin{bmatrix} \mathrm{I} \\ \mathrm{Q} \\ \mathrm{U} \\ \mathrm{V} \end{bmatrix}_H^{\mathrm{S}} = \mathrm{I}_{\mathrm{eff}}^{\mathrm{S}}\mathbf{P}_H^{\mathrm{S}}; \qquad \mathrm{I}_{\mathrm{eff}}^{\mathrm{S}} = \frac{1}{\sqrt{2}}\mathrm{I}_0^{\mathrm{S}} = \frac{\mathrm{I}^{\mathrm{S}}}{\sqrt{2}\cos 2\theta}. \tag{15.13}$$

It should be observed that in case of complete polarization $(2\theta = 45^0)$:

$$\mathrm{I}_{\mathrm{eff}} = \mathrm{I} \tag{15.14}$$

and

$$\mathbf{I}_{\mathrm{eff}H}^{\mathrm{R}} = \mathrm{I}_{\mathrm{eff}}^{\mathrm{R}}\mathbf{P}_H^{\mathrm{R}} = \frac{\mathrm{I}_{\mathrm{eff}}^{\mathrm{R}}}{\sqrt{2}}\begin{bmatrix} 1 \\ \cos 2\gamma \\ \sin 2\gamma\cos 2\delta \\ \sin 2\gamma \sin 2\delta \end{bmatrix}_H^{\mathrm{R}} = \frac{\mathrm{I}^{\mathrm{R}}}{\sqrt{2}}\begin{bmatrix} 1 \\ \mathrm{q} \\ \mathrm{u} \\ \mathrm{v} \end{bmatrix}_H^{\mathrm{R}} \tag{15.15}$$

Equations for power reception by the antenna (or for transmission between two antennas) for arbitrary wave and antenna polarizations (generally both can be partially polarized if considered as independent stochastic processes) presented in terms of the unit Stokes four-vectors are:

$$\begin{aligned} P_r &= \tfrac{1}{2}\tilde{\mathbf{I}}_{0H}^{\mathbf{R}}\mathbf{I}_{0H}^{\mathbf{S}} = \tfrac{1}{2}\mathrm{I}_0^{\mathrm{R}}\mathrm{I}_0^{\mathrm{S}}\,\tilde{\mathbf{P}}_H^{\mathbf{R}}\mathbf{P}_H^{\mathbf{S}} \\ &= \tilde{\mathbf{I}}_{\mathrm{eff}H}^{\mathbf{R}}\mathbf{I}_{\mathrm{eff}H}^{\mathbf{S}} = \mathrm{I}_{\mathrm{eff}}^{\mathrm{R}}\mathrm{I}_{\mathrm{eff}}^{\mathrm{S}}\,\tilde{\mathbf{P}}_H^{\mathbf{R}}\mathbf{P}_H^{\mathbf{S}} \end{aligned} \tag{15.16}$$

where

$$\tilde{\mathbf{P}}_H^{\mathbf{R}}\mathbf{P}_H^{\mathbf{S}} = \cos 2\theta^{\mathrm{R}}\cos 2\theta^{\mathrm{S}} + \sin 2\theta^{\mathrm{R}}\sin 2\theta^{\mathrm{S}}(\mathrm{q}^{\mathrm{R}}\mathrm{q}^{\mathrm{S}} + \mathrm{u}^{\mathrm{R}}\mathrm{u}^{\mathrm{S}} + \mathrm{v}^{\mathrm{R}}\mathrm{v}^{\mathrm{S}})_H. \tag{15.17}$$

Denoting

$$\cos 2\psi^{RS} \equiv \cos(RS) = (q^R q^S + u^R u^S + v^R v^S)_H \tag{15.18}$$

the scalar product of the two unit Stokes four-vectors will be obtained in the form of cosine of an angle on a sphere,

$$\widetilde{\mathbf{P}}_H^R \mathbf{P}_H^S = \cos 2\theta^R \cos 2\theta^S + \sin 2\theta^R \sin 2\theta^S \cos 2\psi^{RS} = \cos 2\Omega^{RS}, \tag{15.19}$$

or in terms of degrees of polarization,

$$\widetilde{\mathbf{P}}_H^R \mathbf{P}_H^S = \cos 2\theta^R \cos 2\theta^S (1 + p^R p^S \cos 2\psi^{RS}), \tag{15.20}$$

thus obtaining

$$\begin{aligned} P_r &= \tfrac{1}{2} I_0^R I_0^S \cos 2\Omega^{RS} = \tfrac{1}{2} I_0^R I_0^S \cos 2\theta^R \cos 2\theta^S (1 + p^R p^S \cos 2\psi^{RS}) \\ &= I_{eff}^R I_{eff}^S \cos 2\Omega^{RS} = I_{eff}^R I_{eff}^S \cos 2\theta^R \cos 2\theta^S (1 + p^R p^S \cos 2\psi^{RS}) \end{aligned} \tag{15.21}$$

In the case of *complete polarization* $(2\theta^R = 2\theta^R = 45^0)$one obtains known result (compare, e.g., with (4.15)):

$$\begin{aligned} \widetilde{\mathbf{P}}_H^R \mathbf{P}_H^S &= \tfrac{1}{2}(1 + \cos 2\psi^{RS}) = \cos^2 \psi^{RS} \\ &= |\widetilde{u}^R u^{S*}|^2, \end{aligned} \tag{15.22}$$

and

$$\begin{aligned} P_r &= \tfrac{1}{2} I_0^R I_0^S \cos^2 \psi^{RS} \\ &= I_{eff}^R I_{eff}^S \cos^2 \psi^{RS}, \end{aligned} \tag{15.23}$$

as well as

$$P_r = A_e S \cos^2 \psi^{RS}, \tag{15.24}$$

which will be obtained when using the usual notation: for the receiving effective area of an antenna,

$$I_{eff}^A = A_e, \tag{15.25a}$$

and for the incident power density,

$$I_{eff}^S = S. \tag{15.25b}$$

15.4. Scattering and transmission power equations for the partially polarized waves

Using similar procedure as in (15.12), the scattering and transmission power equations with the partially depolarizing Kennaugh matrix can be presented as follows (also for partially polarized incident waves, but when the Kennaugh matrix is non-depolarizing, as for a 'point' target [123, Part I-4], or it represents the stochastic process independent of the illumination [123, Part I-5]):

$$K_H \mathbf{P}_H^T = \sigma^T \mathbf{P}_H^S \quad \text{and} \quad P_r = \widetilde{\mathbf{P}}_H^R K_H \mathbf{P}_H^T = \sigma^T \widetilde{\mathbf{P}}_H^R \mathbf{P}_H^S. \tag{15.26}$$

The form of those equations is exactly the same as in the case of nondepolarizing Kennaugh matrices (compare with (7.19)). Only the scattered power σ^T cannot be considered as the square of the distance between the polarization point on the Poincare sphere model of that matrix (now not existing) and an inversion point in it. However, such a scattered power can be treated as the sum of powers scattered by the component 'point' (non-depolarizing) targets corresponding to an incoherent decomposition of that parent matrix (observe an example in Fig. 11.3).

Sometimes it is convenient to present incident and scattered waves by such 'unit' Stokes four-vectors, components of which are Stokes parameters with the unit total power as the first component. So, in cases of completely polarized incident waves and partially polarized scattered waves, by using the 'full unit' Stokes four-vectors instead of formerly defined (efficient) unit vectors:

$$\mathbf{P}_{0H}^{\mathrm{T}} = \sqrt{2}\,\mathbf{P}_{H}^{\mathrm{T}} = \begin{bmatrix} 1 \\ \mathrm{q} \\ \mathrm{u} \\ \mathrm{v} \end{bmatrix}_{H}^{\mathrm{T}} \quad \text{and} \quad \mathbf{P}_{0H}^{\mathrm{S}} = \sqrt{2}\,\mathbf{P}_{H}^{\mathrm{S}} = \sqrt{2}\cos 2\theta^{\mathrm{S}} \begin{bmatrix} 1 \\ \mathrm{pq} \\ \mathrm{pu} \\ \mathrm{pv} \end{bmatrix}_{H}^{\mathrm{T}}, \tag{15.27}$$

the scattered wave may be obtained in the form

$$\mathrm{K}_{H}\mathbf{P}_{0H}^{\mathrm{T}} = \mathrm{K}_{H} \begin{bmatrix} 1 \\ \mathrm{q} \\ \mathrm{u} \\ \mathrm{v} \end{bmatrix}_{H}^{\mathrm{T}} = \sigma^{\mathrm{T}}\,\mathbf{P}_{0H}^{\mathrm{S}} = \sqrt{2}\,\sigma^{\mathrm{T}} \cos 2\theta^{\mathrm{S}} \begin{bmatrix} 1 \\ \mathrm{pq} \\ \mathrm{pu} \\ \mathrm{pv} \end{bmatrix}_{H}^{\mathrm{S}} \tag{15.28}$$

where

$$\mathrm{I}^{\mathrm{S}} = \sqrt{2}\,\sigma^{\mathrm{T}} \cos 2\theta^{\mathrm{S}} \tag{15.29}$$

means the total scattered power for the unit incident total power, and the received power is

$$\begin{aligned} P_r &= \tfrac{1}{2}\widetilde{\mathbf{P}}_{0H}^{\mathrm{R}}\mathrm{K}_{H}\mathbf{P}_{0H}^{\mathrm{T}} = \tfrac{1}{2}\sigma^{\mathrm{T}}\widetilde{\mathbf{P}}_{0H}^{\mathrm{R}}\mathbf{P}_{0H}^{\mathrm{S}} \\ &= \tfrac{1}{2}\begin{bmatrix} 1 & \mathrm{q} & \mathrm{u} & \mathrm{v} \end{bmatrix}_{H}^{\mathrm{R}} \sqrt{2}\,\sigma^{\mathrm{T}} \cos 2\theta^{\mathrm{S}} \begin{bmatrix} 1 \\ \mathrm{pq} \\ \mathrm{pu} \\ \mathrm{pv} \end{bmatrix}_{H}^{\mathrm{S}} \\ &= \tfrac{1}{\sqrt{2}}\sigma^{\mathrm{T}} \cos 2\theta^{\mathrm{S}} [1 + \mathrm{p}^{\mathrm{S}}(\mathrm{q}^{\mathrm{R}}\mathrm{q}^{\mathrm{S}} + \mathrm{u}^{\mathrm{R}}\mathrm{u}^{\mathrm{S}} + \mathrm{v}^{\mathrm{R}}\mathrm{v}^{\mathrm{S}})_{H}] \\ &= \tfrac{1}{\sqrt{2}}\sigma^{\mathrm{T}} \cos 2\theta^{\mathrm{S}} [1 + \mathrm{p}^{\mathrm{S}} \cos 2\psi^{RS}]. \end{aligned} \tag{15.30}$$

15.5. The polarization four-sphere

Following the expressions (15.1), (15.2), (15.3), and (15.19), the (integrated) Stokes four-vector of a partial polarization will be represented by a point on the four-sphere surface (the surface of a sphere in the four-dimensional real space) with a radius equal to the magnitude, I_0, of the 'full' vector. In that case the first of four mutually perpendicular axes of that space will correspond to the total power, I, of the completely polarized, partially polarized, or completely unpolarized wave.

To simplify the drawing of the four-sphere, one of their axes, except of the first one, and sometimes the two axes, will be omitted (see Fig. 15.1 or Fig. 15.2). Such drawings may occur useful to present also the Poincare three-sphere (of complete polarizations) in the four-dimensional space. It takes form of a hypercircle (with a radius equal to I, the total power of the completely polarized wave), being a crossection of the four-sphere surface with a 90^0 hypercone having axis of symmetry coinciding with the I axis

Fig. 15.1 presents also three decompositions of the four-vector into incoherent components: completely polarized (corresponding to elliptical polarizations) and unpolarized (like natural light),

$$\mathbf{I}_0 = \mathbf{I}_{0E} + \mathbf{I}_{0N} = \begin{bmatrix} I_E \\ Q \\ U \\ V \end{bmatrix} + \begin{bmatrix} I_N \\ 0 \\ 0 \\ 0 \end{bmatrix}; \qquad I_E = \sqrt{Q^2 + U^2 + V^2}, \tag{15.31a}$$

two orthogonal, completely polarized,

$$\mathbf{I}_0 = \mathbf{I}_{01} + \mathbf{I}_{02} = \begin{bmatrix} I_1 \\ Q_1 \\ U_1 \\ V_1 \end{bmatrix} + \begin{bmatrix} I_2 \\ Q_2 \\ U_2 \\ V_2 \end{bmatrix}; \qquad I_i = \sqrt{Q_i^2 + U_i^2 + V_i^2}, \quad i = 1,2, \tag{15.31b}$$

and component full Stokes vectors of the total power (intensity) and of pure polarization order,

$$\mathbf{I}_0 = \mathbf{I}_{0I} + \mathbf{I}_{0P} = + \begin{bmatrix} I \\ 0 \\ 0 \\ 0 \end{bmatrix} + \begin{bmatrix} 0 \\ Q \\ U \\ V \end{bmatrix}; \qquad I \geq \sqrt{Q^2 + U^2 + V^2}. \tag{15.31c}$$

It should be mentioned that the last decomposition is also physically realizable but rather in a virtual world. For instance it can be build up in a electronic circuit as a virtual receiving four-vector. In such a virtual space the total power I can take even negative values.

The following mutual relations can be observed between those component vectors:

$$\begin{aligned} \mathbf{I}_{0E} &= \mathbf{I}_{01} - \mathbf{I}_{02}^{\times}, \\ \mathbf{I}_{0N} &= \mathbf{I}_{02} + \mathbf{I}_{02}^{\times}, \end{aligned} \tag{15.32}$$

and their components:

$$\begin{aligned} I &= I_0 \cos 2\theta = I_E + I_N = I_1 + I_2, \\ I_E &= I_0 \sin 2\theta = I_1 - I_2, \\ I_N &= 2I_2, \\ I_0^2 &= 2(I_1^2 + I_2^2), \\ p &= \tan 2\theta = \frac{I_E}{I} = \frac{I_1 - I_2}{I_1 + I_2}. \end{aligned} \tag{15.33}$$

Fig. 15.2. shows the angles $2\theta^{R}$, $2\theta^{S}$, $2\psi^{RS}$ and $2\Omega^{RS}$ as they appeared in the expressions: (15.19), for the scalar product of the (efficient) unit Stokes four-vectors, and in (15.21), for the received scattered power.

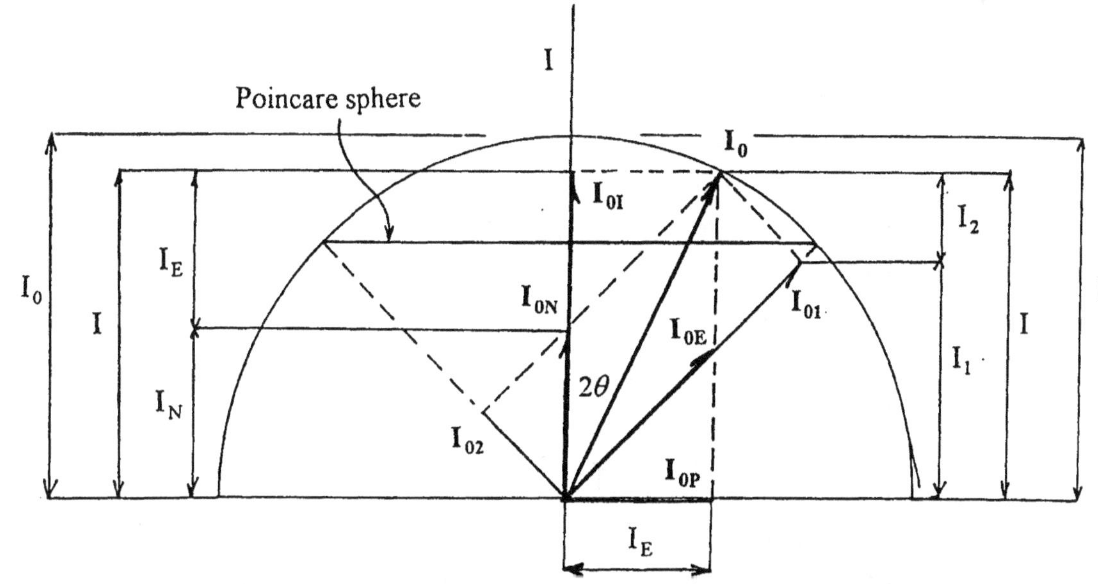

Fig. 15.1. Three decompositions of a full Sokes four-vector into incoherent components

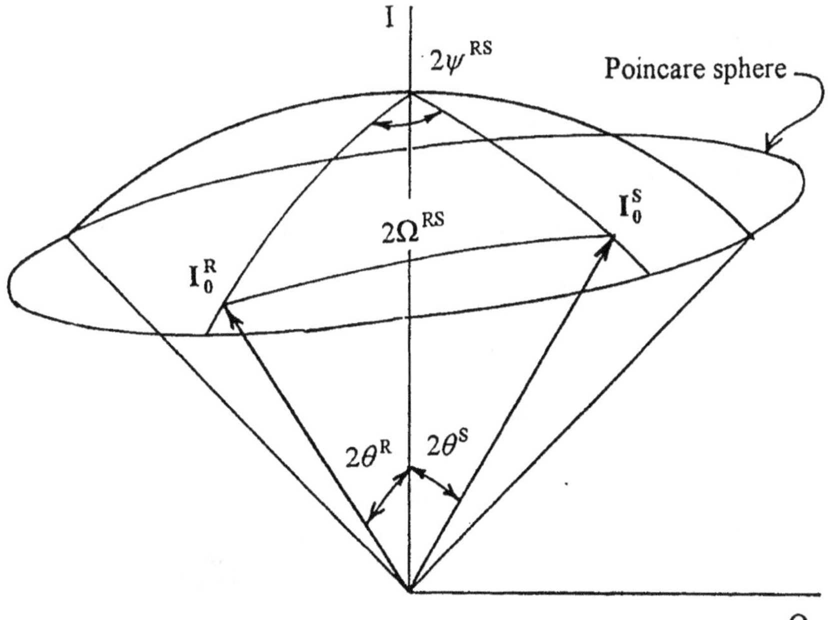

Fig. 15.2. The polarization four-sphere representation of two partially polarized Stokes four-vectors and the angles responsible for their scalar multiplication

15.6. An example of applications. Canceling the partially polarized radar clutter

The concept of formation of the virtual receiving vector perpendicular to the partially polarized radar clutter has been described in the workshop paper [73] and can be explained using the here presented approach based on the polarization four-sphere transformations.

To simplify formulae the following notation will be introduced for the effective Stokes four-vectors:

$$
\begin{array}{ll}
\text{for the power scattered from the target,} & \mathbf{I}^{\mathbf{S}}_{\mathbf{eff}} = \mathbf{S}_{\mathbf{eff}}, \\
\text{for its component perpendicular to the clutter,} & \mathbf{I}^{\mathbf{R}}_{\mathbf{eff}} = \mathbf{R}_{\mathbf{eff}} = \mathrm{R}_{\mathrm{eff}}\mathbf{P}^{\mathbf{R}}, \\
\text{for the desired receiving unit vector,} & \mathbf{P}^{\mathbf{R}}, \\
\text{for the disturbance (source of clutter),} & \mathbf{I}^{\mathbf{D}}_{\mathbf{eff}} = \mathbf{D}_{\mathbf{eff}} = \mathrm{D}_{\mathrm{eff}}\mathbf{P}^{\mathbf{D}}, \\
\text{for the transmitted signal,} & \mathbf{I}^{\mathbf{T}}_{\mathbf{eff}} = \mathbf{T}_{\mathbf{eff}} = \mathrm{T}_{\mathrm{eff}}\mathbf{P}^{\mathbf{T}}, \\
\text{and for the signal plus disturbance,} & \mathbf{W}_{\mathbf{eff}} = \mathbf{T}_{\mathbf{eff}} + \mathbf{D}_{\mathbf{eff}}
\end{array}
\qquad (15.34)
$$

With the desired receiving four-vector perpendicular to the clutter, the received power (see (15.12)) will be

$$P_r = \widetilde{\mathbf{R}}_{\mathbf{eff}}\mathbf{P}^{\mathbf{R}} = \mathrm{R}_{\mathrm{eff}}\widetilde{\mathbf{P}}^{\mathbf{R}}\mathbf{P}^{\mathbf{R}} = \mathrm{R}_{\mathrm{eff}} = |\mathbf{R}_{\mathbf{eff}}| = \mathrm{S}_{\mathrm{eff}}\sin 2\Omega^{\mathrm{DS}}. \qquad (15.35)$$

That is the expression for the received signal which can be extracted from the disturbance. However its value depends also on the transmitted polarization. It will be shown how the optimum transmit polarization can be determined when knowing the Kennaugh matrices for the disturbance, K^{D}, and for the signal plus disturbance, K^{W}.

Having those matrices one can find the four-vectors

$$\mathbf{P}^{\mathbf{D}} = \frac{\mathrm{K}^{\mathrm{D}}\mathbf{P}^{\mathbf{T}}}{|\mathrm{K}^{\mathrm{D}}\mathbf{P}^{\mathbf{T}}|} \quad \text{and} \quad \mathbf{W}_{\mathbf{eff}} = \mathrm{K}^{\mathrm{W}}\mathbf{P}^{\mathbf{T}} \qquad (15.36)$$

The expression for the received power can be found in the form

$$P_r = |\mathbf{R}_{\mathbf{eff}}| = |(\mathbf{1} - \mathbf{P}^{\mathbf{D}}\widetilde{\mathbf{P}}^{\mathbf{D}})\mathbf{W}_{\mathbf{eff}}| \qquad (15.37)$$

where $\mathbf{1}$ is the 4×4 unit matrix. The above expression can be verified as follows:

$$
\begin{aligned}
\mathbf{R}_{\mathbf{eff}} &= (\mathbf{1} - \mathbf{P}^{\mathbf{D}}\widetilde{\mathbf{P}}^{\mathbf{D}})(\mathrm{S}_{\mathrm{eff}}\mathbf{P}^{\mathbf{S}} + \mathrm{D}_{\mathrm{eff}}\mathbf{P}^{\mathbf{D}}) \\
&= \mathrm{S}_{\mathrm{eff}}\mathbf{P}^{\mathbf{S}} + \mathrm{D}_{\mathrm{eff}}\mathbf{P}^{\mathbf{D}} - \mathrm{S}_{\mathrm{eff}}(\widetilde{\mathbf{P}}^{\mathbf{D}}\mathbf{P}^{\mathbf{S}})\mathbf{P}^{\mathbf{D}} - \mathrm{D}_{\mathrm{eff}}(\widetilde{\mathbf{P}}^{\mathbf{D}}\mathbf{P}^{\mathbf{D}})\mathbf{P}^{\mathbf{D}} \\
&= \mathrm{S}_{\mathrm{eff}}(\mathbf{P}^{\mathbf{S}} - (\widetilde{\mathbf{P}}^{\mathbf{D}}\mathbf{P}^{\mathbf{S}})\mathbf{P}^{\mathbf{D}}) = \mathrm{S}_{\mathrm{eff}}(\mathbf{P}^{\mathbf{S}} - \cos 2\Omega^{\mathrm{DS}}\mathbf{P}^{\mathbf{D}}) \\
&= \mathrm{S}_{\mathrm{eff}}\sin 2\Omega^{\mathrm{DS}}\mathbf{P}^{\mathbf{D}}.
\end{aligned}
\qquad (15.38)
$$

Therefore one can write

$$
\begin{aligned}
P_r^2 &= |\mathbf{R}_{\mathbf{eff}}| = |(\mathbf{1} - \mathbf{P}^{\mathbf{D}}\widetilde{\mathbf{P}}^{\mathbf{D}})\mathbf{W}_{\mathbf{eff}}|^2 \\
&= \widetilde{\mathbf{P}}^{\mathbf{T}}\mathrm{K}^{\mathrm{W}}|\mathbf{1} - \frac{\mathrm{K}^{\mathrm{D}}\mathbf{P}^{\mathbf{T}}\widetilde{\mathbf{P}}^{\mathbf{T}}\widetilde{\mathrm{K}}^{\mathrm{D}}}{\widetilde{\mathbf{P}}^{\mathbf{T}}\widetilde{\mathrm{K}}^{\mathrm{D}}\mathrm{K}^{\mathrm{D}}\mathbf{P}^{\mathbf{T}}}|^2\,\mathrm{K}^{\mathrm{W}}\mathbf{P}^{\mathbf{T}},
\end{aligned}
\qquad (15.39)
$$

That expression for the square of the received power is of the form depending directly on the transmit polarization. Its maximization enables one to determine the optimum transmit polarization.

APPENDIX A. USEFUL FORMULAE OF SPHERICAL TRIGONOMETRY

A. 1. Spherical excess

The solid angle subtended by the spherical triangle ABC (see Fig.A.1) is numerically equal to the *spherical excess* of the triangle, *i.e.*, the excess of the sum of its three angles over π.

$$E_{\mathrm{ABC}} = \hat{A} + \hat{B} + \hat{C} - \pi \tag{A.1}$$

The excess is a measure of the *oriented* triangle, what means that

$$E_{\mathrm{ABC}} = -E_{\mathrm{CBA}} \tag{A.2}$$

If Cx be the point diametrically opposite to C (Fig.A.1), then the spherical excess of the triangle CxBA colunar to ABC is

$$E'_{\mathrm{ABC}} \equiv E_{\mathrm{C'BA}} = 2\hat{C} - E_{\mathrm{ABC}} = \pi - (\hat{A} + \hat{B}) + \hat{C} \tag{A.3}$$

because

$$E'_{\mathrm{ABC}} + E_{\mathrm{ABC}} = 2\hat{C}$$

Also

$$\tfrac{1}{2}(E'_{\mathrm{ABC}} - E_{\mathrm{ABC}}) = \pi - (\hat{A} + \hat{B}) \tag{A.4}$$

Consider an A phasor (Fig.A.2), oriented as the CA great circle arc, and then shifted parallel along the three sides of the spherical triangle ABC to the left, without rotation. After one full turn, the phasor shall be rotated versus its original orientation also to the left, by the angle E_{ABC}, numerically equal to the spherical excess of the triangle. Only after the additional rotations by $\pi - \hat{A}$, $\pi - \hat{B}$, and $\pi - \hat{C}$ angles (at the vertices of the triangle, for example), one obtains the total rotation of the phasor equal to 2π.

When shifting the phasor along any circular path (Fig.A.3) considered as a sum of paths around narrow oriented triangles of diminishing width and of summed area equal in the limit to the solid angle subtended by the circle, then, if the shifts along all sides of all triangles be parallel, after the full 2π turn the phasor will be rotated versus its original orientation by the angle numerically equal to the surrounded area. Also direction of the angle will be that of the path. If, when shifting, the phasor be kept parallel to the circular path, then it will undergo an extra rotation in the same direction, to the total of 2π radians. Therefore, in that case, the total angle of rotation can be considered as consisting of two parts: one caused by the true rotation and another one made by surrounding of an area.

A. 2. Other useful formulae

For spherical triangle as in Fig.A.1 and Fig.A.4:

$$\arctan\left[\frac{\cos\frac{a+b}{2}}{\cos\frac{a-b}{2}}\tan\frac{\pm\hat{C}}{2}\right] = \frac{\pm\hat{C} \mp E_{\mathrm{ABC}}}{2} = \pm\frac{\pi - (\hat{A} + \hat{B})}{2} = \pm\frac{E'_{\mathrm{ABC}} - E_{\mathrm{ABC}}}{4} \tag{A.5}$$

$$\cos\frac{E'_{\mathrm{ABC}}}{2} = \frac{\sin^2\frac{1}{2}a + \sin^2\frac{1}{2}b - \sin^2\frac{1}{2}c}{2\sin\frac{1}{2}a\sin\frac{1}{2}b\cos\frac{1}{2}c} = \frac{\cos^2\frac{1}{2}a' + \cos^2\frac{1}{2}b' + \cos^2\frac{1}{2}c - 1}{2\cos\frac{1}{2}a'\cos\frac{1}{2}b'\cos\frac{1}{2}c} = -\cos 2\Delta_B^A \tag{A.6}$$

where

$$2\Delta_B^A = \pi - \hat{C} + \frac{E_{\text{ABC}}}{2} = \pi - \frac{E'_{\text{ABC}}}{2}; \qquad 0 \leq 2\Delta_B^A \leq \pi \quad \text{for} \quad 0 \leq E_{\text{ABC}} \leq 2\pi \tag{A.7}$$

$$\tan\frac{a}{2}\tan\frac{b}{2}\sin\left(\hat{C} - \frac{E_{\text{ABC}}}{2}\right) = \sin\frac{E_{\text{ABC}}}{2}; \qquad \sin\left(\hat{C} - \frac{E_{\text{ABC}}}{2}\right) = \sin 2\Delta_B^A \tag{A.8}$$

$$\cos c = \cos a \, \cos b + \sin a \, \sin b \, \cos\hat{C} \tag{A.9}$$

$$\cot\hat{A} = \frac{\sin b \cot a - \cos b \cos\hat{C}}{\sin\hat{C}}; \qquad \cot\hat{B} = \frac{\sin a \cot b - \cos a \cos\hat{C}}{\sin\hat{C}} \tag{A.10}$$

$$\sin\hat{A} = \frac{\sin a \sin\hat{B}}{\sin b}; \qquad \sin\hat{B} = \frac{\sin b \sin\hat{A}}{\sin a} \tag{A.11}$$

For spherical triangle as in Fig.A.5:

$$\tan(90^\circ - 2\eta) = \cot 2\eta = \frac{\cot 2\delta}{\cos 2\gamma} \tag{A.12}$$

$$E_{\text{ABC}} = 2\delta - 2\eta \tag{A.13}$$

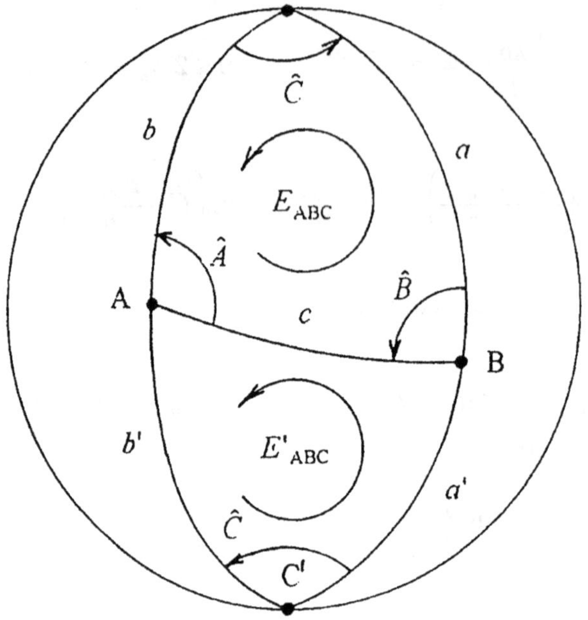

Fig. A. 1.

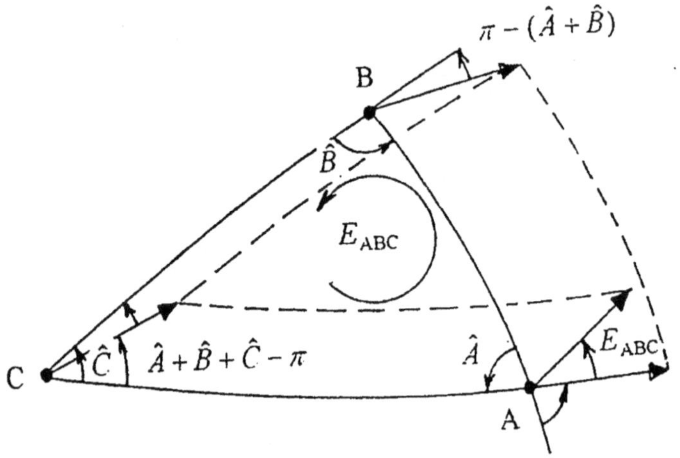

Fig. A. 2.

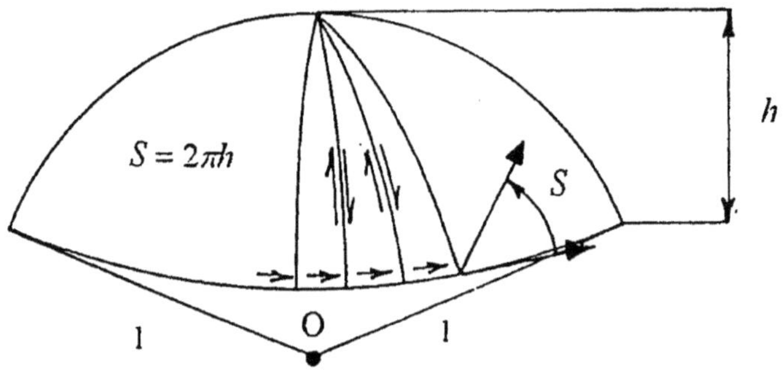

Fig. A. 3.

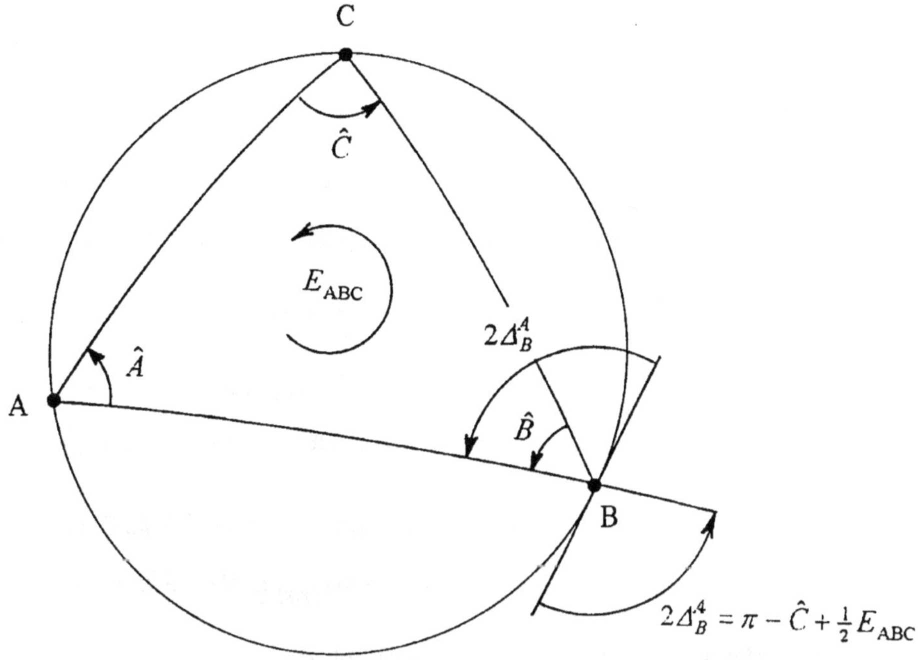

Fig. A. 4.

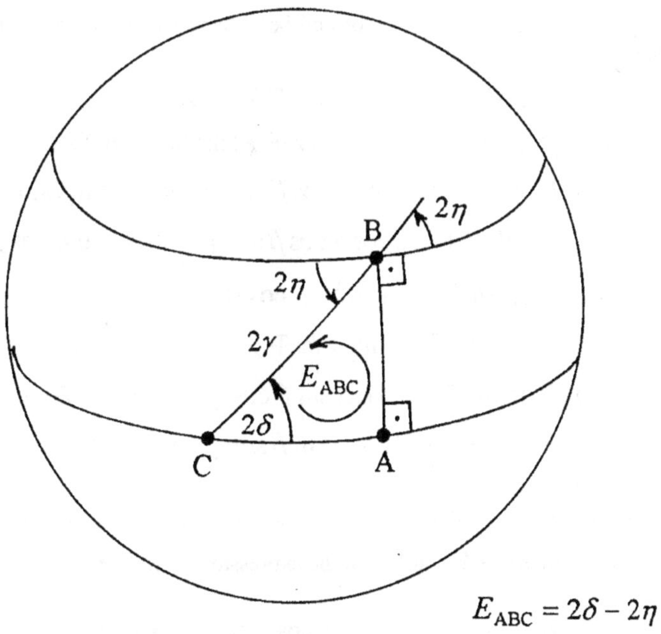

Fig. A. 5.

APPENDIX A1

STOKES FOUR-VECTORS IN TERMS OF ANALYTICAL AND GEOMETRICAL PARAMETERS FOR THE TWO ORDERS OF BASIS VECTORS

Consider 'analytical components' of the elliptically polarized electric vector:

$$\begin{aligned} E_y^{\pm} &= a_y \cos(\omega t \mp kz \mp \nu_y) = a_y[\cos\nu_y \cos(\omega t \mp kz) \pm \sin\nu_y \sin(\omega t \mp kz)] \\ E_x^{\pm} &= a_x \cos(\omega t \mp kz \mp \nu_x) = a_x[\cos\nu_x \cos(\omega t \mp kz) \pm \sin\nu_x \sin(\omega t \mp kz)]. \end{aligned} \tag{A1.1}$$

and its 'geometrical components' in two coordinate systems, $\xi(x,y)\eta(x,y)$ and $\xi(y,x)\eta(y,x)$:

$$\begin{aligned} E_{\xi(x,y)}^{\pm} &= a_0 \cos\alpha_{(x,y)} \cos(\omega t \mp kz \mp \chi) \\ E_{\eta(x,y)}^{\pm} &= \mp a_0 \sin\alpha_{(x,y)} \sin(\omega t \mp kz \mp \chi). \end{aligned} \tag{A1.2}$$

and

$$\begin{aligned} E_{\xi(y,x)}^{\pm} &= \mp a_0 \sin\alpha_{(y,x)} \sin(\omega t \mp kz \mp \chi) \\ E_{\eta(y,x)}^{\pm} &= a_0 \cos\alpha_{(y,x)} \cos(\omega t \mp kz \mp \chi), \end{aligned} \tag{A1.3}$$

with the Monge's circle radius of the polarization ellipse,

$$a_0 = \sqrt{a_x^2 + a_y^2}\ . \tag{A1.4}$$

A1. 1. The case of the (y,x) polarization basis

Analytical components of the electric vector can be obtained in terms of its geometrical components by the following procedure:

$$\begin{aligned} E_y^{\pm} &= E_{\eta(y,x)}^{\pm} \cos\beta_{(y,x)} - E_{\xi(y,x)}^{\pm} \sin\beta_{(y,x)} \\ &= a_0[\cos\alpha\cos\beta\cos(\omega t \mp kz \mp \chi) \pm \sin\alpha\sin\beta\sin(\omega t \mp kz \mp \chi)]_{(y,x)} \\ &= a_0\{\cos(\omega t \mp kz)[\cos\alpha\cos\beta\cos\chi - \sin\alpha\sin\beta\sin\chi]_{(y,x)} \\ &\quad \pm \sin(\omega t \mp kz)[\cos\alpha\cos\beta\sin\chi + \sin\alpha\sin\beta\cos\chi]_{(y,x)}\} \\ E_x^{\pm} &= E_{\eta(y,x)}^{\pm} \sin\beta_{(y,x)} + E_{\xi(y,x)}^{\pm} \cos\beta_{(y,x)} \\ &= a_0[\cos\alpha\sin\beta\cos(\omega t \mp kz \mp \chi) \mp \sin\alpha\cos\beta\sin(\omega t \mp kz \mp \chi)]_{(y,x)} \\ &= a_0\{\cos(\omega t \mp kz)[\cos\alpha\sin\beta\cos\chi + \sin\alpha\cos\beta\sin\chi]_{(y,x)} \\ &\quad \pm \sin(\omega t \mp kz)[\cos\alpha\sin\beta\sin\chi - \sin\alpha\cos\beta\cos\chi]_{(y,x)}\}. \end{aligned} \tag{A1.5}$$

Comparing expressions for analytical components of the electric vector, E_y and E_x, combined with $\cos(\omega t - kz)$ and $\sin(\omega t - kz)$ terms of the expressions (A1.1) we arrive at the equalities

$$\begin{aligned} a_y \cos\nu_y &= a_0(\cos\alpha\cos\beta\cos\chi - \sin\alpha\sin\beta\sin\chi)_{(y,x)} \\ a_y \sin\nu_y &= a_0(\cos\alpha\cos\beta\sin\chi + \sin\alpha\sin\beta\cos\chi)_{(y,x)} \end{aligned} \tag{A1.6}$$

$$a_x \cos v_x = a_0(\cos\alpha \sin\beta \cos\chi + \sin\alpha \cos\beta \sin\chi)_{(y,x)}$$
$$a_x \sin v_x = a_0(\cos\alpha \sin\beta \sin\chi - \sin\alpha \cos\beta \cos\chi)_{(y,x)} \quad \text{(A1.7)}$$

They allow to express

$$a_y^2 = a_0^2(\cos^2\alpha \cos^2\beta + \sin^2\alpha \sin^2\beta)_{(y,x)}$$
$$a_x^2 = a_0^2(\cos^2\alpha \sin^2\beta + \sin^2\alpha \cos^2\beta)_{(y,x)} \quad \text{(A1.8)}$$

and, after simple manipulations,

$$a_y^2 - a_x^2 = a_0^2 \cos 2\alpha_{(y,x)} \cos 2\beta_{(y,x)} \quad \text{(A1.9)}$$

$$2a_y a_x(\cos v_y \cos v_x + \sin v_y \sin v_x) = 2a_y a_x \cos(v_y - v_x)$$
$$= a_0^2 \cos 2\alpha_{(y,x)} \sin 2\beta_{(y,x)} \quad \text{(A1.10)}$$

and

$$2a_y a_x(\sin v_y \cos v_x - \cos v_y \sin v_x) = 2a_y a_x \sin(v_y - v_x)$$
$$= a_0^2 \sin 2\alpha_{(y,x)}. \quad \text{(A1.11)}$$

These equations lead to known expressions for the Stokes four-vector in the (y,x) polarization basis. It formulates the 'intensity representation' of elliptically polarized waves and presents mutual relations between analytical and geometrical parameters of the polarization ellipse in the (y,x) basis:

$$\mathbf{I}_0^{\pm} = \begin{bmatrix} \mathrm{I} \\ \mathrm{Q} \\ \mathrm{U} \\ \mathrm{V} \end{bmatrix}_{(y,x)} = \begin{bmatrix} a_y^2 + a_x^2 \\ a_y^2 - a_x^2 \\ 2a_y a_x \cos(v_y - v_x) \\ 2a_y a_x \sin(v_y - v_x) \end{bmatrix} = a_0^2 \begin{bmatrix} 1 \\ \cos 2\gamma \\ \sin 2\gamma \cos 2\delta \\ \sin 2\gamma \sin 2\delta \end{bmatrix}_{(y,x)} = a_0^2 \begin{bmatrix} 1 \\ \cos 2\alpha \cos 2\beta \\ \cos 2\alpha \sin 2\beta \\ \sin 2\alpha \end{bmatrix}_{(y,x)} \quad \text{(A1.12)}$$

Neglecting the first component, the remaining ones, called 'Stokes parameters', can be interpreted as rectangular coordinates of the polarization point on the Poincare sphere for the elliptically polarized electric vector, components of which are expressed in the (y,x) basis. These Stokes parameters satisfy the equality

$$\mathrm{I}^2 = \mathrm{Q}_{(y,x)}^2 + \mathrm{U}_{(y,x)}^2 + \mathrm{V}_{(y,x)}^2 \quad \text{(A1.13)}$$

and are independent of the direction of wave's propagation or antenna's orientation along the z axis.

A1. 2. The case of the (x,y) polarization basis

The procedure is similar. Again, analytical components of the electric vector in terms of its geometrical components, now in the (x,y) basis, can be obtained as follows:

$$E_x^{\pm} = E_{\xi(x,y)}^{\pm} \cos\beta_{(x,y)} - E_{\eta(x,y)}^{\pm} \sin\beta_{(x,y)}$$
$$= a_0[\cos\alpha \cos\beta \cos(\omega t \mp kz \mp \chi) \pm \sin\alpha \sin\beta \sin(\omega t \mp kz \mp \chi)]_{(x,y)}$$
$$= a_0\{\cos(\omega t \mp kz)[\cos\alpha \cos\beta \cos\chi - \sin\alpha \sin\beta \sin\chi]_{(x,y)}$$
$$\pm \sin(\omega t \mp kz)[\cos\alpha \cos\beta \sin\chi + \sin\alpha \sin\beta \cos\chi]_{(x,y)}\}$$
$$E_y^{\pm} = E_{\xi(x,y)}^{\pm} \sin\beta_{(x,y)} + E_{\eta(x,y)}^{\pm} \cos\beta_{(x,y)} \quad \text{(A1.14)}$$
$$= a_0[\cos\alpha \sin\beta \cos(\omega t \mp kz \mp \chi) \mp \sin\alpha \cos\beta \sin(\omega t \mp kz \mp \chi)]_{(x,y)}$$
$$= a_0\{\cos(\omega t \mp kz)[\cos\alpha \sin\beta \cos\chi + \sin\alpha \cos\beta \sin\chi]_{(x,y)}$$
$$\pm \sin(\omega t \mp kz)[\cos\alpha \sin\beta \sin\chi - \sin\alpha \cos\beta \cos\chi]_{(x,y)}\}.$$

Comparison of expressions (A1.1) and (A1.14) with the same $\cos(\omega t - kz)$ and $\sin(\omega t - kz)$ terms yields:

$$\begin{aligned} a_x \cos \nu_x &= a_0 (\cos\alpha \sin\beta \cos\chi - \sin\alpha \cos\beta \sin\chi)_{(x,y)} \\ a_x \sin \nu_x &= a_0 (\cos\alpha \sin\beta \sin\chi + \sin\alpha \cos\beta \cos\chi)_{(x,y)} \end{aligned} \tag{A1.15}$$

$$\begin{aligned} a_y \cos \nu_y &= a_0 (\cos\alpha \cos\beta \cos\chi + \sin\alpha \sin\beta \sin\chi)_{(x,y)} \\ a_y \sin \nu_y &= a_0 (\cos\alpha \cos\beta \sin\chi - \sin\alpha \sin\beta \cos\chi)_{(x,y)} \end{aligned} \tag{A1.16}$$

That allows to express

$$\begin{aligned} a_x^2 &= a_0^2 (\cos^2\alpha \cos^2\beta + \sin^2\alpha \sin^2\beta)_{(x,y)} \\ a_y^2 &= a_0^2 (\cos^2\alpha \sin^2\beta + \sin^2\alpha \cos^2\beta)_{(x,y)} \end{aligned} \tag{A1.17}$$

and, after simple manipulations,

$$a_x^2 - a_y^2 = a_0^2 \cos 2\alpha_{(x,y)} \cos 2\beta_{(x,y)} \tag{A1.18}$$

$$\begin{aligned} 2a_x a_y (\cos\nu_x \cos\nu_y + \sin\nu_x \sin\nu_y) &= 2a_x a_y \cos(\nu_x - \nu_y) \\ &= a_0^2 \cos 2\alpha_{(x,y)} \sin 2\beta_{(x,y)} \end{aligned} \tag{A1.19}$$

and

$$\begin{aligned} 2a_x a_y (\sin\nu_x \cos\nu_y - \cos\nu_x \sin\nu_y) &= 2a_x a_y \sin(\nu_x - \nu_y) \\ &= a_0^2 \sin 2\alpha_{(x,y)}. \end{aligned} \tag{A1.20}$$

These equations lead to known expressions for Stokes four-vectors, now in the (*x,y*) polarization basis. They present mutual relations between analytical and geometrical parameters of the polarization ellipse in the (*x,y*) basis:

$$\mathbf{I}_0^{\pm} = \begin{bmatrix} \mathrm{I} \\ \mathrm{Q} \\ \mathrm{U} \\ \mathrm{V} \end{bmatrix}_{(x,y)} = \begin{bmatrix} a_x^2 + a_y^2 \\ a_x^2 - a_y^2 \\ 2a_x a_y \cos(\nu_x - \nu_y) \\ 2a_x a_y \sin(\nu_x - \nu_y) \end{bmatrix} = a_0^2 \begin{bmatrix} 1 \\ \cos 2\gamma \\ \sin 2\gamma \cos 2\delta \\ \sin 2\gamma \sin 2\delta \end{bmatrix}_{(x,y)} = a_0^2 \begin{bmatrix} 1 \\ \cos 2\alpha \cos 2\beta \\ \cos 2\alpha \sin 2\beta \\ \sin 2\alpha \end{bmatrix}_{(x,y)} \tag{A1.21}$$

It should be observed an essential result:

$$\begin{aligned} \mathrm{Q}_{(x,y)} &= -\mathrm{Q}_{(y,x)} \\ \mathrm{U}_{(x,y)} &= \mathrm{U}_{(y,x)} \\ \mathrm{V}_{(x,y)} &= -\mathrm{V}_{(y,x)}. \end{aligned} \tag{A1.22}$$

For example, for circular polarizations: $-2\alpha_{(x,y)} = 2\alpha_{(y,x)} = 90^0$. It means that in the (*y,x*) basis the right-handed circular polarization is represented by the upper pole of the Poincare sphere, and in the (*x,y*) basis by its lower pole.

APPENDIX B

DIAMETER OF THE POINCARE SPHERE MODEL OF THE SCATTERING MATRIX

There are two polarization points on the Poincare sphere, M and N, which correspond to maximum and minimum of the scattered power, $\sigma^{M} = \sigma_{max}$ and $\sigma^{N} = \sigma_{min}$, accordingly. These scattered powers correspond to the unit power of the incident wave and can be found immediately from the following scattering equation in any ONP PP basis, the H basis for example,

$$K_H P_{0H}^{M,N} = \begin{bmatrix} a_1 & b_1 & b_3 & b_5 \\ c_1 & a_2 & b_4 & b_6 \\ c_3 & c_4 & a_3 & b_2 \\ c_5 & c_6 & c_2 & a_4 \end{bmatrix}_H \begin{bmatrix} 1 \\ q^{M,N} \\ u^{M,N} \\ v^{M,N} \end{bmatrix}_H = \sigma^{M,N} \begin{bmatrix} 1 \\ q^{M'',N''} \\ u^{M'',N''} \\ v^{M'',N''} \end{bmatrix}_H = \sigma^{M,N} P_{0H}^{M'',N''}. \tag{B.1}$$

The obtained result (in any ONP PP basis) is

$$\begin{aligned} \sigma^{M,N} &= a_1 + b_1 q^{M,N} + b_3 u^{M,N} + b_5 v^{M,N} \\ &= a_1 \pm \left(b_1^2 + b_3^2 + b_5^2\right) / b_0 \\ &= a_1 \pm b_0 \end{aligned} \tag{B.2}$$

where

$$b_0 \equiv \sqrt{b_1^2 + b_3^2 + b_5^2}\,. \tag{B.3}$$

The term b_0 does not depend on the basis, similarly as a_1 which is the half of *span* of the corresponding Sinclair matrix.

The incident wave's unit total power Stokes four-vectors (called also the 'full' unit four-vectors) are accordingly:

$$P_{0H}^{M,N} = \begin{bmatrix} 1 \\ q^{M,N} \\ u^{M,N} \\ v^{M,N} \end{bmatrix}_H = \frac{1}{b_0} \begin{bmatrix} b_0 \\ \pm b_1 \\ \pm b_3 \\ \pm b_5 \end{bmatrix}_H \tag{B.4}$$

what means that they are mutually orthogonal, and their points on the Poincare sphere are antipodal to each other.

Consider now the Poincare sphere of incident waves of unit total power, in any ONP PP basis, and its MN diameter with two points on it: the center of the sphere, O, and an I point (called the inversion point) located somewhere on the ON radius. Let the length of the radius of that sphere be

$$r_0 = (\mathrm{ON})_n = (\mathrm{OM})_n = 1\,. \tag{B.5}$$

Here the subscript 'n' means that the radius of the sphere has been normalized to one. Vectors of those two radii are

$$\left.\begin{array}{l} vec(\mathrm{OM})_n \\ vec(\mathrm{ON})_n \end{array}\right\} = \pm \begin{bmatrix} q^{M} \\ u^{M} \\ v^{M} \end{bmatrix} \tag{B.6}$$

That yields the vector of the $(\mathrm{OI})_n$ segment:

$$vec(\mathrm{OI})_n = \begin{bmatrix} Q_n^{I} \\ U_n^{I} \\ V_n^{I} \end{bmatrix} = -(\mathrm{OI})_n \begin{bmatrix} q^{M} \\ u^{M} \\ v^{M} \end{bmatrix} = \frac{-(\mathrm{OI})_n}{b_0} \begin{bmatrix} b_1 \\ b_3 \\ b_5 \end{bmatrix}. \tag{B.7}$$

and the scalar product

$$vec(\mathrm{OI})_n \bullet vec(\mathrm{OM})_n = \mathrm{Q}_n^{\mathrm{I}}\mathrm{q}^{\mathrm{M}} + \mathrm{U}_n^{\mathrm{I}}\mathrm{u}^{\mathrm{M}} + \mathrm{V}_n^{\mathrm{I}}\mathrm{v}^{\mathrm{M}} = (\mathrm{OI})_n \equiv (\mathrm{OI})_{r_0=1} \leq 1. \qquad \text{(B.8)}$$

Following the equalities

$$\left(\mathrm{q}^{\mathrm{M}}\right)^2 + \left(\mathrm{u}^{\mathrm{M}}\right)^2 + \left(\mathrm{v}^{\mathrm{M}}\right)^2 = 1 \quad \text{and} \quad \left(\mathrm{Q}_n^{\mathrm{I}}\right)^2 + \left(\mathrm{U}_n^{\mathrm{I}}\right)^2 + \left(\mathrm{V}_n^{\mathrm{I}}\right)^2 = (\mathrm{OI})_{r_0=1}^2, \qquad \text{(B.9)}$$

squares of the $(\mathrm{IM})_n$ and $(\mathrm{IN})_n$ segments can be presented as:

$$\left.\begin{matrix}(\mathrm{IM})_n^2 \\ (\mathrm{IN})_n^2\end{matrix}\right\} = \left(\pm\mathrm{q}^{\mathrm{M}} - \mathrm{Q}_n^{\mathrm{I}}\right)^2 + \left(\pm\mathrm{u}^{\mathrm{M}} - \mathrm{U}_n^{\mathrm{I}}\right)^2 + \left(\pm\mathrm{v}^{\mathrm{M}} - \mathrm{V}_n^{\mathrm{I}}\right)^2$$
$$= 1 + (\mathrm{OI})_n^2 \mp 2\left(\mathrm{Q}_n^{\mathrm{I}}\mathrm{q}^{\mathrm{M,N}} + \mathrm{U}_n^{\mathrm{I}}\mathrm{u}^{\mathrm{M,N}} + \mathrm{V}_n^{\mathrm{I}}\mathrm{v}^{\mathrm{M,N}}\right) \qquad \text{(B.10)}$$

That resembles the above presented expressions for the scattered powers. Therefore, denoting after Kennaugh [ ] the radius of the Poincare sphere for scattered waves as

$$r_0 = k \equiv \frac{\sqrt{\sigma_0}}{2} \qquad \text{(B.11)}$$

it is possible to present the following equalities:

$$b_0 = -2k\left(k\mathrm{Q}_n^{\mathrm{I}}\mathrm{q}^{\mathrm{M}} + k\mathrm{U}_n^{\mathrm{I}}\mathrm{u}^{\mathrm{M}} + k\mathrm{V}_n^{\mathrm{I}}\mathrm{v}^{\mathrm{M}}\right) = 2k[k(\mathrm{OI})_{r_0=1}] = 2k(\mathrm{OI})_{r_0=k}$$
$$= \sqrt{\sigma_0}\,(\mathrm{OI})_{r_0=k} \qquad \text{(B.12)}$$

and

$$a_1 = k^2[1 + (\mathrm{OI})_n^2] = k^2 + [k(\mathrm{OI})_n]^2 = k^2 + (\mathrm{OI})_{r_0=k}^2$$
$$= \frac{\sigma_0}{4} + \frac{b_0^2}{\sigma_0} \qquad \text{(B.13)}$$

what yields

$$\sigma^{\mathrm{M,N}} = a_1 \pm b_0 = k^2 + (\mathrm{OI})_{r_0=k}^2 \pm 2k(\mathrm{OI})_{r_0=k} = [k + (\mathrm{OI})_{r_0=k}]^2 = \left\{\begin{matrix}(\mathrm{IM})_{r_0=k}^2 \\ (\mathrm{IN})_{r_0=k}^2\end{matrix}\right. . \qquad \text{(B.14)}$$

Finally, the diameter of the Poincare sphere for scattered waves (the Poincare sphere model of the scattering matrix) is

$$2r_0 = k(\mathrm{IM})_{r_0=1} + k(\mathrm{IN})_{r_0=1} = (\mathrm{IM})_{r_0=k} + (\mathrm{IN})_{r_0=k} = 2k$$
$$= \sqrt{\sigma^{\mathrm{M}}} + \sqrt{\sigma^{\mathrm{N}}} = \sqrt{\sigma_0} \qquad \text{(B.15)}$$

and its square is

$$\sigma_0 = \left(\sqrt{\sigma^{\mathrm{M}}} + \sqrt{\sigma^{\mathrm{N}}}\right)^2$$
$$= SpanA + 2|\det A| \qquad \text{(B.16)}$$

That square of the Poincare sphere diameter is (after Kennaugh [96]) 'proportional to the target's effective crossection (ECS) given by the squared sum of square roots of radar crossections (RCS) for polarizations M and N'. It can be expressed by the span and determinant of the Sinclair matrix A, corresponding to the Kennaugh matrix K, as has been shown above (see also Appendix C).

So, powers are proportional to squares of segments inside the Poincare sphere model of the scattering matrix, while these segments themselves, as well as the diameter of the model, are proportional to amplitude real coefficients

APPENDIX C

OTHER POWER RELATIONS INSIDE THE POINCARE SPHERE MODEL

In the Appendix B developed magnitude of the target's effective crossection (ECS) can be presented by the following sequence of formulae:

$$\begin{aligned}\sigma_0 &= \left(\sqrt{\sigma_{\max}} + \sqrt{\sigma_{\min}}\right)^2 \\ &= \sigma_{\max} + \sigma_{\min} + 2\sqrt{\sigma_{\max}\sigma_{\min}} \\ &= 2\left(a_1 + \sqrt{a_1^2 - b_0^2}\right) \\ &= M_1 + M_2 + M_3 + M_4 + 2\sqrt{M_1M_2 + M_3M_4 - 2\,\mathrm{Re}(A_1A_2A_3{}^*A_4{}^*)} \\ &= SpanA + 2|\det A|\end{aligned} \tag{C.1}$$

Expressing the absolute value of determinant of the Sinclair matrix by

$$a_0 \equiv |\det A| = \sqrt{\sigma_{\max}\sigma_{\min}} = \sqrt{a_1^2 - b_0^2} = \frac{\sigma_0}{2} - a_1 \tag{C.2}$$

one obtains

$$\sigma_0 \equiv (2k)^2 = 2(a_1 + a_0) \tag{C.3}$$

and other useful relations:

$$a_1^2 = a_0^2 + b_0^2, \tag{C.4}$$

$$\begin{aligned}(\mathrm{IM})_n &= 1 + (\mathrm{OI})_n = \sqrt{\sigma_{\max}}/k \\ (\mathrm{IN})_n &= 1 - (\mathrm{OI})_n = \sqrt{\sigma_{\min}}/k \\ (\mathrm{OI})_n &= \frac{\sqrt{\sigma_{\max}} - \sqrt{\sigma_{\min}}}{2k} = \frac{b_0}{2k^2}\end{aligned} \tag{C.5}$$

$$\begin{aligned}1 - (\mathrm{OI})_n^2 = \frac{a_0}{k^2} = a_{0n};\quad 1 + (\mathrm{OI})_n^2 = \frac{a_1}{k^2} = a_{1n} \\ a_{1n,0n} = 1 \pm [(\mathrm{Q}_n^{\mathrm{I}})^2 + (\mathrm{U}_n^{\mathrm{I}})^2 + (\mathrm{V}_n^{\mathrm{I}})^2]\end{aligned} \tag{C.6}$$

$$\begin{bmatrix}\mathrm{Q}\\ \mathrm{U}\\ \mathrm{V}\end{bmatrix}^{\mathrm{I}}_{r_0=1} = \frac{1}{k}\begin{bmatrix}\mathrm{Q}\\ \mathrm{U}\\ \mathrm{V}\end{bmatrix}^{\mathrm{I}}_{r_0=k} = \frac{-1}{2k^2}\begin{bmatrix}b_1\\ b_3\\ b_5\end{bmatrix} = \frac{-2}{\sigma_0}\begin{bmatrix}b_1\\ b_3\\ b_5\end{bmatrix}. \tag{C.7}$$

For any incident polarization point T:

$$\begin{aligned}(\mathrm{IT})_n^2 &= \left(\mathrm{q}^{\mathrm{T}} - \mathrm{Q}_n^{\mathrm{I}}\right)^2 + \left(\mathrm{u}^{\mathrm{T}} - \mathrm{U}_n^{\mathrm{I}}\right)^2 + \left(\mathrm{v}^{\mathrm{T}} - \mathrm{V}_n^{\mathrm{I}}\right)^2 \\ &= 1 + (\mathrm{OI})_n^2 - 2\left(\mathrm{Q}_n^{\mathrm{I}}\mathrm{q}^{\mathrm{T}} + \mathrm{U}_n^{\mathrm{I}}\mathrm{u}^{\mathrm{T}} + \mathrm{V}_n^{\mathrm{I}}\mathrm{v}^{\mathrm{T}}\right) \\ &= \frac{1}{k^2}[a_1 + (b_1\mathrm{q}^{\mathrm{T}} + b_3\mathrm{u}^{\mathrm{T}} + b_5\mathrm{v}^{\mathrm{T}})] \\ &= \frac{1}{k^2}\sigma^{\mathrm{T}}\end{aligned} \tag{C.8}$$

and

$$\sigma^{\mathrm{T}} = (\mathrm{IT})^2_{r_0=k\equiv\sqrt{\sigma_0}/2}. \tag{C.9}$$

APPENDIX D

GEOMETRICAL REPRESENTATION OF THE CO-POLARIZED RECEIVED POWER FOR THE BISTATIC SCATTERING (NONSYMMETRICAL) MATRIX

Following the original derivation by Kennaugh [95], and using the here proposed notation, the co-polarized received voltage will be presented by the amplitude transmission equation in the characteristic ONP PP basis K, in the form:

$$V_c = \begin{bmatrix} a & b \end{bmatrix}_K^T \begin{bmatrix} A_2 & B_1 + jB_2 \\ -B_1 - jB_2 & A_1 \end{bmatrix}_{CCS} \begin{bmatrix} a \\ b \end{bmatrix}_K^T = A_{2CCS}(a_K^T)^2 + A_{1CCS}(b_K^T)^2 \tag{D.1}$$

where the exponent term, $\exp\{j\mu\}$, of the A_K scattering matrix has been omitted, leaving the diagonal elements of that matrix,

$$A_2 \equiv A_{2CCS} \geq A_1 \equiv A_{1CCS}, \tag{D.2}$$

nonnegative real. It can be immediately seen that the problem is almost exactly like for the symmetrical matrix. The only difference will be in the value of the effective crossection, $\sigma_0' = (A_1 + A_2)^2 < \sigma_0$, representing the square of diameter of the Poincare sphere model for that matrix.
When using the known formulae for the Cayley-Klein parameters in terms of the analytical parameters,

$$\begin{aligned} a &= \cos\gamma \exp\{-j(\delta + \varepsilon)\} \\ b &= \sin\gamma \exp\{j(\delta - \varepsilon)\} \end{aligned}, \tag{D.3}$$

the co-polarized received power can be presented as follows,

$$P_c = |V_c|^2 = (A_2 \cos^2\gamma_K^T + A_1 \sin^2\gamma_K^T)^2 - (\sqrt{A_1 A_2} \sin 2\gamma_K^T \sin 2\delta_K^T)^2. \tag{D.4}$$

For the CO-POL Null points, O_1 and O_2, in the CCS we have

$$\cos 2\gamma_K^{O_{1,2}} = \frac{A_1 - A_2}{A_1 + A_2} \leq 0, \quad \sin 2\gamma_K^{O_{1,2}} = \frac{2\sqrt{A_1 A_2}}{A_1 + A_2} \geq 0. \tag{D.5}$$

Denoting

$$2\gamma_K^{O_1} = 2\gamma_K^{O_2} = 2\gamma_K^{O} \tag{D.6}$$

we obtain

$$A_2 \cos^2\gamma_K^T + A_1 \sin^2\gamma_K^T = \tfrac{1}{2}(A_1 + A_2)(1 - \cos 2\gamma_K^O \cos 2\gamma_K^T) \tag{D.7}$$

and

$$\sqrt{A_1 A_2} \sin 2\gamma_K^T \sin 2\delta_K^T = -\tfrac{1}{2}(A_1 + A_2) \sin 2\gamma_K^O \sin 2\gamma_K^T \cos(90^0 + 2\delta_K^T). \tag{D.8}$$

Presenting the received power as a product of the sum and difference of the above expressions, applying the known dependence from spherical trigonometry,

$$\cos 2\psi_{1,2} = \cos 2\gamma_K^O \cos 2\gamma - \sin 2\gamma_K^O \sin 2\gamma \cos(90^0 \pm 2\delta), \tag{D.9}$$

and the equality

$$\tfrac{1}{2}(1 - \cos 2\psi) = \sin^2\psi, \tag{D.10}$$

we arrive at the formula

$$P_c = (A_1 + A_2)^2 \sin^2\psi_1 \sin^2\psi_2. \tag{D.11}$$

In view of the equality

$$\sin\psi_{1,2} = \tfrac{1}{2}(\mathrm{O}_{1,2}\mathrm{T})_{r_0=1} = \tfrac{1}{\sqrt{\sigma_0'}}(\mathrm{O}_{1,2}\mathrm{T})_{2r_0=\sqrt{\sigma_0'}}, \quad \text{with} \quad \sqrt{\sigma_0'} = A_1 + A_2, \tag{D.12}$$

the equivalent geometrical dependence for the co-polarized received power takes the form,

$$\begin{aligned} P_c &= \frac{(A_1 + A_2)^2}{16}[(\mathrm{O}_1\mathrm{T})^2 \times (\mathrm{O}_2\mathrm{T})^2]_{r_0=1} \\ &= \frac{[(\mathrm{O}_1\mathrm{T})^2 \times (\mathrm{O}_2\mathrm{T})^2]_{2r_0=A_1+A_2=\sqrt{\sigma_0'}}}{(A_1 + A_2)^2} \\ &= [(\mathrm{O}_1\mathrm{T})^2 \times (\mathrm{O}_2\mathrm{T})^2]_{2r_0=\sqrt{A_1+A_2}=\sqrt[4]{\sigma_0'}} \\ &= \frac{\sigma_0'}{(2r_0)^4}[(\mathrm{O}_1\mathrm{T})^2 \times (\mathrm{O}_2\mathrm{T})^2] \end{aligned} \tag{D.13}$$

where

$$\begin{aligned} \sigma_0' &\equiv (2k')^2 = (A_1 + A_2)^2 \\ &\le \sigma_0 \equiv (2k)^2 = A_2^2 + A_1^2 + 2(B_1^2 + B_2^2) + 2\sqrt{(A_1A_2 + B_1^2 - B_2^2)^2 + 4B_1^2B_2^2}\,. \end{aligned} \tag{D14.}$$

APPENDIX E

RELATIONS BETWEEN ELEMENTS OF SINCLAIR AND KENNAUGH MATRICES. CONDITIONS FOR PRESERVATION OF THE COMPLETE POLARIZATION

E 1. Bistatic scattering. General form of matrices

The following two notations will be used for presenting the Kennaugh matrix in the ONP horizontal/vertical linear basis: after Perrin [126]/van de Hulst [82] (originally being applied to the FAA, but here to the BSA), and after Huynen/Cloude (Cloude [32] extended the Huynen's notation, originally introduced for symmetrical matrices, to the bistatic scattering case),

$$\mathrm{K}_H = \begin{bmatrix} \mathrm{a}_1 & \mathrm{b}_1 & \mathrm{b}_3 & \mathrm{b}_5 \\ \mathrm{c}_1 & \mathrm{a}_2 & \mathrm{b}_4 & \mathrm{b}_6 \\ \mathrm{c}_3 & \mathrm{c}_4 & \mathrm{a}_3 & \mathrm{b}_2 \\ \mathrm{c}_5 & \mathrm{c}_6 & \mathrm{c}_2 & \mathrm{a}_4 \end{bmatrix}_H = \begin{bmatrix} A_0+B_0 & C+N & H+L & F+I \\ C-N & A+B & E+J & G+K \\ H-L & E-J & A-B & D+M \\ F-I & G-K & D-M & -A_0+B_0 \end{bmatrix}_H . \tag{E.1}$$

That Kennaugh matrix corresponds to the following Sinclair matrix in the appropriate notations

$$A_H = \begin{bmatrix} A_2 & A_3 \\ A_4 & A_1 \end{bmatrix}_H = \frac{e^{j\phi}}{\sqrt{2(A_0+A)}} \begin{bmatrix} A_0+A+(C+jD) & H-jG-j(I+jJ) \\ H-jG+j(I+jJ) & A_0+A-(C+jD) \end{bmatrix}_H \tag{E.2}$$

and can be obtained according to the formula

$$\mathrm{K}_H = \tilde{\mathrm{U}}(A_H \otimes A_H{}^*)\mathrm{U}; \qquad \mathrm{U} = \frac{1}{\sqrt{2}} \begin{bmatrix} 1 & 1 & 0 & 0 \\ 0 & 0 & 1 & -j \\ 0 & 0 & 1 & j \\ 1 & -1 & 0 & 0 \end{bmatrix} . \tag{E.3}$$

It should be observed that matrices in the above explained Perrin/van de Hulst notation, and all further presented formulae in that notation, can be related to any other ONP PP basis, not only to the H basis. They are equally valid for the natural and reversed order of basis vectors, but only for the time-convention $\exp(+j\omega t)$, and for propagation z-axes directed towards the target for the incident and scattered waves.

Expressions for elements of the Kennaugh matrix will be presented with the use of the additional auxiliary notation:

$$\begin{gathered} M_k = A_k A_k{}^* \\ S_{ki} = S_{ik} = \tfrac{1}{2}(A_i A_k{}^* + A_k A_i{}^*) = \mathrm{Re}(A_k A_i{}^*) = \mathrm{Re}(A_i A_k{}^*) \\ -D_{ki} = D_{ik} = \tfrac{j}{2}(A_i A_k{}^* - A_k A_i{}^*) = \mathrm{Im}(A_k A_i{}^*) = -\mathrm{Im}(A_i A_k{}^*) \end{gathered} \tag{E.4}$$

The resulting formulae are

$$\begin{aligned} \mathrm{a}_1 &= \tfrac{1}{2}(M_2 + M_3 + M_4 + M_1) \\ \mathrm{a}_2 &= \tfrac{1}{2}(M_2 - M_3 - M_4 + M_1) \\ \mathrm{b}_1 &= \tfrac{1}{2}(M_2 - M_3 + M_4 - M_1) \\ \mathrm{c}_1 &= \tfrac{1}{2}(M_2 + M_3 - M_4 - M_1) \end{aligned} \tag{E.5a}$$

$$\begin{aligned} \mathrm{a}_3 &= S_{34} + S_{12} = \mathrm{Re}(A_4 A_3{}^* + A_2 A_1{}^*) \\ \mathrm{a}_4 &= S_{34} - S_{12} = \mathrm{Re}(A_4 A_3{}^* - A_2 A_1{}^*) \end{aligned} \tag{E.5b}$$

$$b_2 = D_{12} + D_{34} = \mathrm{Im}(A_2 A_1{}^* + A_4 A_3{}^*)$$
$$c_2 = D_{12} - D_{34} = \mathrm{Im}(A_2 A_1{}^* - A_4 A_3{}^*) \tag{E.5c}$$

$$b_3 = S_{32} + S_{14} = \mathrm{Re}(A_2 A_3{}^* + A_4 A_1{}^*)$$
$$b_4 = S_{32} - S_{14} = \mathrm{Re}(A_2 A_3{}^* - A_4 A_1{}^*)$$
$$b_5 = D_{32} + D_{14} = \mathrm{Im}(A_2 A_3{}^* + A_4 A_1{}^*) \tag{E.5d}$$
$$b_6 = D_{32} - D_{14} = \mathrm{Im}(A_2 A_3{}^* - A_4 A_1{}^*)$$

$$c_3 = S_{42} + S_{13} = \mathrm{Re}(A_2 A_4{}^* + A_3 A_1{}^*)$$
$$c_4 = S_{42} - S_{13} = \mathrm{Re}(A_2 A_4{}^* - A_3 A_1{}^*)$$
$$c_5 = D_{42} + D_{13} = \mathrm{Im}(A_2 A_4{}^* + A_3 A_1{}^*) \tag{E.5e}$$
$$c_6 = D_{42} - D_{13} = \mathrm{Im}(A_2 A_4{}^* - A_3 A_1{}^*)$$

For nondepolarizing (or 'point') target , only 7 independent elements of the Kennaugh matrix may exist. Therefore the following mutual relations between elements of that matrix can be found. They are known also as conditions for preservation of the complete polarization:

$$\left.\begin{aligned} a_1b_1 - a_2c_1 &= c_3c_4 + c_5c_6 \\ a_1c_1 - a_2b_1 &= b_3b_4 + b_5b_6 \\ a_3b_1 + a_4c_1 &= b_3c_4 + c_5b_6 \\ a_3c_1 + a_4b_1 &= c_3b_4 + b_5c_6 \end{aligned}\right\} \Leftrightarrow \left\{\begin{aligned} C(B_0 - B) &= EH + FG \\ C(A_0 - A) &= IK + JL \\ N(B_0 + B) &= -EL - FK \\ N(A_0 + A) &= -GI - HJ \end{aligned}\right. \tag{E.6a}$$

$$\left.\begin{aligned} a_1b_2 - a_2c_2 &= c_3b_5 - b_4c_6 \\ a_1c_2 - a_2b_2 &= b_3c_5 - c_4b_6 \\ a_3b_2 + a_4c_2 &= b_3b_5 - b_4b_6 \\ a_3c_2 + a_4b_2 &= c_3c_5 - c_4c_6 \end{aligned}\right\} \Leftrightarrow \left\{\begin{aligned} D(B_0 - B) &= -EG + FH \\ D(A_0 - A) &= -IL + JK \\ M(B_0 + B) &= EK - FL \\ M(A_0 + A) &= -GJ + HI \end{aligned}\right. \tag{E.6b}$$

$$\left.\begin{aligned} a_1b_4 - a_4c_4 &= c_1b_3 - b_2c_6 \\ a_1c_4 - a_4b_4 &= b_1c_3 - c_2b_6 \\ a_2b_4 + a_3c_4 &= b_1b_3 - c_2c_6 \\ a_2c_4 + a_3b_4 &= c_1c_3 - b_2b_6 \end{aligned}\right\} \Leftrightarrow \left\{\begin{aligned} E(A_0 + A) &= CH - DG \\ E(A_0 - A) &= KM - LN \\ J(B_0 + B) &= CL + DK \\ J(B_0 - B) &= -GM - HN \end{aligned}\right. \tag{E.6c}$$

$$\left.\begin{aligned} a_1b_5 - a_4c_5 &= c_1b_6 + b_2c_3 \\ a_1c_5 - a_4b_5 &= b_1c_6 + c_2b_3 \\ a_2b_5 + a_3c_5 &= b_1b_6 + c_2c_3 \\ a_2c_5 + a_3b_5 &= c_1c_6 + b_2b_3 \end{aligned}\right\} \Leftrightarrow \left\{\begin{aligned} F(A_0 + A) &= CG + DH \\ F(A_0 - A) &= -KN - LM \\ I(B_0 + B) &= CK - DL \\ I(B_0 - B) &= -GN + HM \end{aligned}\right. \tag{E.6d}$$

$$\left.\begin{aligned} a_1b_6 - a_3c_6 &= c_1b_5 - c_2c_4 \\ a_1c_6 - a_3b_6 &= b_1c_5 - b_2b_4 \\ a_2b_6 + a_4c_6 &= b_1b_5 - b_2c_4 \\ a_2c_6 + a_4b_6 &= c_1c_5 - c_2b_4 \end{aligned}\right\} \Leftrightarrow \left\{\begin{aligned} G(B_0 + B) &= CF - DE \\ G(A_0 - A) &= -IN - JM \\ K(A_0 + A) &= CI + DJ \\ K(B_0 - B) &= EM - FN \end{aligned}\right. \tag{E.6e}$$

$$\left.\begin{aligned} a_1b_3 - a_3c_3 &= c_1b_4 + c_2c_5 \\ a_1c_3 - a_3b_3 &= b_1c_4 + b_2b_5 \\ a_2b_3 + a_4c_3 &= b_1b_4 + b_2c_5 \\ a_2c_3 + a_4b_3 &= c_1c_4 + c_2b_5 \end{aligned}\right\} \Leftrightarrow \begin{cases} H(B_0 + B) = CE + DF \\ H(A_0 - A) = IM - JN \\ L(A_0 + A) = CJ - DI \\ L(B_0 - B) = -EN - FM \end{cases} \tag{E.6f}$$

$$\begin{aligned} b_0^2 = b_1^2 + b_3^2 + b_5^2 = c_1^2 + c_3^2 + c_5^2 &\Leftrightarrow CN + FI + HL = 0 \\ b_1^2 + b_4^2 + b_6^2 = c_1^2 + c_4^2 + c_6^2 &\Leftrightarrow CN + EJ + GK = 0 \\ b_3^2 - b_4^2 + b_2^2 = c_3^2 - c_4^2 + c_2^2 &\Leftrightarrow DM - HJ + HL = 0 \\ b_5^2 - b_6^2 - b_2^2 = c_5^2 - c_6^2 - c_2^2 &\Leftrightarrow DM - FI + GK = 0 \end{aligned} \tag{E.6g}$$

$$a_1a_2 + a_3a_4 = b_1c_1 + b_2c_2 \Leftrightarrow \begin{cases} C^2 + D^2 = (A_0 + A)(B_0 + B) \\ M^2 + N^2 = (A_0 - A)(B_0 - B) \end{cases} \tag{E.6h}$$

$$a_1a_3 + a_2a_4 = b_3c_3 + b_6c_6 \Leftrightarrow \begin{cases} G^2 + H^2 = (A_0 + A)(B_0 - B) \\ K^2 + L^2 = (A_0 - A)(B_0 + B) \end{cases} \tag{E.6i}$$

$$a_1a_4 + a_2a_3 = b_5c_5 + b_4c_4 \Leftrightarrow \begin{cases} E^2 + F^2 = B_0^2 - B^2 \\ I^2 + J^2 = A_0^2 - A^2 \end{cases} \tag{E.6j}$$

Other relations:

$$\begin{aligned} SpanK &= 4a_1^2 = (SpanA)^2 \\ \det K &= -a_0^4 = -|\det A|^4 . \end{aligned} \tag{E.7}$$

Inverse relations (for A_2 positive real):

$$\begin{aligned} A_2 &= \sqrt{\tfrac{1}{2}(a_1 + a_2 + b_1 + c_1)} \\ A_3 &= [b_3 + b_4 - j(b_5 + b_6)]/(2A_2) \\ A_4 &= [c_3 + c_4 - j(c_5 + c_6)]/(2A_2) \\ A_1 &= [a_3 - a_4 - j(b_2 + c_2)]/(2A_2) \end{aligned} \tag{E.8}$$

E 2. Monostatic scattering. General form of matrices

For monostatic scattering:

$$\begin{aligned} &A_4 = A_3 \\ &c_i = b_i, \quad i = 1,\ldots,6 \\ &I = J = K = L = M = N = 0, \quad A = A_0 \end{aligned} \tag{E.9}$$

Thus, Kennaugh and Sinclair matrices take the forms

$$K_H = \begin{bmatrix} a_1 & b_1 & b_3 & b_5 \\ b_1 & a_2 & b_4 & b_6 \\ b_3 & b_4 & a_3 & b_2 \\ b_5 & b_6 & b_2 & a_4 \end{bmatrix}_H = \begin{bmatrix} A_0 + B_0 & C & H & F \\ C & A_0 + B & E & G \\ H & E & A_0 - B & D \\ F & G & D & -A_0 + B_0 \end{bmatrix}_H , \tag{E.10}$$

$$A_H = \begin{bmatrix} A_2 & A_3 \\ A_3 & A_1 \end{bmatrix}_H = \frac{e^{j\phi}}{2\sqrt{A_0}} \begin{bmatrix} 2A_0 + (C + jD) & H - jG \\ H - jG & 2A_0 - (C + jD) \end{bmatrix}_H . \tag{E.11}$$

Elements of the Kennaugh matrix are

$$\begin{aligned} a_1 &= \tfrac{1}{2}(M_2 + 2M_3 + M_1) \\ a_2 &= \tfrac{1}{2}(M_2 - 2M_3 + M_1) \\ b_1 &= \tfrac{1}{2}(M_2 - M_1) \end{aligned} \tag{E.12a}$$

$$\begin{aligned} a_3 &= M_3 + S_{12} = M_3 + \mathrm{Re}(A_2 A_1^*) \\ a_4 &= M_3 - S_{12} = M_3 - \mathrm{Re}(A_2 A_1^*) \end{aligned} \tag{E.12b}$$

$$b_2 = D_{12} = \mathrm{Im}(A_2 A_1^*) \tag{E.12c}$$

$$\begin{aligned} b_3 &= S_{32} + S_{13} = \mathrm{Re}[A_3^*(A_2 + A_1)] \\ b_4 &= S_{32} - S_{13} = \mathrm{Re}[A_3^*(A_2 - A_1)] \\ b_5 &= D_{32} + D_{13} = \mathrm{Im}[A_3^*(A_2 - A_1)] \\ b_6 &= D_{32} - D_{13} = \mathrm{Im}[A_3^*(A_2 + A_1)] \end{aligned} \tag{E.12d}$$

For nondepolarizing (or 'point') target , only 5 independent elements of the Kennaugh matrix may exist. Therefore the following mutual relations between elements of that matrix can be found. They are known also as conditions for preservation of the complete polarization:

$$a_1 = a_2 + a_3 + a_4 \tag{E.13}$$

$$\begin{aligned} (a_1 - a_4)(a_1 - a_3) &= a_1 a_2 + a_3 a_4 = b_1^2 + b_2^2 \Leftrightarrow C^2 + D^2 = 2A_0(B_0 + B) \\ (a_1 - a_4)(a_1 - a_2) &= a_1 a_3 + a_2 a_4 = b_3^2 + b_6^2 \Leftrightarrow H^2 + G^2 = 2A_0(B_0 - B) \\ (a_1 - a_2)(a_1 - a_3) &= a_1 a_4 + a_2 a_3 = b_5^2 + b_4^2 \Leftrightarrow F^2 + E^2 = B_0^2 - B^2 \end{aligned} \tag{E.14a}$$

$$\begin{aligned} a_1^2 - a_2^2 &= b_3^2 + b_4^2 + b_5^2 + b_6^2 \\ a_1^2 - a_3^2 &= b_1^2 + b_2^2 + b_4^2 + b_5^2 \\ a_1^2 - a_4^2 &= b_1^2 + b_2^2 + b_3^2 + b_6^2 \end{aligned} \tag{E.14b}$$

$$b_1 b_3 b_5 = b_1 b_4 b_6 + b_2 b_3 b_4 + b_2 b_5 b_6 \tag{E.14c}$$

$$\begin{aligned} b_1(a_1 - a_2) &= b_3 b_4 + b_5 b_6 \Leftrightarrow C(B_0 - B) = EH + FG \\ b_2(a_1 - a_2) &= b_3 b_5 - b_4 b_6 \Leftrightarrow D(B_0 - B) = -EG + FH \\ b_3(a_1 - a_3) &= b_1 b_4 + b_2 b_5 \Leftrightarrow H(B_0 + B) = CE + DF \\ b_4(a_1 - a_4) &= b_1 b_3 - b_2 b_6 \Leftrightarrow 2A_0 E = CH - DG \\ b_5(a_1 - a_4) &= b_1 b_6 + b_2 b_3 \Leftrightarrow 2A_0 F = CG + DH \\ b_6(a_1 - a_3) &= b_1 b_5 - b_2 b_4 \Leftrightarrow G(B_0 + B) = CF - DE \end{aligned} \tag{E.14d}$$

E 3. Bistatic scattering in the characteristic basis

In the characteristic coordinate system (CCS) of Stokes parameters, corresponding to the characteristic basis, K, of the orthogonal null-phase (ONP) polarization and phase (PP) basis vectors, u^K and $u^L = u^{K\times}$, we obtain

$$\mathrm{K}_{CCS} \equiv \mathrm{K}_K = \begin{bmatrix} \mathrm{a}_1 & \mathrm{b}_1 & \mathrm{b}_3 & \mathrm{b}_5 \\ \mathrm{b}_1 & \mathrm{a}_2 & \mathrm{b}_4 & \mathrm{b}_6 \\ -\mathrm{b}_3 & -\mathrm{b}_4 & \mathrm{a}_3 & 0 \\ -\mathrm{b}_5 & -\mathrm{b}_6 & 0 & \mathrm{a}_4 \end{bmatrix}_K = \tilde{\mathrm{D}}_H^K \mathrm{K}_H \mathrm{D}_H^K = \tilde{\mathrm{U}}(A_K \otimes A_K{}^*)\mathrm{U} \tag{E.15}$$

for

$$\mathrm{D}_H^K = \tilde{\mathrm{U}}^*(C_H^K \otimes C_H^K{}^*)\mathrm{U} \tag{E.16}$$

and

$$A_K = \begin{bmatrix} A_2 & A_3 \\ -A_3 & A_1 \end{bmatrix}_K = \begin{bmatrix} A_2 & B_1 + jB_2 \\ -B_1 + jB_2 & A_1 \end{bmatrix}_{CCS} e^{j\mu} = \tilde{C}_H^K A_H C_H^K \tag{E.17}$$

where

$$A_2 \equiv A_{2CCS} = A_{2K}\, e^{-j\mu} \geq A_1 \equiv A_{1CCS} = A_{1K}\, e^{-j\mu} \geq 0 \tag{E.18a}$$

$$B_2 \geq 0 \text{ with } B_1 \geq 0 \text{ if } B_2 = 0. \tag{E.18b}$$

and

$$C_H^K = \begin{bmatrix} a & -b^* \\ b & a^* \end{bmatrix}_H^K. \tag{E.19}$$

Components of the K matrix in the K basis are (lower indices K have been omitted for simplicity of notation):

$$\begin{aligned} &\mathrm{a}_1 = \tfrac{1}{2}\left(A_2^2 + A_1^2\right) + B_1^2 + B_2^2 \geq 0, && \mathrm{b}_3 = B_1\left(A_2 - A_1\right), \\ &\mathrm{a}_2 = \tfrac{1}{2}\left(A_2^2 + A_1^2\right) - B_1^2 - B_2^2, && \mathrm{b}_4 = B_1\left(A_2 + A_1\right), \\ &\mathrm{a}_3 = A_1 A_2 - B_1^2 - B_2^2, && \mathrm{b}_5 = -B_2\left(A_2 + A_1\right) \leq 0, \\ &\mathrm{a}_4 = -A_1 A_2 - B_1^2 - B_2^2 \leq 0, && \mathrm{b}_6 = -B_2\left(A_2 - A_1\right) \leq 0. \\ &\mathrm{b}_1 = \tfrac{1}{2}\left(A_2^2 - A_1^2\right) \geq 0, && \mathrm{b}_2 = 0 \end{aligned} \tag{E.20}$$

They satisfy the linear equation:

$$\mathrm{a}_2 = \mathrm{a}_1 + \mathrm{a}_3 + \mathrm{a}_4. \tag{E.21}$$

The corresponding conditions for preservation of complete polarization are:

$$\begin{aligned} (\mathrm{a}_2 - \mathrm{a}_4)(\mathrm{a}_2 - \mathrm{a}_3) = (\mathrm{a}_1 + \mathrm{a}_4)(\mathrm{a}_1 + \mathrm{a}_3) = \mathrm{a}_1\mathrm{a}_2 + \mathrm{a}_3\mathrm{a}_4 = \mathrm{b}_1^2 \\ (\mathrm{a}_1 + \mathrm{a}_4)(\mathrm{a}_3 + \mathrm{a}_4) = \mathrm{a}_1\mathrm{a}_3 + \mathrm{a}_2\mathrm{a}_4 = -\mathrm{b}_3^2 - \mathrm{b}_6^2 \\ (\mathrm{a}_1 + \mathrm{a}_3)(\mathrm{a}_3 + \mathrm{a}_4) = \mathrm{a}_1\mathrm{a}_4 + \mathrm{a}_2\mathrm{a}_3 = -\mathrm{b}_5^2 - \mathrm{b}_4^2 \end{aligned} \tag{E.22a}$$

$$\begin{aligned} \mathrm{a}_1^2 - \mathrm{a}_2^2 = \mathrm{b}_3^2 + \mathrm{b}_4^2 + \mathrm{b}_5^2 + \mathrm{b}_6^2 \\ \mathrm{a}_1^2 - \mathrm{a}_3^2 = \mathrm{b}_1^2 + \mathrm{b}_4^2 + \mathrm{b}_5^2 \\ \mathrm{a}_1^2 - \mathrm{a}_4^2 = \mathrm{b}_1^2 + \mathrm{b}_3^2 + \mathrm{b}_6^2 \end{aligned} \tag{E.22b}$$

and

$$b_1(a_1 - a_2) = b_3 b_4 + b_5 b_6$$
$$0 = b_3 b_5 - b_4 b_6 \qquad \text{(E.22c)}$$
$$b_3(a_1 + a_3) = b_1 b_4$$
$$b_4(a_1 + a_4) = b_1 b_3$$
$$b_5(a_1 + a_4) = b_1 b_6 \qquad \text{(E.22d)}$$
$$b_6(a_1 + a_3) = b_1 b_5$$

E 4. Monostatic scattering in the characteristic coordinate system

For matrices

$$K_{CCS} \equiv K_K = \begin{bmatrix} a_1 & b_1 & 0 & 0 \\ b_1 & a_1 & 0 & 0 \\ 0 & 0 & a_3 & 0 \\ 0 & 0 & 0 & -a_3 \end{bmatrix}_K = \widetilde{D}_H^K K_H D_H^K = \widetilde{U}(A_K \otimes A_K{}^*)U \qquad \text{(E.23)}$$

and

$$A_K = \begin{bmatrix} A_2 & 0 \\ 0 & A_1 \end{bmatrix}_K = \begin{bmatrix} A_2 & 0 \\ 0 & A_1 \end{bmatrix}_{CCS} e^{j\mu} = \widetilde{C}_H^K A_H C_H^K \qquad \text{(E.24)}$$

we obtain values of the Kennaugh matrix components

$$a_1 = \tfrac{1}{2}(A_2^2 + A_1^2) = \tfrac{1}{2} SpanA$$
$$b_1 = \tfrac{1}{2}(A_2^2 - A_1^2) = b_0 \qquad \text{(E.25)}$$
$$a_3 = A_1 A_2 = a_0 = |\det A| \geq 0.$$

The one only condition for preservation of complete polarization is

$$a_1^2 = a_3^2 + b_1^2. \qquad \text{(E.26)}$$

APPENDIX F
COVARIANCE AND COHERENCE MATRICES

F 1. Definitions

For Sinclair or Jones matrices,

$$A = \begin{bmatrix} A_2 & A_3 \\ A_4 & A_1 \end{bmatrix}, \qquad A^0 = \begin{bmatrix} A_2^0 & A_3^0 \\ A_4^0 & A_1^0 \end{bmatrix} \qquad \text{(F.1)}$$

and the corresponding amplitude complex four-vectors,

$$k_L \equiv vecA = \begin{bmatrix} A_2 \\ A_3 \\ A_4 \\ A_1 \end{bmatrix}, \qquad k_L{}^0 \equiv vecA^0 = \begin{bmatrix} A_2^0 \\ A_3^0 \\ A_4^0 \\ A_1^0 \end{bmatrix}, \qquad \text{(F.2)}$$

or

$$k_P = \widetilde{U}^* k_L, \qquad k_P^0 = \widetilde{U}^* k_P^0, \qquad \text{(F.3)}$$

using the unitary matrix with columns presenting Pauli matrices in vector forms,

$$
\mathrm{U}=\frac{1}{\sqrt{2}}\begin{bmatrix} 1 & 1 & 0 & 0 \\ 0 & 0 & 1 & -j \\ 0 & 0 & 1 & j \\ 1 & -1 & 0 & 0 \end{bmatrix}, \tag{F.4}
$$

we define, in the ONP PP basis H, two covariance matrices, Σ and Σ^0, and two coherence matrices, T and T^0, with mutual relations between them:

$$
\Sigma_H = k_{LH}\widetilde{k}_{LH}{}^* = \mathrm{U}T_H\widetilde{\mathrm{U}}^*, \qquad \Sigma_H^0 = k_{LH}^0\widetilde{k}_{LH}^0{}^* = \mathrm{U}T_H^0\widetilde{\mathrm{U}}^* \tag{F.5}
$$

$$
T_H = k_{PH}\widetilde{k}_{PH}{}^* = \widetilde{\mathrm{U}}^*\Sigma_H\mathrm{U}, \qquad T_H^0 = k_{PH}^0\widetilde{k}_{PH}^0{}^* = \widetilde{\mathrm{U}}^*\Sigma_H^0\mathrm{U} \tag{F.6}
$$

Explicitly (see notation in Appendix E, formulae (E.1)), the covariance matrix is:

$$
\Sigma_H = \frac{1}{2}\begin{bmatrix}
a_1+a_2+b_1+c_1 & b_3+b_4+j(b_5+b_6) & c_3+c_4+j(c_5+c_6) & a_3-a_4+j(b_2+c_2) \\
b_3+b_4-j(b_5+b_6) & a_1-a_2-b_1+c_1 & a_3+a_4-j(b_2-c_2) & c_3-c_4+j(c_5-c_6) \\
c_3+c_4-j(c_5+c_6) & a_3+a_4+j(b_2-c_2) & a_1-a_2+b_1-c_1 & b_3-b_4+j(b_5-b_6) \\
a_3-a_4-j(b_2+c_2) & c_3-c_4-j(c_5-c_6) & b_3-b_4-j(b_5-b_6) & a_1+a_2-b_1-c_1
\end{bmatrix}_H
$$

$$
=\frac{1}{2}\begin{bmatrix}
A_0+A+B_0+B+2C & H+E+J+L+j(F+G+I+K) & H+E-J-L+j(F+G-I-K) & A_0+A-B_0-B+j2D \\
H+E+J+L-j(F+G+I+K) & B_0-B+A_0-A-2N & B_0-B-A_0+A-j2M & H-E+J-L+j(F-G-I+K) \\
H+E-J-L-j(F+G-I-K) & B_0-B-A_0+A+j2M & B_0-B+A_0-A+2N & H-E-J+L+j(F-G+I-K) \\
A_0+A-B_0-B-j2D & H-E+J-L-j(F-G-I+K) & H-E-J+L-j(F-G+I-K) & A_0+A+B_0+B-2C
\end{bmatrix}_H
$$

The coherence matrix is:

$$
T_H = \frac{1}{2}\begin{bmatrix}
a_1+a_2+a_3-a_4 & b_1+c_1-j(b_2+c_2) & b_3+c_3+j(b_6+c_6) & b_5-c_5-j(b_4-c_4) \\
b_1+c_1+j(b_2+c_2) & a_1+a_2-a_3+a_4 & b_4+c_4+j(b_5+c_5) & b_6-c_6-j(b_3-c_3) \\
b_3+c_3-j(b_6+c_6) & b_4+c_4-j(b_5+c_5) & a_1-a_2+a_3+a_4 & b_2-c_2+j(b_1-c_1) \\
b_5-c_5+j(b_4-c_4) & b_6-c_6+j(b_3-c_3) & b_2-c_2-j(b_1-c_1) & a_1-a_2-a_3-a_4
\end{bmatrix}_H
$$

$$
=\begin{bmatrix}
A_0+A & C-jD & H+jG & I-jJ \\
C+JD & B_0+B & E+jF & K-jL \\
H-JG & E-jF & B_0-B & M+jN \\
I+jJ & K+jL & M-jN & A_0-A
\end{bmatrix}_H \tag{F.7}
$$

For the monostatic scattering, elements of the last column and row of that matrix are equal to null.

F 2. The spatial reversal transformation

Starting from the known equation with the Sinclair matrix in any ONP PP basis B we immediately obtain the corresponding relation for the amplitude Sinclair four-vector:

$$A_B^0 = C_B^0 * A_B; \qquad C_B^0 = \widetilde{C}_H^B * \begin{bmatrix} -1 & 0 \\ 0 & 1 \end{bmatrix} C_H^B = \begin{bmatrix} -w & u \\ u & w* \end{bmatrix}_H^B, \qquad C_B^0 * C_B^0 = \begin{bmatrix} 1 & 0 \\ 0 & 1 \end{bmatrix}$$

$$k_{LB}^0 = \mathrm{U}_B^0 * k_{LB}; \quad \mathrm{U}_B^0 * = (C_B^0 * \otimes \begin{bmatrix} 1 & 0 \\ 0 & 1 \end{bmatrix}) = \begin{bmatrix} -w* & 0 & u & 0 \\ 0 & -w* & 0 & u \\ u & 0 & w & 0 \\ 0 & u & 0 & w \end{bmatrix}_H^B, \quad \mathrm{U}_B^0 * \mathrm{U}_B^0 = diag\{1,\ 1,\ 1,\ 1\}$$

with (F.8)

$$\mathrm{u}_H^B = \sin 2\gamma_H^B \cos 2\delta_H^B \tag{F.9}$$

and

$$\begin{aligned} \mathrm{w}_H^B &= \cos 2\gamma_H^B \cos 2\delta_H^B \cos 2\varepsilon_H^B - \sin 2\delta_H^B \sin 2\varepsilon_H^B \\ &\quad -j\left(\cos 2\gamma_H^B \cos 2\delta_H^B \sin 2\varepsilon_H^B + \sin 2\delta_H^B \cos 2\varepsilon_H^B\right) \end{aligned} \tag{F.10}$$

The corresponding relations for covariance and coherency matrices are

$$\begin{aligned} \Sigma_B^0 &= k_{LB}^0 \widetilde{k}_{LB}^0 * \\ &= \mathrm{U}_B^0 * k_{LB} \widetilde{k}_{LB} * \mathrm{U}_B^0 \\ &= \mathrm{U}_B^0 * \Sigma_B \mathrm{U}_B^0 \end{aligned} \tag{F.11}$$

and

$$T_B^0 = \widetilde{\mathrm{U}} * \mathrm{U}_B^0 * \mathrm{U} T_B \widetilde{\mathrm{U}} * \mathrm{U}_B^0 \mathrm{U} \tag{F.12}$$

because

$$\Sigma_B^0 = \mathrm{U} T_B^0 \widetilde{\mathrm{U}} * = \mathrm{U}_B^0 * \Sigma_B \mathrm{U}_B^0 = \mathrm{U}_B^0 * \mathrm{U} T_B \widetilde{\mathrm{U}} * \mathrm{U}_B^0. \tag{F.13}$$

F 3. The transmission equations

The received voltage can be presented in terms of the Sinclair and Jones amplitude complex four-vectors in any ONP PP basis B as follows:

$$\begin{aligned} V_r &= \widetilde{u}_B^R A_B u_B^T = (\widetilde{u}_B^R \otimes \widetilde{u}_B^T) k_{LB} = \widetilde{s}_B k_{PB} \\ &= \widetilde{u}_B^{R0} * A_B^0 u_B^T = (\widetilde{u}_B^{R0} * \otimes \widetilde{u}_B^T) k_{LB}^0 = \widetilde{s}_B^0 k_{PB}^0 \end{aligned} \tag{F.14}$$

where

$$\begin{aligned} s_B &= \widetilde{\mathrm{U}}(u_B^R \otimes u_B^T) \\ s_B^0 &= \widetilde{\mathrm{U}}(u_B^{R0} * \otimes u_B^T) \\ &= \widetilde{\mathrm{U}}(C_B^0 u_B^R \otimes u_B^T) = \widetilde{\mathrm{U}}(C_B^0 \otimes \begin{bmatrix} 1 & 0 \\ 0 & 1 \end{bmatrix}) \mathrm{U} * \widetilde{\mathrm{U}}(u_B^R \otimes u_B^T) \\ &= \widetilde{\mathrm{U}} \mathrm{U}_B^0 \mathrm{U} * s_B \end{aligned} \tag{F.15}$$

and the received power in terms of the corresponding Kennaugh and Mueller coherency matrices

$$
\begin{aligned}
P_r &= V_r\widetilde{V}_r{}^* \\
&= \widetilde{s}_H k_{PH}\widetilde{k}_{PH}{}^* s_H{}^* = \widetilde{s}_H T_H s_H{}^* \\
&= \widetilde{s}_H^0 k_{PH}^0\widetilde{k}_{PH}^0{}^* s_H^0{}^* = \widetilde{s}_H^0 T_H^0 s_H^0{}^*.
\end{aligned}
\tag{F.16}
$$

F 4. The change of basis transformations

Again, starting from known equations for amplitude Sinclair and Jones matrices we can immediately obtain the corresponding relations for the covariance matrices in the new ONP PP basis. Taking

$$
A_B = \widetilde{C}_H^B A_H C_H^B, \qquad A_B^0 = \widetilde{C}_H^B{}^* A_H^0 C_H^B
\tag{F.17}
$$

we find

$$
k_{LB} = \begin{bmatrix} A_2 \\ A_3 \\ A_4 \\ A_1 \end{bmatrix}_B = (\widetilde{C}_H^B \otimes \widetilde{C}_H^B)k_{LH}, \qquad k_{LB}^0 = \begin{bmatrix} A_2^0 \\ A_3^0 \\ A_4^0 \\ A_1^0 \end{bmatrix}_B = (\widetilde{C}_H^B{}^* \otimes \widetilde{C}_H^B)k_{LH}^0
\tag{F.18}
$$

and

$$
\begin{aligned}
\Sigma_B &= k_{LB}\widetilde{k}_{LB}{}^* \\
&= (\widetilde{C}_H^B \otimes \widetilde{C}_H^B)k_{LH}\widetilde{k}_{LH}{}^*(C_H^B{}^* \otimes C^*{}_H^B) \\
&= (\widetilde{C}_H^B \otimes \widetilde{C}_H^B)\Sigma_H(C_H^B{}^* \otimes C^*{}_H^B) = \widetilde{\Sigma}_B{}^*,
\end{aligned}
\tag{F.19}
$$

$$
\begin{aligned}
\Sigma_B^0 &= k_{LB}^0\widetilde{k}_{LB}^0{}^* \\
&= (\widetilde{C}_H^B{}^* \otimes \widetilde{C}_H^B)k_{LH}^0\widetilde{k}_{LH}^0{}^*(C_H^B \otimes C^*{}_H^B) \\
&= (\widetilde{C}_H^B{}^* \otimes \widetilde{C}_H^B)\Sigma_H^0(C_H^B \otimes C^*{}_H^B) = \widetilde{\Sigma}_B^0{}^*,
\end{aligned}
\tag{F.20}
$$

as well as

$$
\begin{aligned}
T_B &= k_{PB}\widetilde{k}_{PB}{}^* \\
&= \widetilde{\mathrm{U}}^* k_{LB}\widetilde{k}_{LB}{}^*\mathrm{U} = \widetilde{\mathrm{U}}^*\Sigma_B\mathrm{U} \\
&= \widetilde{\mathrm{U}}^*(\widetilde{C}_H^B \otimes \widetilde{C}_H^B)\Sigma_H(C_H^B{}^* \otimes C^*{}_H^B)\mathrm{U} \\
&= \widetilde{\mathrm{U}}^*(\widetilde{C}_H^B \otimes \widetilde{C}_H^B)\mathrm{U}T_H\widetilde{\mathrm{U}}^*(C_H^B{}^* \otimes C^*{}_H^B)\mathrm{U} = \widetilde{T}_B{}^*,
\end{aligned}
\tag{F.21}
$$

$$
\begin{aligned}
T_B^0 &= k_{PB}^0\widetilde{k}_{PB}^0{}^* \\
&= \widetilde{\mathrm{U}}^* k_{LB}^0\widetilde{k}_{LB}^0{}^*\mathrm{U} = \widetilde{\mathrm{U}}^*\Sigma_B^0\mathrm{U} \\
&= \widetilde{\mathrm{U}}^*(\widetilde{C}_H^B{}^* \otimes \widetilde{C}_H^B)\Sigma_H^0(C_H^B \otimes C^*{}_H^B)\mathrm{U} \\
&= \widetilde{\mathrm{U}}^*(\widetilde{C}_H^B{}^* \otimes \widetilde{C}_H^B)\mathrm{U}T_H^0\widetilde{\mathrm{U}}^*(C_H^B \otimes C^*{}_H^B)\mathrm{U} = \widetilde{\mathrm{D}}_H^B T_H^0 \mathrm{D}_H^B = \widetilde{T}_B^0{}^*.
\end{aligned}
\tag{F.22}
$$

It is interesting to compare the just obtained equalities with the corresponding change of basis equations for the Kennaugh and Mueller matrices,

$$
\begin{aligned}
\mathrm{K}_B &= \widetilde{\mathrm{D}}_H^B\mathrm{K}_H\mathrm{D}_H^B = \widetilde{\mathrm{U}}^*(\widetilde{C}_H^B{}^* \otimes \widetilde{C}_H^B)\mathrm{U}\mathrm{K}_H\widetilde{\mathrm{U}}^*(C_H^B \otimes C_H^B{}^*)\mathrm{U}, \\
\mathrm{K}_B^0 &= \widetilde{\mathrm{D}}_H^B\mathrm{K}_H^0\mathrm{D}_H^B = \widetilde{\mathrm{U}}^*(\widetilde{C}_H^B{}^* \otimes \widetilde{C}_H^B)\mathrm{U}\mathrm{K}_H^0\widetilde{\mathrm{U}}^*(C_H^B \otimes C_H^B{}^*)\mathrm{U}.
\end{aligned}
\tag{F.23}
$$

APPENDIX G
THE CHARACTERISTIC POLARIZATION RATIO AND THE SINCLAIR MATRIX IN THE CHARACTERISTIC BASIS

The two-step procedure of obtaining the Sinclair matrix in the characteristic polarization basis uses as the first step the transformation

$$A_{K'} = \widetilde{C}_H^{K'} A_H C_H^{K'} = \begin{bmatrix} A_2 & A_3 \\ A_4 & A_1 \end{bmatrix}_{K'} \equiv \begin{bmatrix} A_2' & A_3' \\ -A_3' & A_1' \end{bmatrix} \tag{G.1}$$

with the transformation matrix

$$\begin{aligned} C_H^{K'} &= \begin{bmatrix} e^{-j\delta} & 0 \\ 0 & e^{j\delta} \end{bmatrix}_H^{K'} \begin{bmatrix} \cos\gamma & -\sin\gamma \\ \sin\gamma & \cos\gamma \end{bmatrix}_H^{K'} \begin{bmatrix} e^{j\delta} & 0 \\ 0 & e^{-j\delta} \end{bmatrix}_H^{K'} \\ &= \begin{bmatrix} \cos\gamma & -\sin\gamma e^{-j2\delta} \\ \sin\gamma e^{j2\delta} & \cos\gamma \end{bmatrix}_H^{K'} \\ &= \frac{1}{\sqrt{1+\rho\rho^*}} \begin{bmatrix} 1 & -\rho^* \\ \rho & 1 \end{bmatrix}_H^{K'} \end{aligned} \tag{G.2}$$

The second-step transformation will change only phases of diagonal elements. Using the transformation matrix

$$C_{K'}^{K} = \begin{bmatrix} e^{-j(\delta+\varepsilon)} & 0 \\ 0 & e^{j(\delta+\varepsilon)} \end{bmatrix}_H^{K} \tag{G.3}$$

we obtain

$$A_K = \widetilde{C}_{K'}^{K} A_{K'} C_{K'}^{K} = \begin{bmatrix} A'_2 e^{-j2(\delta+\varepsilon)} & A'_3 \\ A'_4 & A'_1 e^{j2(\delta+\varepsilon)} \end{bmatrix} = \begin{bmatrix} A_2 & A_3 \\ -A_3 & A_1 \end{bmatrix}_K . \tag{G.4}$$

In the characteristic basis we want to have equal phases of diagonal elements. It can be achieved when taking

$$2\varepsilon_H^K = \tfrac{1}{2}(\arg A'_2 - \arg A'_1) - 2\delta_H^K \tag{G.5}$$

These elements will become real when excluding from the matrix the exponent term, $\exp\{j\mu\}$, with the phase argument

$$\mu = \tfrac{1}{2}(\arg A'_2 + \arg A'_1). \tag{G.6}$$

The problem remains of obtaining the characteristic polarization ratio

$$\rho = \rho_H^K = \rho_H^{K'} \tag{G.7}$$

for known elements, $A_{2H}, A_{3H}, A_{4H}, A_{1H}$, of the A_H matrix. Such polarization ratio should fulfill the equality

$$A_{3K} + A_{4K} = A_{3K'} + A_{4K'} = 0. \tag{G.8}$$

The first of the two transformations yields expressions

$$\begin{aligned} A'_3 &= [-A_{2H}\rho^* + A_{3H} - A_{4H}\rho\rho^* + A_{1H}\rho]/(1+\rho\rho^*) \\ A'_4 &= [-A_{2H}\rho^* - A_{3H}\rho\rho^* + A_{4H} + A_{1H}\rho]/(1+\rho\rho^*) \end{aligned} \tag{G.9}$$

Neglecting denominators we will add these two expressions equalizing the sum to null, and take also its conjugate value, thus obtaining the two equations:

$$
\begin{aligned}
&-2A_{2H}\rho^* + 2A_{1H}\rho + (A_{3H} + A_{4H})(1-\rho\rho^*) = 0 \\
&2A_{1H}{}^*\rho^* - 2A_{2H}{}^*\rho + (A_{3H}{}^* + A_{4H}{}^*)(1-\rho\rho^*) = 0
\end{aligned} \tag{G.10}
$$

Multiplying the first equation by $A_{2H}{}^*$, the second one by A_{1H}, and taking their sum we obtain

$$
-2(A_{2H}A_{2H}{}^* - A_{1H}A_{1H}{}^*)\rho^* + [A_{1H}(A_{3H}{}^* + A_{4H}{}^*) + A_{2H}{}^*(A_{3H} + A_{4H})](1-\rho\rho^*) = 0. \tag{G.11}
$$

Then, denoting

$$
\begin{aligned}
R_1 &= A_{2H}A_{2H}{}^* - A_{1H}A_{1H}{}^* \\
R_2 &= -A_{1H}(A_{3H}{}^* + A_{4H}{}^*) - A_{2H}{}^*(A_{3H} + A_{4H})
\end{aligned} \tag{G.12}
$$

we arrive at the equation

$$
-2R_1\rho^* - R_2 + R_2\rho\rho^* = 0 \tag{G.13}
$$

which after multiplication by ρ/ρ^* takes the form

$$
R_2\rho^2 - 2R_1\rho - R_2\frac{\rho}{\rho^*} = 0. \tag{G.14}
$$

Solution of that equation is

$$
\rho = \frac{R_1 - \sqrt{R_1^2 + R_2R_2{}^*}}{R_2} = \rho_H^K \tag{G.15}
$$

because then

$$
\frac{\rho}{\rho^*} = \frac{R_2{}^*}{R_2} \quad \text{or} \quad R_2\frac{\rho}{\rho^*} = R_2{}^* \tag{G.15}
$$

and our equation becomes

$$
R_2\rho^2 - 2R_1\rho - R_2{}^* = 0. \tag{G.16}
$$

That equation exactly agrees with the proposed solution and determines the two lacking arguments

$$
\begin{aligned}
2\delta_H^K &= \arg \rho_H^K \\
2\gamma_H^K &= 2\arctan|\rho_H^K|.
\end{aligned} \tag{G.17}
$$

Now, all elements of the Sinclair matrix can be found from the equation

$$
A_K e^{-j\mu} = \begin{bmatrix} A_2 & B_1 + jB_2 \\ -B_1 - jB_2 & A_1 \end{bmatrix}_{CCS} = \begin{bmatrix} |A_2'| & A_3' e^{-j\mu} \\ -A_3' e^{-j\mu} & |A_1'| \end{bmatrix}. \tag{G.18}
$$

They are

$$
\begin{aligned}
A_2 &\equiv A_{2CCS} = |A'_2|, \\
A_1 &\equiv A_{1CCS} = |A'_1|, \\
B_1 + jB_2 &\equiv (B_1 + jB_2)_{CCS} = A'_3 e^{-j\mu}
\end{aligned} \tag{G.19}
$$

where

$$
\begin{aligned}
A'_2 &= [A_{2H} + (A_{3H} + A_{4H})\rho + A_{1H}\rho^2]/(1+\rho\rho^*) \\
A'_3 &= [-A_{2H}\rho^* + A_{3H} - A_{4H}\rho\rho^* + A_{1H}\rho]/(1+\rho\rho^*) \\
A'_1 &= [A_{2H}\rho^{*2} - (A_{3H} + A_{4H})\rho^* + A_{1H}]/(1+\rho\rho^*).
\end{aligned} \tag{G.20}
$$

APPENDIX H

THE ALLOWED REGIONS FOR THE INVERSION POINT INSIDE THE POINCARE SPHERE OF UNIT RADIUS

H1. Boundary surfaces of the allowed regions

Polarization properties of the Sinclair scattering matrix, such as the scattered power formation, or the Poincare sphere inversion (before its rotation, in the process of polarization transformation when scattering), depend on location of the inversion point I inside the Poincare sphere model of that matrix. Also reconstruction of the whole Sinclair or Kennaugh scattering matrix is possible for known coordinates of that I point in the characteristic coordinate system (CCS), when the sphere diameter is known. However, location of the I point in the CCS cannot be arbitrary, and in some regions two solutions for reconstruction the matrix exist. Therefore, it is of special importance to find boundary surfaces of the allowed regions for that point inside the sphere in the CCS. Location of the I point on those surfaces, or between them, may also serve to classify scattering matrices independently of their polarization bases because the geometry of the model is basis invariant.

To simplify considerations and notation the sphere radius will be chosen equal to one ($k = 1$), and the upper and lower indices of coordinates of the I point will be omitted by taking:

$$Q \equiv Q_n^I, \quad U \equiv U_n^I, \quad V \equiv V_n^I. \tag{H.1}$$

One boundary surface corresponds to common solution, I and II, satisfying the equation

$$\begin{aligned} |R| &= |R_n| \equiv |\mathrm{Re}(\det A_{CCSn})| \\ &= \sqrt{[1-(Q^2+U^2+V^2)]^2 - [2UV/Q]^2} = 0 \end{aligned} \tag{H.2}$$

In the Q=const crossections of that surface, V as a function of U presents two symmetrical hyperbolic branches, for U<0 and U>0,

$$V = \frac{1}{|Q|}\sqrt{(Q^2+U^2)(1-Q^2)} - |U|, \tag{H.3}$$

determined in the range $|U| \le V$, with the common point at $U = 0$. A part of the allowed region for the inversion point is below that surface but for $V \ge 0$.

Another boundary surface can be found when considering the CCS conditions $A_2 \ge A_1 \ge 0$ in the two solutions. From (9.38a) and (9.39a) we have

$$b_1 S \ge 2(b_1^2 + b_3^2) \quad \text{for solution I} \tag{H.4a}$$

and

$$2(b_1^2 + b_5^2) \ge b_1 S \quad \text{for solution II.} \tag{H.4b}$$

For $k = 1$, and nonpositive Q, we obtain from (9.41)

$$b_1 = 2|Q|, \quad b_3^2 = 4U^2, \quad b_5^2 = 4V^2 \tag{H.5}$$

and according to (9.35)

$$\begin{aligned} S = S_n &= 2\left(1 + Q^2 + U^2 + V^2 + \sqrt{\left[1-\left(Q^2+U^2+V^2\right)\right]^2 - \left(2UV/Q\right)^2}\right) \\ &= 2(a_1 + |R|) \ge 0 \end{aligned} \tag{H.6}$$

with

$$a_1 = a_{1n} = 1 + Q^2 + U^2 + V^2 \tag{H.7}$$

and

$$|R| = |R_n| = +\sqrt{a_1^2 - \frac{1}{b_1^2}(b_1^2 + b_3^2)(b_1^2 + b_3^2)}$$
$$= +\sqrt{a_0^2 - (b_3 b_5 / b_1)^2} \tag{H.8}$$

with

$$a_0 = a_{0n} \equiv |\det(A_{CCSn})|$$
$$= 1 - \left(Q^2 + U^2 + V^2\right) \geq 0. \tag{H.9}$$

The new boundary surface corresponds to limiting cases of the two conditions, with inequalities being changed for equalities. That results in solution I:

$$2|Q|\left(1 + Q^2 + U^2 + V^2 + \sqrt{\left[1 - \left(Q^2 + U^2 + V^2\right)\right]^2 - \left(2UV/Q\right)^2}\right) = 4(Q^2 + U^2), \tag{H.10}$$

what after rearranging and squaring yields

$$(Q^2 + U^2 + V^2 + Q)(1 + Q)(Q^2 + U^2) = 0. \tag{H.11}$$

Taking the null value of the first term in brackets we arrive at the equation of the so-called 'small sphere',

$$(|Q| - \tfrac{1}{2})^2 + U^2 + V^2 = \tfrac{1}{4}, \tag{H.12}$$

with its center of coordinates $Q = -0.5$, $U = V = 0$, and the radius $r = 0.5$. Only the upper part of that sphere (for $V \geq 0$) can be considered as a part of the allowed region for the inversion point. It appears that the two just obtained boundary surfaces are tangent to each other along the circles formed by their crossections with the half-planes

$$\pm U + V = 0: \qquad V \geq 0. \tag{H.13}$$

However, for the solution I, only a part of that surface, on and below those half-planes, is a true boundary above which the $|R|$ term becomes imaginary. The remaining part, above those half-planes, is the boundary surface for the solution II only. The I points for that solution can be subtended above that surface and beneath the boundary surface for common solution. In that region the two solutions exist. That can be checked by inspection of the corresponding equation for the solution II,

$$4(Q^2 + V^2) = 2|Q|\left(1 + Q^2 + U^2 + V^2 + \sqrt{\left[1 - \left(Q^2 + U^2 + V^2\right)\right]^2 - \left(2UV/Q\right)^2}\right), \tag{H.14}$$

leading to the same equation of the 'small sphere'.

H2. The allowed regions for the inversion point

That way the allowed regions have been found in which the inversion point I can be located. Outside those regions elements of the Kennaugh matrix would become complex because of the imaginary R values. Coordinates of the I point inside the sphere of the unit radius, in the ČCS, are subtended in the ranges:

$$-1 \leq Q \leq 0$$
$$-\sqrt{|Q| - Q^2} \leq U \leq \sqrt{|Q| - Q^2} \tag{H.15}$$

$$0 \leq V \leq \begin{cases} \dfrac{1}{|Q|}\left[\sqrt{\left(Q^2 + U^2\right)\left(1 - Q^2\right)} - |U|\right] & \text{for } V \geq |U| \\ \sqrt{|Q| - Q^2 - U^2} & \text{for } V \leq |U| \end{cases} \tag{H.16}$$

Such location of the I point allows to obtain the solution I. The solution II is also possible but only for $V \geq |U|$ and above the 'small sphere', for

$$V \geq \sqrt{|Q|-Q^2 - U^2} \tag{H.17}$$

(see Fig. H1).

H3. The *S* and *R* parameter dependence on the A_{CCS} matrix elements

It is interesting to examine how the *S* and *R* parameters depend on the elements of the A_{CCS} matrix belonging to the two solutions. Using equalities (7.25) (see also Section E 3 in the Appendix E) we find that

$$a_0^2 = (A_2 A_1 + B_1^2 - B_2^2)^2 + 4B_1^2 B_2^2 \quad \text{and} \quad b_3 b_5 / b_1 = -2B_1 B_2 \tag{H.18}$$

thus obtaining

$$|R| = +\sqrt{(A_2 A_1 + B_1^2 - B_2^2)^2}\,. \tag{H.19}$$

Denoting in turn

$$S^{\mathrm{I}} = (A_2 + A_1)^2 + 4B_1^2 \tag{H.20a}$$

and

$$S^{\mathrm{II}} = (A_2 - A_1)^2 + 4B_2^2 \tag{H.20b}$$

we can show that

$$S = \tfrac{1}{2}[S^{\mathrm{I}} + S^{\mathrm{II}} + \sqrt{(S^{\mathrm{I}} - S^{\mathrm{II}})^2}\,] = \max(S^{\mathrm{I}}, S^{\mathrm{II}}). \tag{H.21}$$

It can be shown also that

$$b_1 S^{\mathrm{I}} - 2(b_1^2 + b_3^2) = A_1 (A_2 - A_1) S^{\mathrm{I}} \geq 0 \tag{H.22a}$$

and

$$2(b_1^2 + b_5^2) - b_1 S^{\mathrm{II}} = A_1 (A_2 + A_1) S^{\mathrm{II}} \geq 0 \tag{H.22b}$$

what means that S^{I} and S^{II} are the *S* parameters for the solutions I and II, accordingly. Also, having

$$|R| = \tfrac{1}{4}\sqrt{(S^{\mathrm{I}} - S^{\mathrm{II}})^2} \tag{H.23}$$

we obtain conditions:

$$R = A_2 A_1 + B_1^2 - B_2^2 \geq 0\,, \quad \text{for the solution I, when } S = S^{\mathrm{I}} \geq S^{\mathrm{II}}, \tag{H.24a}$$

and

$$R = A_2 A_1 + B_1^2 - B_2^2 \leq 0\,, \quad \text{for the solution II, when } S = S^{\mathrm{II}} \geq S^{\mathrm{I}}. \tag{H.24b}$$

H4. The *S* and *R* parameter dependence on the I point location

However, if we will choose $A_2^{\mathrm{I}}, A_1^{\mathrm{I}}, B_2^{\mathrm{I}}, B_1^{\mathrm{I}}$ and $A_2^{\mathrm{II}}, A_1^{\mathrm{II}}, B_2^{\mathrm{II}}, B_1^{\mathrm{II}}$ corresponding to different solutions but for the same Q, U and V coordinates of the I point, then in virtue of the equality

$$|R| = \sqrt{\left[1 - \left(Q^2 + U^2 + V^2\right)\right]^2 - \left(2UV/Q\right)^2} \tag{H.25}$$

we obtain

$$|R| = |R^{\rm I}| \equiv |A_2^{\rm I} A_1^{\rm I} + (B_1^{\rm I})^2 - (B_2^{\rm I})^2| = |R^{\rm II}| \equiv |A_2^{\rm II} A_1^{\rm II} + (B_1^{\rm II})^2 - (B_2^{\rm II})^2|, \tag{H.26}$$

with

$$R^{\rm I} = -R^{\rm II} \geq 0 \tag{H.27}$$

and, what follows,

$$S = 2(a_1 + |R|) = S^{\rm I} = S^{\rm II}. \tag{H.28}$$

Having these results it is interesting to observe that starting upwards with the I point from the V = 0 plane in its region inside the circle

$$(|\mathrm{Q}| - \tfrac{1}{2})^2 + \mathrm{U}^2 = \tfrac{1}{4} \tag{H.29}$$

we first begin with the maximum positive R value,

$$R = \mathrm{a}_0 = 1 - \mathrm{Q}^2 - \mathrm{U}^2, \tag{H.30}$$

corresponding to the solution I, until reaching the boundary surface $R = 0$ for common I and II solution. Then, applying the solution II, we can move with the I point downwards having negative R values, until reaching the boundary surface for that solution corresponding to the minimum R value,

$$R = -1 + |\mathrm{Q}| + 2\frac{\mathrm{U}^2}{|\mathrm{Q}|}. \tag{H.31}$$

APPENDIX I

EXAMPLES OF SCATTERING MATRICES AND THEIR POINCARE SPHERE MODELS FOR SPECIAL LOCATIONS OF THE INVERSION I POINT

General remark. All models are normalized to the sphere radius $k = 1$. The resulting square of the sphere diameter is $\sigma_{0n} = 4$. The I point coordinates are expressed in the characteristic coordinate system (CCS) and have been denoted as:

$$Q \equiv Q^{I}_{Kn}, \quad U \equiv U^{I}_{Kn}, \quad V \equiv V^{I}_{Kn}. \tag{I.1}$$

The V coordinate should be nonnegative because for $V < 0$ it is always possible to rotate the polarization sphere by 180^0 about the KL (OQ_K) axis preserving the condition $A_{4K} = -A_{3K}$ required for the characteristic ONP PP basis, known also as the CCS in the Stokes' parameters domain (see also Appendix K).

Other parameters are:

$$t = U/Q = b_3/b_1 \tag{I.2}$$

$$e = -Q/V = -b_1/b_5 \geq 0 \tag{I.3}$$

$$a_{0n,1n} = 1 \mp (Q^2 + U^2 + V^2); \quad a_{0,1} = k^2 a_{0n,1n}, \tag{I.4}$$

$$\sigma_{0n} = 2(a_{1n} + a_{0n}) = 4 \tag{I.5}$$

$$\begin{aligned} |R_n| &= \sqrt{[1-(Q^2+U^2+V^2)]^2 - [2UV/Q]^2} \\ &= +\sqrt{a_0^2 - (b_3 b_5/b_1)^2}; \quad |R| = k^2 |R_n|, \end{aligned} \tag{I.6}$$

$$\begin{aligned} S_n &= 2(a_{1n} + |R_n|) \\ &= 2\left(1 + Q^2 + U^2 + V^2 + \sqrt{\left[1-\left(Q^2+U^2+V^2\right)\right]^2 - \left(2UV/Q\right)^2}\right); \quad S = k^2 S_n \end{aligned} \tag{I.7}$$

Scattering matrices have been presented in the form (for comparison see (7.21b) and (7.20b)):

$$A_{CCSn} = \begin{bmatrix} A_2 & B_1 + jB_2 \\ -B_1 - jB_2 & A_1 \end{bmatrix}_{CCSn}, \quad K_{Kn} = \begin{bmatrix} a_1 & b_1 & b_3 & b_5 \\ b_1 & a_2 & b_4 & b_6 \\ -b_3 & -b_4 & a_3 & 0 \\ -b_5 & -b_6 & 0 & a_4 \end{bmatrix}_{Kn} \tag{I.8}$$

Elements of matrices correspond to formulae (9.38-39) and (9.43). They depend on the Q, U, V coordinates of the inversion point in the CCS and their functions: ratios t and e, parameters R and S, and elements of the first raw of the K_K matrix. Sometimes two solutions are possible satisfying conditions (9.40).

Axis and angle of rotation after inversion have been presented after the formulae (9.45-46) with upper or lower signs related to the solution I or II, accordingly. The axis of rotation after inversion has been expressed, after (9.45), by its tilt angle, the ratio of its unit vector components (n_2 along the U_K axis, and n_3 along the V_K axis) and in terms of all the above defined parameters as follows:

$$\tan 2\delta^P_K = -\frac{n_2}{n_3} = -\frac{a_{1n} \mp |R_n|}{a_{1n} \pm |R_n|} \cdot \frac{2t + (1/t)(a_{0n} \mp |R_n|)}{(2/e) + e(a_{0n} \pm |R_n|)} \tag{I.9a}$$

$$= \frac{b_6}{b_4} \cdot \frac{a_3 - a_o}{a_4 - a_o}; \quad n_3 \geq 0. \tag{I.9b}$$

The angle of rotation about that axis, 2ϕ, is after (9.46):

$$\cos 2\gamma_K^P = \cos 2\phi$$
$$= \frac{2a_{1n}t^2 - e^2(|R_n|^2 + a_{1n}a_{0n}) \mp 2(1-e^2t^2)|R_n|}{2a_{0n}(1+t^2)(e^2+1)} \tag{I.10}$$
$$= \frac{a_{3n}a_{4n} - a_{0n}a_{2n}}{2a_{0n}} = \frac{b_{1n}^2 - 2a_{2n}}{2a_{0n}}$$

with

$$\left.\begin{array}{ll} 2\phi = 2\gamma_K^P & \text{for } U \le 0 \\ 2\phi = 2\pi - 2\gamma_K^P & \text{for } U \ge 0 \end{array}\right\} \text{for } 0 \le 2\gamma_K^P \le \pi, \text{ with } 0 \le 2\phi \le 2\pi. \tag{I.11}$$

According to the above formulae, the rotation after inversion axis is always situated in the $U_K V_K$ coordinate plane with its n_3 component always nonnegative. For the I point in the $Q_K V_K$ plane it is always vertical ($n_3 = 1$) for the solution I and always horizontal ($n_2 = \pm 1$) for the solution II. However, the U component of the I point and the n_2 component of the rotation axis are always of opposite sign. Therefore, when the I point penetrates the $Q_K V_K$ plane, the axis of rotation after inversion *corresponding to the solution II* reverses its direction in order to maintain the value $n_3 \ge 0$. Simultaneous change of the angle of rotation after inversion from $2\phi > 0$ to $2\pi - 2\phi$ keeps the rotation angle about that axis positive. It must reverse the phase of the scattering matrix expressed in the CCS and rotating, after inversion, the *two-folded* Riemann surface of the PP sphere. However, it has no influence on the transformation of polarization.

Beneath, scattering matrices and their Poincare sphere models will be presented in both solutions for the inversion point I located on: two coordinate planes, boundary surfaces, some circular lines, coordinate axes, and special polarization points in the CCS.

I. 1. I point in the QU plane of the CCS

Here, for V = 0, the remaining allowed coordinates of the inversion point being considered are:

$$-1 < Q < 0, \qquad -\sqrt{|Q| - Q^2} < U < 0 \tag{I.12}$$

Only negative U values (posirive B_1) are possible when V = 0 (see Appendix K). Values of other essential parameters are:

$$\begin{gathered} 0 < t < \sqrt{\frac{1}{|Q|} - 1}, \\ 1/e = 0, \\ a_{0n,1n} = 1 \mp (Q^2 + U^2) \\ R_n = a_{0n} \\ S_n = 4 = \sigma_{0n} \end{gathered} \tag{I.13}$$

There exists the solution I only. Scattering matrices are of the form:

$$A_{CCSn} = \begin{bmatrix} A_2 & B_1 \\ -B_1 & A_1 \end{bmatrix}_{CCSn}, \qquad K_{Kn} = \begin{bmatrix} a_1 & b_1 & b_3 & 0 \\ b_1 & a_2 & b_4 & 0 \\ -b_3 & -b_4 & a_3 & 0 \\ 0 & 0 & 0 & a_4 \end{bmatrix}_{Kn}. \tag{I.14}$$

Their elements and parameters of the rotation axis and angle are:

$$A_{2n,1n} = \frac{1 \pm |Q|(1+t^2)}{\sqrt{1+t^2}}, \quad B_{1n} = \frac{t}{\sqrt{1+t^2}} > 0, \tag{I.15}$$

$$a_{2n} = \frac{a_{1n} - t^2 a_{0n}}{1+t^2}, \quad a_{3n} = \frac{-a_{1n} t^2 + a_{0n}}{1+t^2}, \quad a_{4n} = -a_{0n}, \tag{I.16a}$$

$$b_{1n} = 2|Q|, \quad b_{3n} = -2U, \quad b_{4n} = \frac{2t}{1+t^2} = \sin 2\gamma_K^P, \tag{I.16b}$$

and

$$\cos 2\gamma_K^P = \frac{t^2 - 1}{t^2 + 1}, \quad 2\delta_K^P = \pi. \tag{I.17}$$

The last equality means that the rotation after inversion axis is oriented along the OV_K coordinate axis.

I. 2. I point in the QV plane of the CCS

For U = 0 the two solutions can exist in the allowing regions:

$$1 < Q < 0, \quad \sqrt{1-Q^2} > V > \begin{cases} 0 & \text{for the solution I} \\ \sqrt{|Q|-Q^2} & \text{for the solution II} \end{cases} \tag{I.18}$$

The remaining parameters are:

$$t = 0$$

$$\frac{|Q|}{\sqrt{1-Q^2}} < e < \begin{cases} \infty & \text{for the solution I} \\ \sqrt{\frac{|Q|}{1-|Q|}} & \text{for the solution II} \end{cases} \tag{I.19}$$

$$a_{0n,1n} = 1 \mp (Q^2 + V^2)$$
$$|R_n| = a_{0n} \tag{I.20}$$
$$S_n = 4 = \sigma_{0n}$$

Both matrices have the same form for the two solutions:

$$A_{CCSn} = \begin{bmatrix} A_2 & jB_2 \\ -jB_2 & A_1 \end{bmatrix}_{CCSn}, \quad K_{Kn} = \begin{bmatrix} a_1 & b_1 & 0 & b_5 \\ b_1 & a_2 & 0 & b_6 \\ 0 & 0 & a_3 & 0 \\ -b_5 & -b_6 & 0 & a_4 \end{bmatrix}_{Kn} \tag{I.21}$$

but their elements are different:

$$A_{2n,1n} = 1 \pm |Q|, \quad B_{2n} = V \quad \text{for the solution I}$$

$$A_{2n,1n} = \frac{V(e^2+1) \pm e}{\sqrt{e^2+1}}, \quad B_{2n} = \frac{1}{\sqrt{e^2+1}} \quad \text{for the solution II} \tag{I.22}$$

and

$$a_{2n} = \frac{a_{1n} e^2 \pm a_{0n}}{e^2+1}, \quad a_{3n} = \pm a_{0n}, \quad a_{4n} = \frac{-a_{1n} \mp e^2 a_{0n}}{e^2+1},$$

$$b_{1n} = 2|Q|, \quad b_{5n} = -2V, \quad b_{6n} = -e\frac{-a_{1n} \mp a_{0n}}{e^2+1} \tag{I.23}$$

Angle and axis for the solution I are determined by:

$$2\gamma_K^P = 2\delta_K^P = \pi\,, \tag{I.24a}$$

and for the solution II by:

$$\cos 2\gamma_K^P = -\frac{e^2-1}{e^2+1}, \quad 2\delta_K^P = \pi/2 \text{ for } 2\phi = 2\gamma_K^P < \pi\,. \tag{I.24b}$$

The last equality means that the rotation after inversion axis is oriented along the OU_K coordinate axis.

I. 3. I point on the 'small sphere' surface

The region under consideration is for V > 0, given by the equation

$$(|Q|-\tfrac{1}{2})^2 + U^2 + V^2 = \tfrac{1}{4}, \qquad \text{or equivalently} \qquad Q^2 + U^2 + V^2 = |Q|\,. \tag{I.25}$$

Mutual dependence between parameters t and e is

$$t^2 + (1/e^2) = (1/|Q|) - 1 \tag{I.26}$$

and other parameters are:

$$a_{0n,1n} = 1 \mp |Q|$$

$$|R_n| = \frac{U^2 - V^2}{|Q|} \tag{I.27a,b,c}$$

$$S_n = \begin{cases} 4|Q|(1+t^2) = 4B_{1n}^2 + A_{2n}^2 & \text{for the solution I and for } |U| \geq V \text{ only} \\ 4V\left(e+\dfrac{1}{e}\right) = 4B_{2n}^2 + A_{2n}^2 & \text{for } |U| \leq V \text{ but for the solution II only} \end{cases}$$

The two matrices:

$$A_{CCSn} = \begin{bmatrix} A_2 & B_1 + jB_2 \\ -B_1 - jB_2 & 0 \end{bmatrix}_{CCSn}, \quad K_{Kn} = \begin{bmatrix} a_1 & b_1 & b_3 & b_5 \\ b_1 & a_2 & b_3 & b_5 \\ -b_3 & -b_3 & a_3 & 0 \\ -b_5 & -b_5 & 0 & a_3 \end{bmatrix}_{Kn} \tag{I.28}$$

have the same elements for both solutions:

$$A_{2n} = 2\sqrt{|Q|}, \quad A_{1n} = 0, \quad B_{1n} = \frac{-U}{\sqrt{|Q|}}, \quad B_{2n} = \frac{V}{\sqrt{|Q|}} \tag{I.29}$$

and

$$a_{2n} = 3|Q| - 1, \quad a_{3n} = a_{0n}, \quad b_{1n} = 2|Q|, \quad b_{3n} = -2U, \quad b_{5n} = -2V\,. \tag{I.30}$$

Angular functions are also identical for the two solutions:

$$\cos 2\gamma_K^P = 1 - 2|Q|, \qquad \tan 2\delta_K^P = \frac{V}{U}\,. \tag{I.31}$$

The last equality means that the axis of rotation after inversion is parallel to the straight line tangent to the circle of the small sphere surface crossection by the plane $Q_K = Q$, at the I point.

What should be observed in that interesting case is: vanishing of the A_1 element, the existence of one double CO-POL NULL polarization point, Q = –1, and formation of a great eigencircle crossing the OQ_K axis, and located in the plane containing the I point (see Fig.J.4). That plane is inclined versus the coordinate $Q_K U_K$ plane at an angle $\arctan(B_2 / B_1)$.

I. 4. I point on the surface of common solutions ($R = 0$)

That boundary surface can be presented for $|U| \leq V$ only, in the form

$$Q^2 + U^2 + V^2 = 1 - 2|t|V. \tag{I.32}$$

If the inversion point would be located on such a surface for $|U| > V$, it would mean that to arrive at the uniquely defined ONP characteristic coordinate system its K phasor should be rotated in its orientation by plus or minus 90 degrees.

On that $R = 0$ surface the t and e parameters are mutually dependent:

$$|t| = \frac{1}{|Q|e}[\sqrt{(1-Q^2)(1+e^2)} - 1], \tag{I.33}$$

$$e = |Q|\frac{\sqrt{(1-Q^2)(1+t^2)} - |t|}{Q^2(1+t^2) - 1}. \tag{I.34}$$

The remaining parameters are:

$$\begin{aligned} a_{0n} &= 2|t|V \\ a_{1n} &= 1 + Q^2 + U^2 + V^2 = 2(1-|t|V) \\ R_n &= 0 \\ S_n &= 2a_{1n} = 4(1-|t|V) \end{aligned} \tag{I.35}$$

Elements of scattering matrices are the same for the two solutions:

$$B_{1n} = B_{1n}^{\mathrm{I}} = t\sqrt{\frac{1-|t|V}{1+t^2}} = B_{1n}^{\mathrm{II}} = tV\sqrt{\frac{e^2+1}{1-|t|V}} \tag{I.36a}$$

$$B_{2n} = B_{2n}^{\mathrm{I}} = V\sqrt{\frac{1+t^2}{1-|t|V}} = B_{2n}^{\mathrm{II}} = tV\sqrt{\frac{1-|t|V}{e^2+1}} \tag{I.36b}$$

with

$$B_{1n}B_{2n} = tV \tag{I.37}$$

and

$$A_{2n,1n} = \frac{V}{B_{2n}} \pm \frac{|Q|B_{2n}}{V} = \sqrt{\frac{1-|t|V}{1+t^2}} \pm |Q|\sqrt{\frac{1+t^2}{1-|t|V}}, \tag{I.38}$$

$$b_{1n} = 2|Q|, \quad b_{3n} = -2U, \quad b_{5n} = -2V, \tag{I.39}$$

$$\begin{aligned} a_{2n} = a_{1n}\frac{e^2 - t^2}{(1+t^2)(e^2+1)}, \quad a_{3n} = -a_{1n}\frac{t^2}{1+t^2}, \quad a_{4n} = -a_{1n}\frac{1}{e^2+1}, \\ b_{4n} = a_1\frac{t}{1+t^2}, \quad b_{6n} = -a_{1n}\frac{e}{e^2+1}. \end{aligned} \tag{I.40}$$

Angular parameters satisfy the equalities:

$$\begin{aligned} \cos 2\gamma_K^P = \cos\phi &= \frac{a_{1n}(2t^2 - e^2 a_{0n})}{2a_{0n}(1+t^2)(e^2+1)} \\ &= \frac{(1-|t|V)(|t| - e^2 V)}{V(1+t^2)(e^2+1)} = -\frac{|U|V - |Q|^3}{|U|V - |Q|} \end{aligned} \tag{I.41}$$

and

$$\tan 2\delta_K^P = -\frac{2t + (a_{0n}/t)}{(2/e) + ea_{0n}} = \frac{U^2 + |QU|V}{U(V + |QU|)} \tag{I.42}$$
$$= (\operatorname{sgn} U)\frac{|U| + |Q|V}{V + |Q||U|}.$$

For $t \to 0$ (when I point tends to the surface of the polarization sphere), these functions become

$$\cos 2\gamma_K^P = -Q^2, \qquad \tan 2\delta_K^P = \mp|Q| \quad \text{(for } U<0 \text{ or } U>0, \text{ accordingly)} \tag{I.43}$$

and for V=|U| (another limiting case),

$$\cos 2\gamma_K^P = 1 - 2|Q|, \qquad \tan 2\delta_K^P = \mp 1 \quad \text{(for } U<0 \text{ or } U>0, \text{ accordingly)} \tag{I.44}$$

Generally,

$$\frac{\partial V}{\partial |U|} = -\frac{1}{|Q|} + |t|\sqrt{\frac{1-Q^2}{Q^2+U^2}} \tag{I.45}$$
$$= -\frac{V + |Q||U|}{|U| + |Q|V} = \frac{\partial V}{\partial U}\operatorname{sgn} U < 0$$

where

$$\frac{\partial V}{\partial U} = -\cot 2\delta_K^P = -\operatorname{sgn} U \frac{V + |Q||U|}{|U| + |Q|V} = \frac{-U(V + |Q||U|)}{U^2 + |Q||U|V}. \tag{I.46}$$

I. 5. I point on the OQ_K coordinate axis

This is the case of symmetrical matrices, corresponding to the monostatic scattering. For the whole range

$$-1 < Q < 0 \tag{I.47a}$$

the solution I only exists. Other parameters are:

$$U = V = t = 1/e = 0, \quad a_{0n,1n} = 1 \mp Q^2, \quad R_n = a_{0n}, \quad S_n = 4. \tag{I.47b}$$

Scattering matrices have the following elements:

$$A_{CCSn} = \begin{bmatrix} 1+|Q| & 0 \\ 0 & 1-|Q| \end{bmatrix}_{CCSn}, \qquad K_{Kn} = \begin{bmatrix} 1+Q^2 & 2|Q| & 0 & 0 \\ 2|Q| & 1+Q^2 & 0 & 0 \\ 0 & 0 & 1-Q^2 & 0 \\ 0 & 0 & 0 & -1+Q^2 \end{bmatrix}_{Kn} \tag{I.48}$$

and the angles are:

$$2\gamma_K^P = 2\delta_K^P = \pi. \tag{I.49}$$

That is the only case with two common points: K, M, E_2 and L, N, E_1 (Fig.).

I. 6. I point on the OV_K coordinate axis

There are two solutions in the whole range

$$0 < V < 1 \tag{I.50}$$

with Q = U = 0. Other parameters are:

$$0 < |t| < \frac{1-V^2}{2V},$$

$$e = 0,$$
$$a_{0n,1n} = 1 \mp V^2,$$
$$|R_n| = \sqrt{(1-V^2)^2 - 4t^2V^2},$$
$$S_n = 2(1 + V^2 + \sqrt{(1-V^2)^2 - 4t^2V^2}). \tag{I.51}$$

Matrices are:

$$A_{CCSn} = \begin{bmatrix} A_2 & B_1 + jB_2 \\ -B_1 - jB_2 & A_2 \end{bmatrix}_{CCSn}, \quad K_{Kn} = \begin{bmatrix} a_1 & 0 & 0 & b_5 \\ 0 & a_2 & b_4 & 0 \\ 0 & -b_4 & a_2 & 0 \\ -b_5 & 0 & 0 & -a_1 \end{bmatrix}_{Kn} \tag{I.52}$$

with elements for solution I:

$$A_{2n} = \frac{1}{2}\sqrt{\frac{S_n}{1+t^2}}, \quad B_{1n} = \frac{t}{2}\sqrt{\frac{S_n}{1+t^2}}, \quad B_{2n} = 2V\sqrt{\frac{1+t^2}{S_n}} \tag{I.53a}$$

for solution II:

$$A_{2n} = \frac{2V}{\sqrt{S_n}}, \quad B_{1n} = \frac{2tV}{\sqrt{S_n}}, \quad B_{2n} = \frac{\sqrt{S_n}}{2} \tag{I.53b}$$

and for both solutions:

$$a_{2n} = \frac{-a_{1n}t^2 \pm |R_n|}{1+t^2}, \quad b_{4n} = t\frac{a_{1n} \pm |R_n|}{1+t^2}, \quad b_{5n} = -2V. \tag{I.54}$$
$$\cos 2\gamma_K^P = -a_{2n} / a_{0n}, \quad 2\delta_K^P = \pi.$$

Special property of that case should be observed:

$$t = \frac{B_{1n}}{B_{2n}} . \tag{I.55}$$

That value is independent of the I point location and can be chosen arbitrarily in its allowed range, thus generating a continuum number of solutions I and II which for each V and t values have two different rotation angles satisfying the equation

$$\cos 2\gamma_K^P = \frac{(1+V^2)t^2 \mp \sqrt{(1-V^2)^2 - 4t^2V^2}}{(1+t^2)(1-V^2)}. \tag{I.56}$$

Greater of those rotation angles corresponds to the solution I, and smaller to the solution II. Between those values is the common solution I and II with

$$\cos 2\gamma_K^P = \frac{1-V^2}{1+V^2}, \quad \text{for} \quad t = t_{\max} = \frac{1-V^2}{2V}. \tag{I.57}$$

In the other limit case, for $t = 0$,

$$2\gamma_K^P = \begin{cases} \pi, & \text{for the solution I} \\ 0, & \text{for the solution II.} \end{cases} \tag{I.58}$$

For all solutions the axis of rotation after inversion coincides with the coordinate axis OV_K. Therefore all eigencircles are in planes perpendicular to that axis. However, eigenpolarizations exist only for the Solution I and $t = 0$ because only then the rotation angle is π and all points of the eigencircle are eigenpolarizations. The solution II for $t = 0$ is also interesting because of the simple rule of transformation of the incident polarization (no rotation after inversion). Also it is interesting to observe that the M and O_1 points coincide (at the lower pole), what cannot be the case for other models.

I. 7. I point on the semicircle $(-U)_{max}$ in the CCS

There exists the solution I only. For the I point location,

$$-1 < Q < 0, \quad -U = \sqrt{|Q| - Q^2}, \quad V = 0, \tag{I.59}$$

other parameters are

$$0 < t < \sqrt{\frac{1}{|Q|} - 1},$$
$$1/e = 0, \tag{I.60}$$
$$a_{0n,1n} = 1 \mp |Q|, \quad |R_n| = a_0, \quad S_n = 4$$

yielding matrices

$$A_{CCSn} = \begin{bmatrix} A_2 & B_1 \\ -B_1 & 0 \end{bmatrix}_{CCSn}, \quad K_{Kn} = \begin{bmatrix} a_1 & b_1 & b_3 & 0 \\ b_1 & a_2 & b_3 & 0 \\ -b_3 & -b_3 & a_3 & 0 \\ 0 & 0 & 0 & a_3 \end{bmatrix}_{Kn} \tag{I.61}$$

with elements

$$A_2 = 2\sqrt{|Q|}, \quad B_1 = \sqrt{1 - |Q|} > 0,$$
$$a_2 = -1 + 3|Q|, \quad a_3 = -1 + |Q|, \quad b_1 = 2|Q|, \quad b_3 = -2U. \tag{I.62}$$

For the angles one obtains

$$\cos 2\gamma_K^P = 1 - 2|Q|, \quad 2\delta_K^P = \pi. \tag{I.63}$$

I. 8. I point on the surface of the polarization sphere

There exists one double, I and II, solution. The I point is located on a quarter of the great circle arc. Its coordinates and other parameters necessary to compute the elements of matrices are

$$-1 < Q < 0, \quad U = 0, \quad V = \sqrt{1 - Q^2} > 0, \tag{I.64}$$

$$t = 0, \quad e = \sqrt{\frac{Q^2}{1 - Q^2}},$$
$$a_{0n} = 0, \quad a_{1n} = 2, \tag{I.65}$$
$$|R_n| = 0, \quad S_n = 4.$$

The matrices are

$$A_{CCSn} = \begin{bmatrix} A_2 & jB_2 \\ -jB_2 & A_1 \end{bmatrix}_{CCSn}, \quad K_{Kn} = \begin{bmatrix} a_1 & b_1 & 0 & b_5 \\ b_1 & a_2 & 0 & b_6 \\ 0 & 0 & 0 & 0 \\ -b_5 & -b_6 & 0 & a_4 \end{bmatrix}_{Kn}. \tag{I.66}$$

with elements

$$A_2 = 1 + |Q|, \quad A_1 = 1 - |Q|, \quad B_2 = \sqrt{1 - Q^2} = V,$$
$$a_2 = 2Q^2, \quad a_4 = -2(1 - Q^2), \quad b_1 = 2|Q|, \quad b_5 = -2V, \quad b_6 = -2|Q|\sqrt{1 - Q^2} = -2|Q|V. \tag{I.67}$$

To find the axes and angles of rotation after inversion it is possible to investigate continua of solutions as in the case of the I point on the OV_K axis (here as functions of the $\partial U / \partial V$ parameter, what has been analyzed in [18]). However, it is recommended to use one solution only,

$$2\gamma_K^P = 2\delta_K^P = \pi, \tag{I.68}$$

because all other lead to the same result for the I point on the Poincare sphere surface.

I. 9. I point on the crossection of the small hemisphere with planes through the Q_K axis inclined at $+45^0$ and -45^0 angles versus the $Q_K U_K$ plane.

Coordinates of the inversion I point satisfy the equalities

$$Q^2 + 2U^2 + Q = 0, \quad |U| = V \geq 0. \tag{I.25'}$$

There exists common solution I and II which can be considered as a limiting case of solution I or II in I.3 for |U|=V resulting in $B_1 = B_2$. Scattering matrices are as in I.3 but with $B_1 = B_2$. Their geometrical model is also as in Fig. J.4.

I. 10. I point on the polarization sphere equator

This is the limit case for the I point location, at Q = −1. And again, the one only solution is recommended, corresponding to the limit case for the inversion point on the OQ_K axis, with following parameters:

$$t = 1/e = 0, \quad a_{0n} = 0, \quad a_{1n} = 2, \quad |R_n| = 0, \quad S_n = 4, \tag{I.69}$$

corresponding matrices:

$$A_{CCSn} = \begin{bmatrix} 2 & 0 \\ 0 & 0 \end{bmatrix}_{CCSn}, \quad K_{Kn} = \begin{bmatrix} 2 & 2 & 0 & 0 \\ 2 & 2 & 0 & 0 \\ 0 & 0 & 0 & 0 \\ 0 & 0 & 0 & 0 \end{bmatrix}_{Kn} \tag{I.70}$$

and angles:

$$2\gamma_K^P = 2\delta_K^P = \pi. \tag{I.71}$$

The unique properties of that model are: coincidence of the K, M, M" and E_2 points being the only scattered polarization point, and no scattered power for the E_1 incident polarization coinciding with I and L points. Though for each I point on the Poincare sphere surface there is one only, E_2, scattered polarization point but, beyond Q = −1 coordinate of the I point, it was never the M" point.

I. 11. I point at the (upper) pole of the polarization sphere

That location of the inversion point, V = 1, is recommended for consideration as the limiting case for the I point on the OV_K coordinate axis. The corresponding parameters and the yielded matrices are:

$$t = e = 0, \quad a_{0n} = 0, \quad a_{1n} = 2, \quad |R_n| = 0, \quad S_n = 4, \tag{I.72}$$

$$A_{CCSn} = \begin{bmatrix} 1 & j \\ -j & 1 \end{bmatrix}_{CCSn}, \quad K_{Kn} = \begin{bmatrix} 2 & 0 & 0 & -2 \\ 0 & 0 & 0 & 0 \\ 0 & 0 & 0 & 0 \\ 2 & 0 & 0 & -2 \end{bmatrix}_{Kn}. \tag{I.73}$$

The rotation after inversion angle,

$$2\gamma_K^P = \pi, \tag{I.74}$$

has no influence on the scattered polarization point, located always at the pole of the sphere, because of the rotation axis coming through that point. Only the scattered wave's phasor will be rotated accordingly.

A special property of that model is the coincidence of points I and O_2. Therefore the incident polarization point, corresponding to the maximum scattered power, is located at the lower pole of the sphere and produces the orthogonal scattered polarization (the M and O_1 points coincide). No matrix of monostatic scattering may exhibit such a property.

I. 12. I point in the center of the polarization sphere.

Two solutions, I and II, exist for the I point of coordinates Q = U = V = 0. There is a continuum of solution I versions for t parameter in the range

$$-\infty \le t \le \infty . \tag{I.75}$$

Taking into account:

$$e = 0, \quad a_{0n} = a_{1n} = 1, \quad |R_n| = 1, \quad S_n = 4, \tag{I.76}$$

the following matrices will be obtained for the solution I:

$$A_{CCSn} = \begin{bmatrix} A_2 & B_1 \\ -B_1 & A_2 \end{bmatrix}_{CCSn}, \quad K_{Kn} = \begin{bmatrix} a_1 & 0 & 0 & b_5 \\ 0 & a_2 & b_4 & 0 \\ 0 & -b_4 & a_2 & 0 \\ -b_5 & 0 & 0 & -a_1 \end{bmatrix}_{Kn} \tag{I.77}$$

with elements for solution I:

$$A_{2n} = \frac{1}{\sqrt{1+t^2}} = \sin 2\gamma_K^P, \quad B_{1n} = \frac{t}{\sqrt{1+t^2}} = \cos 2\gamma_K^P, \tag{I.78}$$

$$a_{2n} = -\frac{t^2 - 1|}{t^2 + 1} = -\cos 2\gamma_K^P, \quad b_{4n} = \frac{2t}{1+t^2} = \sin 2\gamma_K^P . \tag{I.79}$$

For the whole range of the t parameter we obtain

$$0 \le 2\gamma_K^P \le 0, \quad 2\delta_K^P = \pi . \tag{I.80}$$

A special property of that model is independence of the scattered power from the incident polarization. Eigenpolarizations appear only for $t = 0$. Then $2\gamma_K^P = \pi$ and all points of the polarization sphere equator are eigenpolarizations. That model corresponds to the scattering by a sphere.

An important conclusion which can be drawn from that example is that there is no scatterer preserving each incident polarization.

Of a special interest is the case of $t = \pm\infty$. The corresponding matrices are:

$$A_{CCSn} = \begin{bmatrix} 0 & 1 \\ -1 & 0 \end{bmatrix}_{CCSn}, \quad K_{Kn} = \begin{bmatrix} 1 & 0 & 0 & 0 \\ 0 & -1 & 0 & 0 \\ 0 & 0 & -1 & 0 \\ 0 & 0 & 0 & -1 \end{bmatrix}_{Kn} . \tag{I.81}$$

Similar result exhibits the solution II. For the same parameters as above we obtain matrices

$$A_{CCSn} = \begin{bmatrix} 0 & j \\ -j & 0 \end{bmatrix}_{CCSn}, \quad K_{Kn} = \begin{bmatrix} 1 & 0 & 0 & 0 \\ 0 & -1 & 0 & 0 \\ 0 & 0 & -1 & 0 \\ 0 & 0 & 0 & -1 \end{bmatrix}_{Kn} \tag{I.82}$$

In both cases the rotation after inversion angle is

$$2\gamma_K^P = 0 . \tag{I.83}$$

The obtained model is an orthogonalizer which converts any incident polarization point into the antipodal point of the scattered wave without changing the wave's intensity. So, any incident polarization point is the M and O_1 point.

An extra phase change can be arbitrary. Usually for the orthogonalizer (represented by the C^X matrix as in (7.II) or (6.4)) we choose $B_1 = -1$ and $B_2 = 0$, while for the 'invertor' through the polarization sphere center (represented by the A_{0K}^{INV} matrix as in (8.30) with Q=U=V=0 and σ_0=4) we apply $B_1 = 1$ and $B_2 = 0$. The obtained phase difference between results of both those transformations is π because the inversion for the two cases can be interpreted as rotation by the $\pm\pi$ angle (in opposite directions) about an axis through the center of the sphere and perpendicular to the incident polarization phasor. Speaking precisely, the two transformations do not change the phase, shifting phasors parallel along the great semicircles of the polarization sphere. Then, however, the resulting phasors should be interpreted as 'oppositely' (not orthogonally !) polarized what interprets the obtained phase difference.

APPENDIX J

GEOMETRICAL CONSTRUCTIONS OF THE POINCARE SPHERE MODELS OF SCATTERING MATRICES IN THE CCS

J. 1. Mutual locations of special polarization points of the Poincare sphere models of scattering matrices

Having coordinates of special polarization points in the CCS as presented in Chapter 9, Section 9.8 (formulae (9.47) - (9.57)), it is possible to distinguish their four groups. Points of each group are located on a common plane through the K L diameter along the OQ coordinate axis of the CCS (see Fig. J. 1).

Apart from K and L, the following points belonge to those four groups:

1. O_1 and O_2,
2. M, N, M", N" and I
3. E_1 and E_2, and
4. P.

(J.1)

The P point is such an inverted point of an incident polarization which by rotation after inversion, about an axis in the UV plane of the CCS, becomes the characteristic point K of the Poincare sphere model of the scattering matrix.

Moreover, also the fifth group of points can be specified. It forms the plane of the so-called eigencircle, not necessarily coming through K and L points (in the cases of 'small eigencircles'). The plane of the eigencircle is always perpendicular to the axis of rotation after inversion and contains the inversion I point as well as the eigenpolarizations, E_1 and E_2, if they exist for the scattering matrix under consideration. A special property of the eigencircle is that its points remain on that circle after the inversion and rotation, even if the eigenpolarizations do not exist. In case of a great eigencircle it contains also the P point, thus belonging to the fourth of the above defined groups.

The four groups can be determined by the angles of inclination of their planes versus the QU coordinate plane in the CCS as follows:

$$\tan 2\delta_K^{O_2} = +\infty$$

$$\tan 2\delta_K^{N} = \frac{b_5}{b_3} = \frac{-B_2(A_2 + A_1)}{B_1(A_2 - A_1)} \qquad (J.2)$$

$$\tan 2\delta_K^{E} = \frac{b_6}{b_4} = \frac{-B_2(A_2 - A_1)}{B_1(A_2 + A_1)}$$

$$\tan 2\delta_K^{P} = \frac{b_6(a_0 - a_3)}{b_4(a_0 - a_4)} = \tan 2\delta_K^{E} \frac{\sigma_0 - 4r_1^2}{\sigma_0 - 4r_2^2} = \frac{-n_2}{n_3}; \quad n_3 \geq 0. \qquad (J.3)$$

where

$$\sigma_0 = 2(a_1 + a_0) = A_2^2 + A_1^2 + 2(B_2^2 + B_1^2) + 2\sqrt{(A_1A_2 + B_1^2 - B_2^2)^2 + 4B_1^2B_2^2}, \qquad (J.4)$$

$$2r_1 = A_2 + A_1, \quad 2r_2 = A_2 - A_1.$$

What is interesting to observe, it is mutual orientation of those planes,

$$180^0 \geq 2\delta_K^{P} \geq 2\delta_K^{E} \geq 2\delta_K^{N} \geq 2\delta_K^{O_2} = 90^0 \quad \text{for} \quad U_K^I \leq 0 \qquad (J.5a)$$

and

$$0^0 \leq 2\delta_K^{P} \leq 2\delta_K^{E} \leq 2\delta_K^{N} \leq 2\delta_K^{O_2} = 90^0 \quad \text{for} \quad U_K^I \geq 0. \qquad (J.5b)$$

J. 2. The cases of great eigencircles (Poincare sphere models of scattering matrices for the I point in the QV and QU planes, and on the 'small sphere' surface)

The simplest constructions of the Poincare sphere models of scattering matrices can be presented in cases when great eigencircles exist, also when there are no eigenpolarizations. Three kinds of such constructions can be distinguished for scattering matrices resulting from the I point location (in the CCS):

1. In the QV plane $(\mathrm{U}_K^{\mathrm{I}} = 0)$ for the solution II (see Fig. J. 2),
2. In the QU plane $(\mathrm{V}_K^{\mathrm{I}} = 0)$ for the solution I (Fig. J.3),
3. On the 'small sphere' surface and solution I, for $|\mathrm{U}| \geq \mathrm{V}$, or solution II, for $|\mathrm{U}| \leq \mathrm{V}$ (Fig. J.4),

including limiting cases for location of the inversion point on boundaries of its allowed regions.

In the cases of the first kind (Fig. J.2): $B_1 = 0$, $B_2 = \mathrm{V}_K^{\mathrm{N}}$ (the V coordinate of the illuminating polarization corresponding to the minimum scattered power) and the great eigencircle is in the QV plane. The polarization sphere radius is then

$$r_0 \equiv \frac{\sqrt{\sigma_0}}{2} = \sqrt{r_2^2 + B_2^2}; \quad r_2 \equiv \frac{A_2 - A_1}{2}. \tag{J.6}$$

The I point should be located beyond, or on the circle,

$$(|\mathrm{Q}|-(\sqrt{\sigma_0}/4))^2 + \mathrm{V}^2 = (\sqrt{\sigma_0}/4)^2, \tag{J.7}$$

which is a crossection of the 'small sphere' with the QV plane. For the existence of eigenpolarizations the distance of the I point from the center of the sphere should be

$$r_1 \equiv \frac{A_2 + A_1}{2} \geq B_2 = \frac{\sqrt{\sigma_0}}{2}\cos\gamma_K^P \tag{J.8}$$

where the equality corresponds to the one double eigenpolarization point at the upper pole of the sphere.

The model shows how location of the I point determines: A_1, A_2 and B_2 parameters of the A_{CCS} matrix, the M, N, M", N", $\mathrm{O_1}$, $\mathrm{O_2}$, $\mathrm{E_1}$, and $\mathrm{E_2}$ points, and the rotation after inversion angle , $2\phi = 2\gamma_K^P$, corresponding to the rotation axis directed along the OU coordinate axis. It should be observed that $\mathrm{E_1}$, and $\mathrm{O_2}$ polarizations can be found as the end points of radii through the crossection of a circle of the (OI) radius and straight lines through the N point, parallel to the Q and V axes, accordingly.

Dependences between elements of the A_{CCS} scattering matrix for the limiting cases of the I point locations are:

$$B_2 = \sqrt{A_1 A_2}, \text{ and } \quad A_1 = 0. \tag{J.9}$$

They correspond to the inversion point on the Poincare sphere surface and on the 'small sphere' surface, accordingly. In the first case there is one only the eigenpolarization point, $\mathrm{E_2}$, because incident polarizations $\mathrm{E_1}$, $\mathrm{O_2}$ and I meet in one point giving no scattered power. In the second limiting case, $A_1 = 0$, the double $\mathrm{O_{1,2}}$ point occurs. That explains geometrically why closer location of the I point to the center of the polarization sphere is impossible for the solution II.

Constructions of the second kind present great eigencircles in the UV plane. The I point is contained inside the circular region

$$(|\mathrm{Q}|-(\sqrt{\sigma_0}/4))^2 + \mathrm{U}^2 = (\sqrt{\sigma_0}/4)^2. \tag{J.10}$$

As can be seen from the Fig. J. 3, the radius of the polarization sphere is now

$$r_0 \equiv \frac{\sqrt{\sigma_0}}{2} = \sqrt{r_1^2 + B_1^2}; \quad r_1 \equiv \frac{A_2 + A_1}{2}. \tag{J.11}$$

The eigenpolarizations exist for the I point distance from the center of the sphere

$$r_2 \equiv \frac{A_2 - A_1}{2} \geq B_1 = \frac{\sqrt{\sigma_0}}{2} \cos \gamma_K^P \tag{J.12}$$

where the equality corresponds to one double eigenpolarization point at |U| = 1. Projection of the O_1 and O_2 points on the QU plane is determined by crossection, with the Q axis, of a segment of the straight line through the point I and perpendicular to the ON radius.

Constructions of great eigencircles are possible also in any plane through the OQ axis but only for I points on the 'small sphere' surface (Fig. J. 4). Then $A_1 = 0$ and one double $O_{1,2}$ point exists. The radius of the Poincare sphere for that third kind of models is

$$r_0 \equiv \frac{\sqrt{\sigma_0}}{2} = \sqrt{(A_2 / 2)^2 + B_1^2 + B_2^2} \, . \tag{J.13}$$

The eigenpolarizations exist when

$$r_1 = r_2 = \frac{A_2}{2} \geq \sqrt{B_1^2 + B_2^2} = \frac{\sqrt{\sigma_0}}{2} \cos \gamma_K^P . \tag{J.14}$$

The rotation after inversion angle is $2\phi = 2\gamma_K^P$, for $U_K^I \leq 0$, or $2\phi = 2\pi - 2\gamma_K^P$, for $U_K^I \geq 0$). As usually, the axis of rotation after inversion is perpendicular to the plane of the eigencircle and directed upwards versus the QU coordinate plane of the CCS.

J. 3. The cases of small eigencircles

The case of the solution I for the inversion point in the QV plane can serve as a classical example of the small eigencircle formation. The corresponding model of the scattering matrix has been shown in the Fig. J. 5. Here, the small eigencircle in represented by the line E_1 -E_2 . The axis of rotation after inversion is the V_K coordinate axis and the angle of rotation is 180^0. The model shows its construction based on the I point location.

However, in general case of the scattering matrix in the CCS, the rotation after inversion matrix is inclined versus the V_K coordinate axis by an angle $180^0 - 2\delta_K^P$ but remains in the UV coordinate plane. Therefore, in order to present the eigencircle in its plane, it is necessary to rotate the model about the OQ axis by that angle in the opposite direction to obtain the new V' axis perpendicular to the small eigencircle plane (see Fig.J. 6). The following transformation procedure should be applied:

$$\begin{bmatrix} U' \\ V' \end{bmatrix} = \begin{bmatrix} -\cos 2\delta_K^P & -\sin 2\delta_K^P \\ \sin 2\delta_K^P & -\cos 2\delta_K^P \end{bmatrix} \begin{bmatrix} U \\ V \end{bmatrix} . \tag{J.15}$$

For the Poincare sphere model of diameter $2r_0 = \sqrt{\sigma_0} = 2k$, the radius of an eigencircle is

$$\begin{aligned} r_E &= \sqrt{\frac{\sigma_0}{4} - (V'^I_K)^2} = \sqrt{\frac{\sigma_0}{4} - (U_K^I \sin 2\delta_K^P - V_K^I \cos 2\delta_K^P)^2} \\ &= k \sqrt{1 - \frac{B_1^2 B_2^2 [r_2^2 (k^2 - r_1^2) - r_1^2 (k^2 - r_2^2)]^2}{k^4 [B_1^2 r_1^2 (k^2 - r_2^2)^2 + B_2^2 r_2^2 (k^2 - r_1^2)^2]}} . \end{aligned} \tag{J.16}$$

The above result can be obtained by introducing the formerly derived expression for $\tan 2\delta_k^P$, substituting:

$$\sigma_0 = 4k^2, \quad A_2 + A_1 = 2r_1, \quad A_2 - A_1 = 2r_2,$$
$$\mathrm{U}_K^{\mathrm{I}} = \frac{-\mathrm{b}_3}{2k} = \frac{-2B_1 r_2}{2k}, \quad \mathrm{V}_K^{\mathrm{I}} = \frac{-\mathrm{b}_5}{2k} = \frac{2B_2 r_1}{2k}, \tag{J.17}$$

and making use of known trigonometric formulae:

$$\sin 2\delta_K^P = \frac{\mp \tan 2\delta_K^P}{\sqrt{1+\tan^2 2\delta_K^P}}, \quad \cos 2\delta_K^P = \frac{\mp 1}{\sqrt{1+\tan^2 2\delta_K^P}}. \tag{J.18}$$

The choice of sign depends on the trigonometric quarter of the $2\delta_K^P$ angle. The upper sign corresponds to the second and the lower to the first quarter.

Other similarly obtainable useful formulae (see Fig. J. 6) are:

$$\tan \gamma_K^P = \frac{\mathrm{Q}_K^{\mathrm{I}}}{\mathrm{U'}_K^{\mathrm{I}}} = \frac{\sqrt{B_1^2 r_1^2 (k^2 - r_2^2)^2 + B_2^2 r_2^2 (k^2 - r_1^2)^2}}{B_1^2 (k^2 - r_2^2) + B_2^2 (k^2 - r_1^2)}, \tag{J.19}$$

and

$$r_{\mathrm{E}} \cos \gamma_K^P = \sqrt{(\mathrm{Q}_K^{\mathrm{I}})^2 + (\mathrm{U'}_K^{\mathrm{I}})^2}\,\frac{-\mathrm{U'}_K^{\mathrm{E}}}{r_{\mathrm{E}}}. \tag{J.20}$$

They justify construction of the small eigencircle of Fig. J. 6. and location of eigenpolarization points E_1 and E_2 resembling models of Fig. J. 2-4.

J. 4. Location of the CO-POL-NULL points in general case of the scattering matrix model

Two constructions are possible depending on the solution chosen. Therefore it is advisable a quick recognizing when which solution may exist in the CCS. To do that two simple methods can be applied.

In the first method, direct application of formulae (9.40) locates the inversion point of coordinates (9.29):

$$\begin{bmatrix} \mathrm{Q} \\ \mathrm{U} \\ \mathrm{V} \end{bmatrix}_{K;\, 2r_0=\sqrt{\sigma_0}}^{\mathrm{I}} = \frac{-1}{\sqrt{\sigma_0}} \begin{bmatrix} \mathrm{b}_1 \\ \mathrm{b}_3 \\ \mathrm{b}_5 \end{bmatrix}_K = \frac{1}{-2k} \begin{bmatrix} \mathrm{b}_1 \\ \mathrm{b}_3 \\ \mathrm{b}_5 \end{bmatrix}_K \tag{J.21}$$

in the $\mathrm{Q}_K^{\mathrm{I}} = \mathrm{const.}$ crossection of the Poincare sphere model of unit radius (see Fig. J 7) in the allowed regions: OABCGHO for the solution I, and in DBCGD for the solution II. It should be observed that the whole region of the allowed solution II is contained inside the region for the solution I.

Another method is based on computation of the (OI) distance according to the same formula (9.29) and elongating them when multiplying by the ratio of σ_0 / S. If the end point , I', of projection of such an elongated (OI) distance on the QU plane is inside the shadowed 'small circle' of Fig. J. 8, then the solution I exists. If its projection, I", on the QV plane is in the shadowed region outside another 'small circle', then also the solution II exists. To explain those conditions (again see Fig.J. 8) it can be observed that the postulated requirements are:

$$\begin{aligned} &2k \cos 2\alpha \geq (\mathrm{OI'}) \quad \text{for solution I, and} \\ &2k \sin 2\beta \leq (\mathrm{OI''}) \quad \text{for solution II.} \end{aligned} \tag{J.22}$$

They correspond to the previously found requirements, for the solution I:

$$\mathrm{b}_1 S \geq 2(\mathrm{b}_1^2 + \mathrm{b}_3^2) \Rightarrow \frac{2k\,\mathrm{b}_1}{\sqrt{\mathrm{b}_1^2 + \mathrm{b}_3^2}} = 2k \cos\alpha \geq \frac{2k}{S}\sqrt{\mathrm{b}_1^2 + \mathrm{b}_3^2} = \frac{\sigma_0}{S}\sqrt{(\mathrm{Q}_K^{\mathrm{I}})^2 + (\mathrm{U}_K^{\mathrm{I}})^2}, \tag{J.23a}$$

and for the solution II:

$$b_1 S \le 2(b_1^2 + b_5^2) \Rightarrow \frac{2k\,b_1}{\sqrt{b_1^2 + b_5^2}} = 2k\cos\beta \le \frac{2k}{S}\sqrt{b_1^2 + b_5^2} = \frac{\sigma_0}{S}\sqrt{(Q_K^I)^2 + (V_K^I)^2}\,. \quad \text{(J.23b)}$$

The same σ_0 / S ratio can be used to constructional determination of the CO-POL-NULL points, O_1 and O_2. Similar considerations, taking advantage of formulae (9.50), (9.38a) and (9.39a), lead to the following results:

$$\begin{aligned} q_K^{O_{1,2}} &= -\frac{A_2 - A_1}{A_2 + A_1} \\ &= -\frac{2(b_1^2 + b_3^2)}{b_1 S} = -\frac{\sigma_0}{S}\frac{(Q_K^I)^2 + (U_K^I)^2}{k|Q_K^I|} \quad \text{for the solution I} \\ &= -\frac{b_1 S}{2(b_1^2 + b_5^2)} = -\frac{S}{\sigma_0}\frac{k|Q_K^I|}{(Q_K^I)^2 + (V_K^I)^2} \quad \text{for the solution II.} \end{aligned} \quad \text{(J.24)}$$

Geometrically, for $Q_K^{O_{1,2}} = kq_K^{O_{1,2}}$, it means

$$\frac{|Q_K^{O_{1,2}}|}{(\sigma_0 / S)(OI)} = \frac{(OI)}{|Q_K^I|} = \frac{1}{\cos\alpha} \quad \text{for the solution I, and} \quad \text{(J.25a)}$$

$$k\frac{|Q_K^I|}{(OI)} = k\cos\beta = \frac{\sigma_0}{S}(OI)\frac{|Q_K^{O_{1,2}}|}{k} \quad \text{for the solution II.} \quad \text{(J.25b)}$$

So, simple constructions are obtainable especially in cases of simple relations between σ_0 and S. For instance, the last equation describes construction determining the O_2 point location for the inversion point I in the QV plane of the CCS (where $S = \sigma_0$), for the solution II (Fig, J. 2).

Other figures illustrate special behavior of polarizations being transformed along the eigencircles: when the eigenpolarizations do not exist, Fig. J.9, how one eigenpolarization 'attracts' the polarization point when scattering and the other 'repels' them, Fig. J.10, and how eigenpolarizations are transforming into themselves, Fig. J.11.

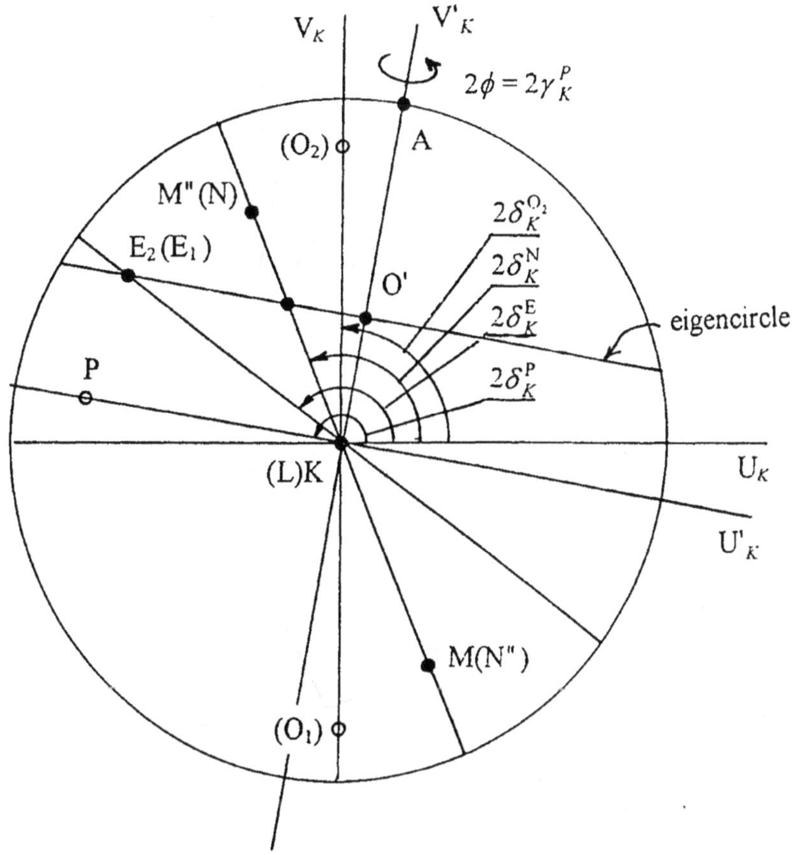

Fig. J.1. Projection of special polarization points onto the UV plane in the CCS.

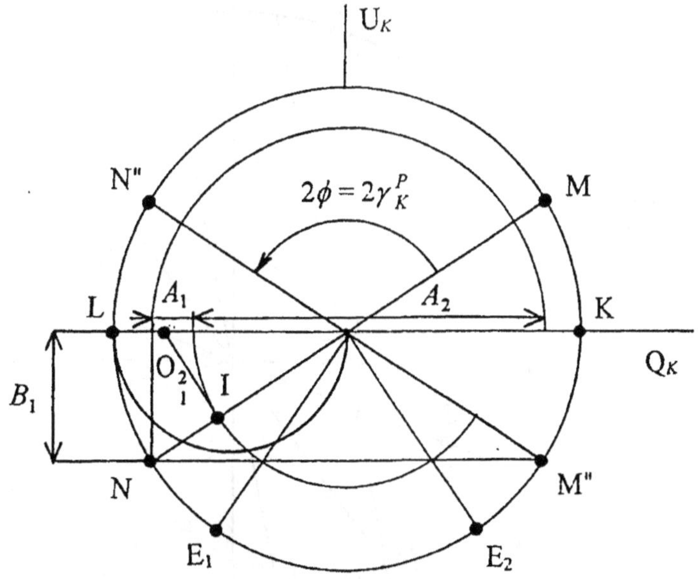

Fig. J.3. I Point in the QU plane in the CCS and the great eigencircle in that plane (solution I is here the only possible).

PROJECTION OF SPECIAL POLARIZATION POINTS ON THE $V_K U_K$ PLANE FOR TWO *I* POINT LOCATIONS

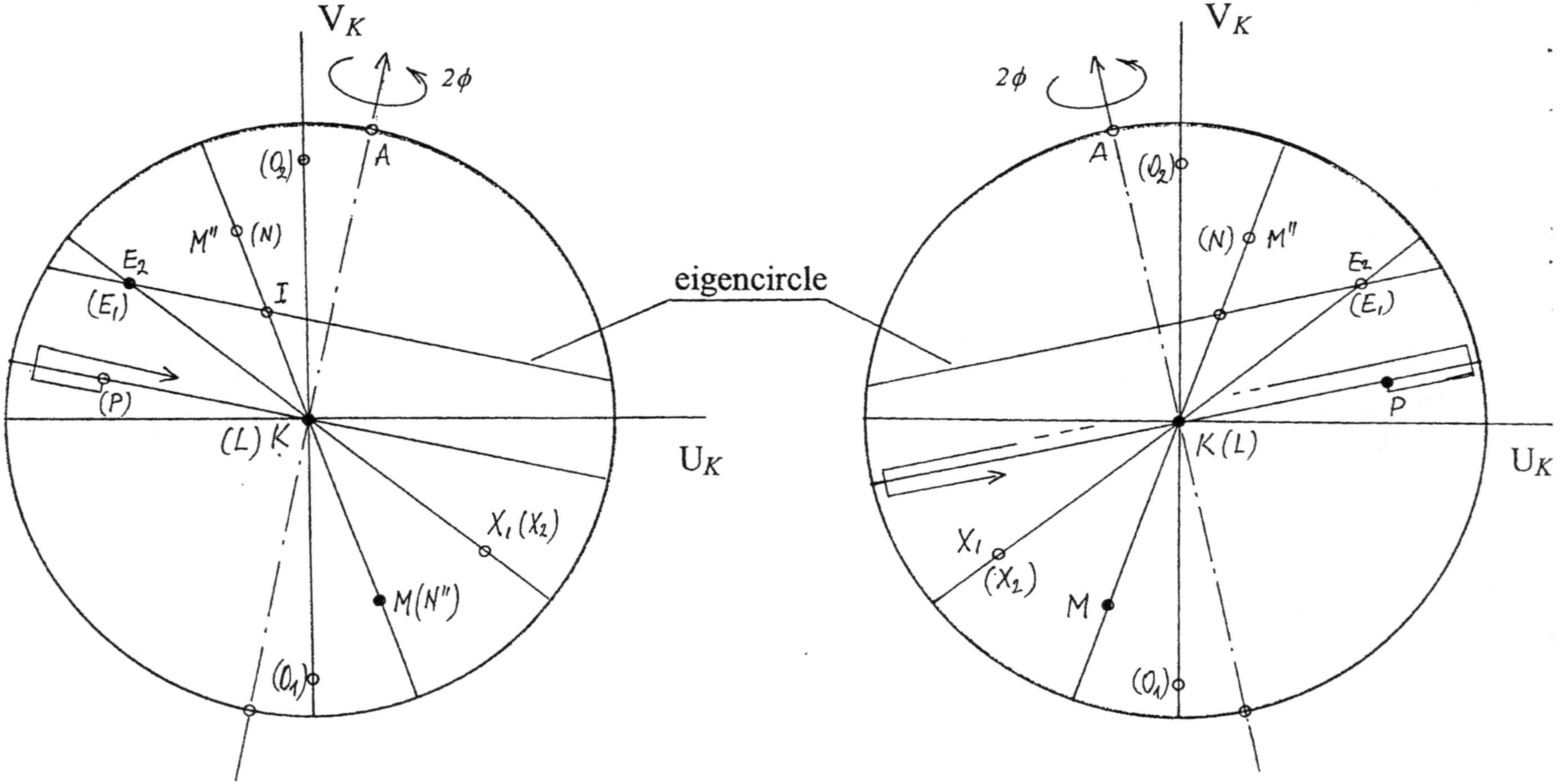

Fig. J.1a

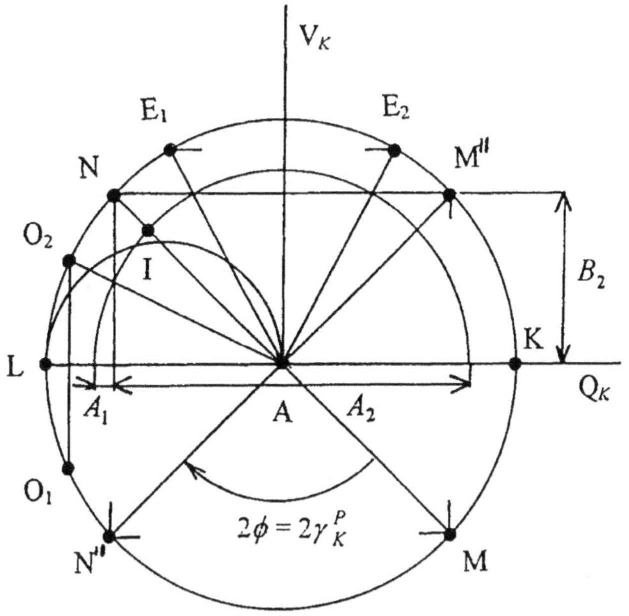

Fig. J.2. I point in the QV plane of the CCS - solution II.
Great eigencircle in that plane. The case of $U^I_K = 0$.

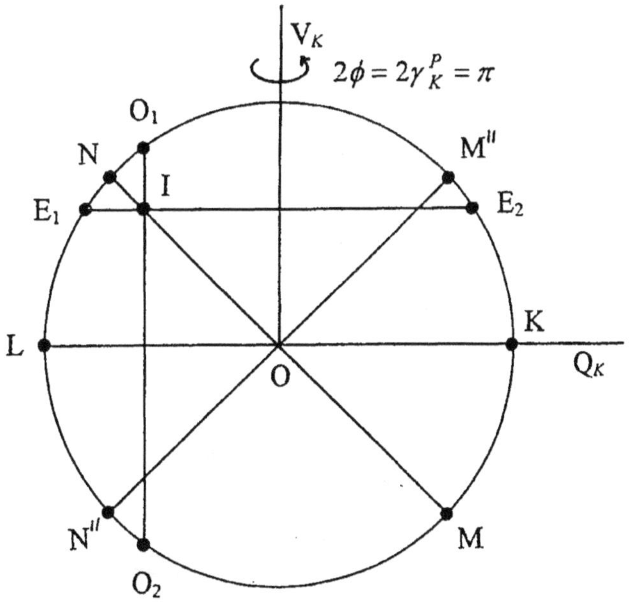

Fig. J.5. I point in the QV plane of the CCS - solution I.
Small eigencircle in the plane perpendicular to
the OV axis. The case of $U^I_K = 0$.

$$2\phi = 2\gamma_K^P \text{ for } U_K^I < 0$$
$$2\phi = 2\pi - 2\gamma_K^P \text{ for } U_K^I > 0$$

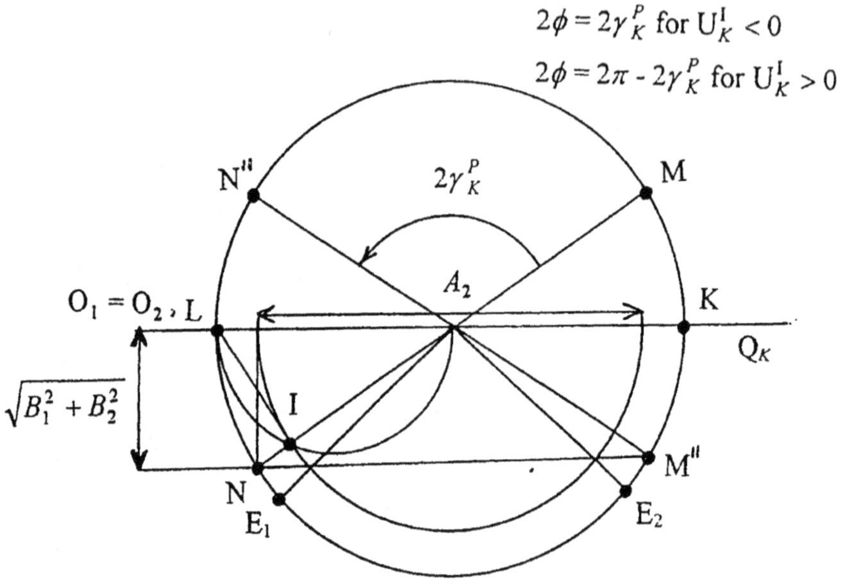

Fig. J.4. I point on the 'small sphere' surface -
solution I for the $|U_K^I| > V_K$ and solution II for the $|U_K^I| < V_K$.
The great eigencircle is inclined at an angle of arc $\tan(B_2/B_1)$
versus the QU plane in the CCS.

$$2\phi = 2\gamma_K^P \text{ for } U_K^I < 0$$
$$2\phi = 2\pi - 2\gamma_K^P \text{ for } U_K^I > 0$$

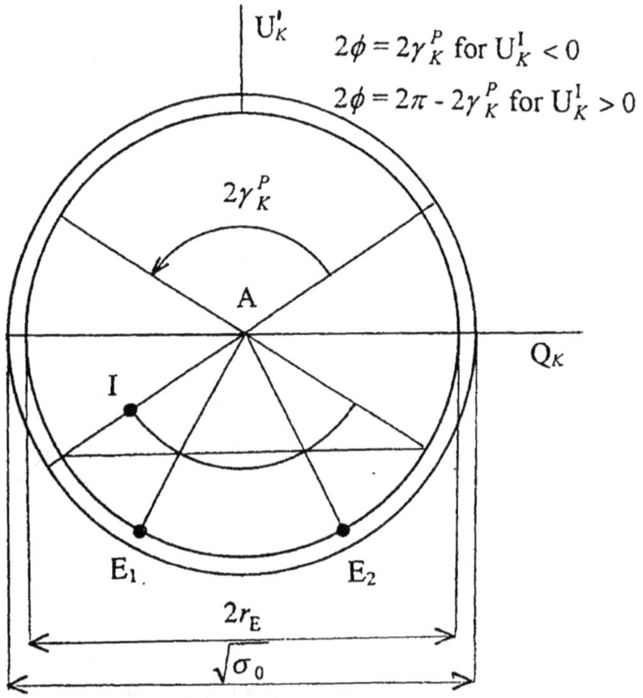

Fig. J.6. General case of the small eigencircle.

ALLOWED REGIONS FOR *I* POINT IN THE CCS

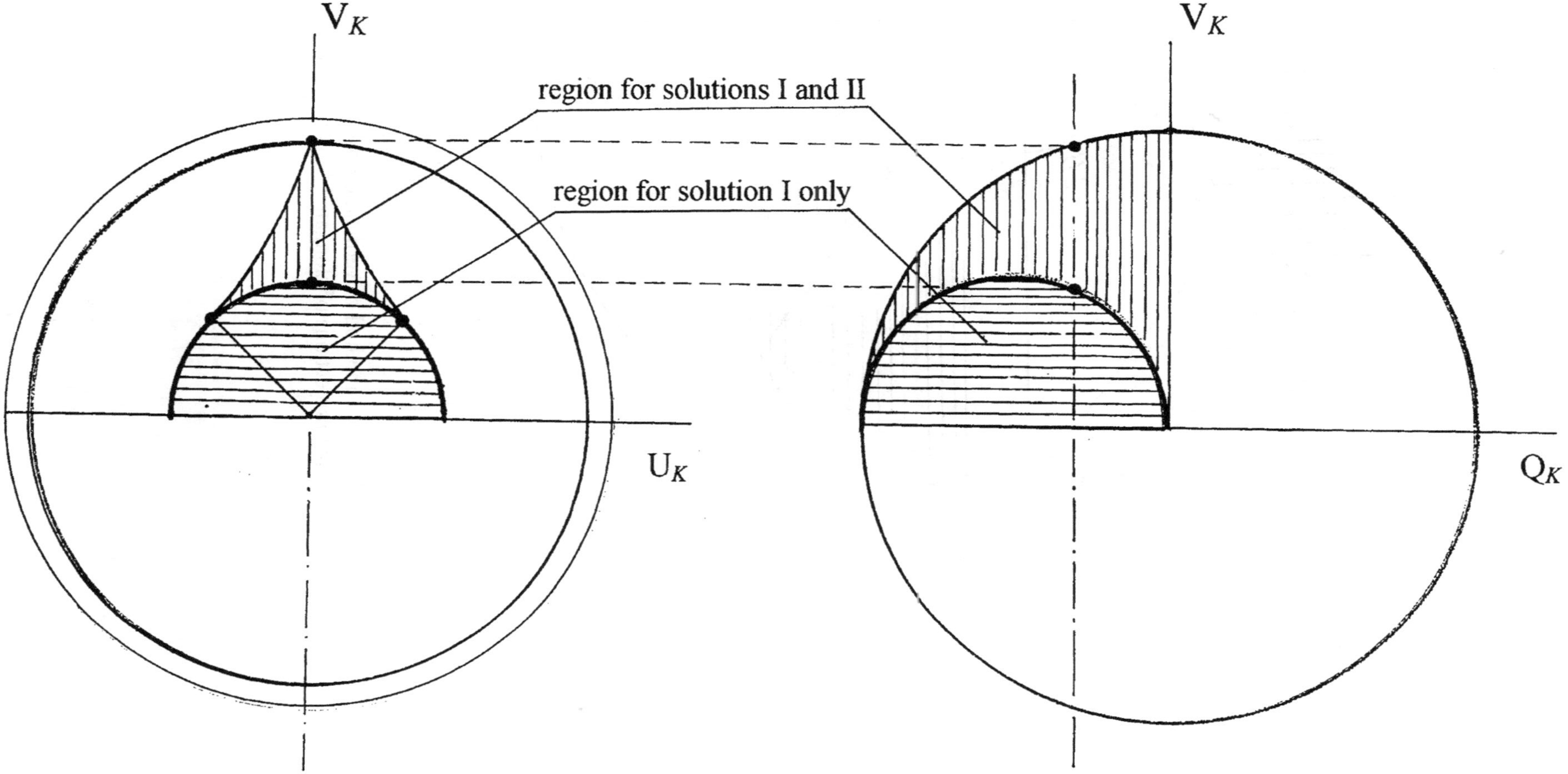

Fig. J.7

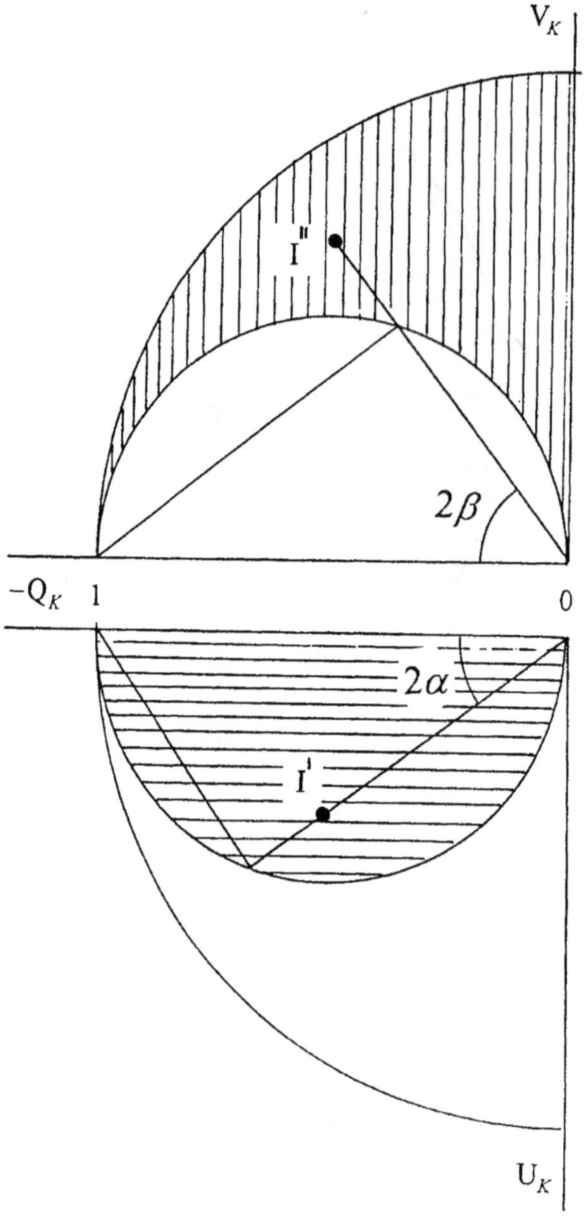

Fig. J.8. To the method of verification solutions I and II using projections of elongated (OI) distances

CHANGE OF POLARIZATION ALONG THE EIGENCIRCLE (TOP VIEW)
THE CASE OF NO EIGENPOLARIZATIONS

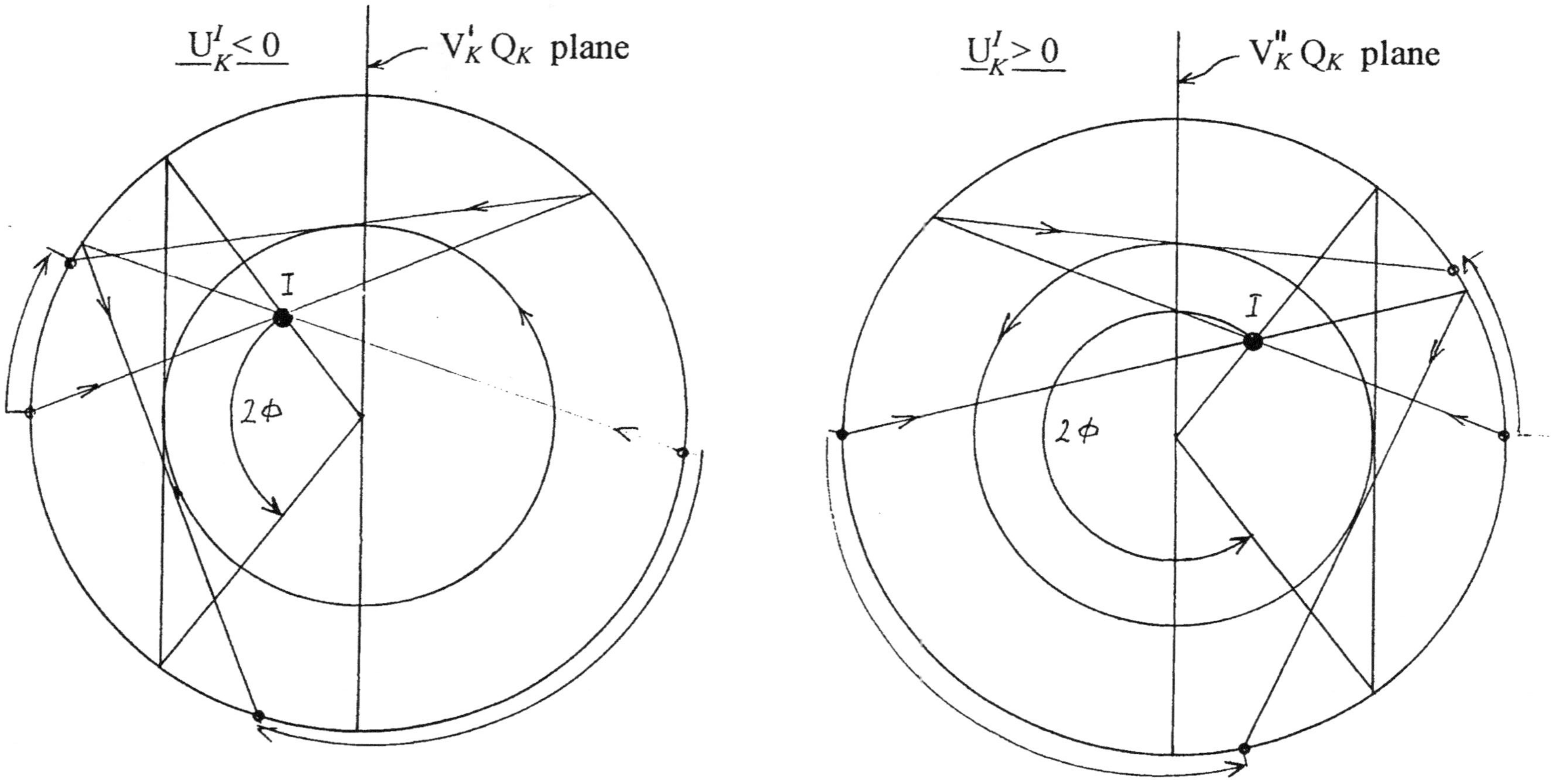

Fig. J.9

ATTRACTION OF POLARIZATION BY E_2 AND REPULSION BY E_1 EIGENPOLARIZATIONS

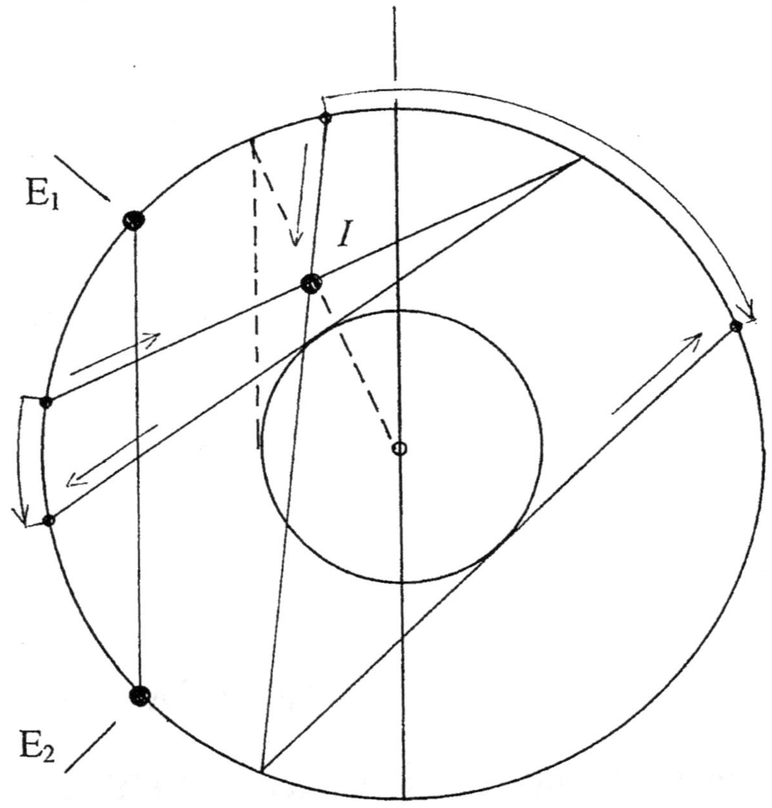

Fig. J.10

TRANSFORMATION OF EIGENPOLARIZATIONS ON THEMSELVES BY INVERSION AND ROTATION IN THE EIGENCIRCLE PLANE

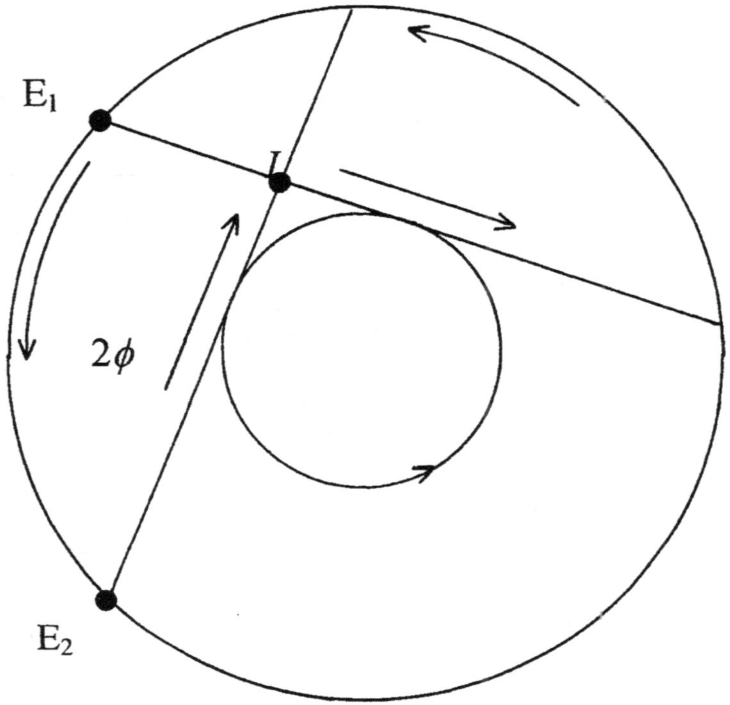

Fig. J.11

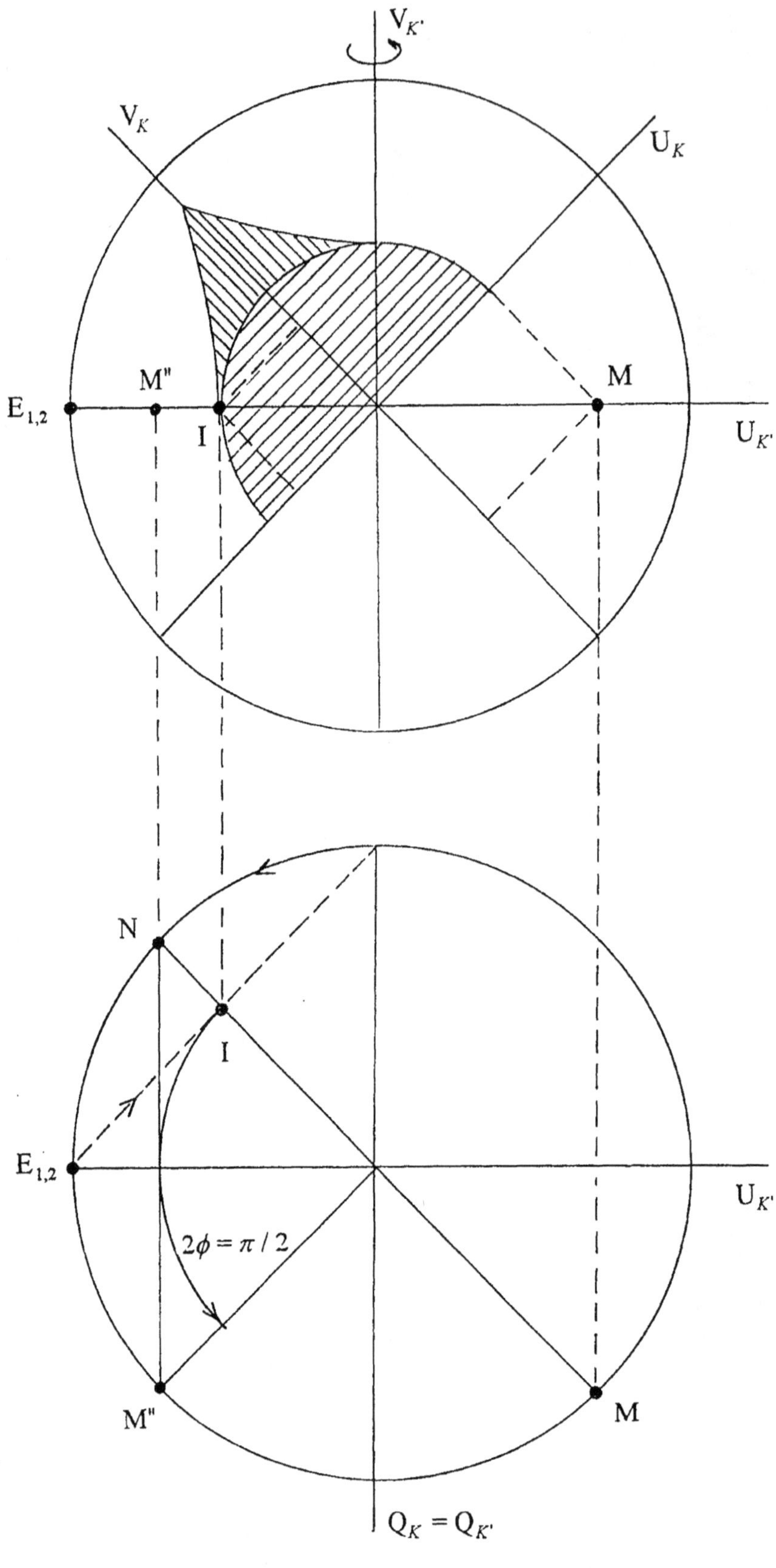

Fig. J.12. The great eigencircle and double eigenpolarization for $Q^I = -0.5$, $U^I = V^I = \frac{-1}{2\sqrt{2}}$.

APPENDIX K

ABOUT UNIQUENESS OF THE CHARACTERISTIC ONP PP BASIS *K*

The proposed form of the Sinclair matrix in the characteristic orthogonal null-phase (ONP) polarization and phase (PP) basis is

$$A_K = \begin{bmatrix} A_2 & B_1 + jB_2 \\ -B_1 - jB_2 & A_1 \end{bmatrix}_{CCS} e^{j\mu}. \tag{K.1}$$

That 'canonical' matrix, of an especially simple form, is determined by location and orientation of its characteristic tangential phasor *K* uniquely represented by the characteristic PP vector (of the form independent of the ONP PP basis order):

$$u_H^K = \begin{bmatrix} \cos\gamma e^{-j(\delta+\varepsilon)} \\ \sin\gamma e^{j(\delta-\varepsilon)} \end{bmatrix}_H^K \tag{K.2}$$

and corresponding to the characteristic coordinate system (CCS) of the three Stokes' parameters Q, U and V.
To obtain the uniqueness of the *K* basis the following requirements has been stated regarding the ranges for values of real elements of the Sinclair matrix in that basis:

$$\begin{aligned} &A_2 \geq A_1 \geq 0 \\ &B_2 \geq 0 \\ &B_1 \geq 0 \ \text{ if } \ B_2 = 0 \end{aligned} \tag{K.3}$$

The first of those conditions determines the only non-zero Q component of the inversion I point to be contained in the range

$$-r \leq Q_K^{\mathrm{I}} \leq 0\,; \qquad r = \sqrt{\sigma_0}\,/\,2\,, \tag{K.4}$$

with

$$\sigma_0 = A_2^2 + A_1^2 + 2(B_2^2 + B_1^2) + 2\sqrt{(A_2A_1 + B_1^2 + B_2^2)^2 + 4B_1^2B_2^2} \tag{K.5}$$

The second condition, $B_2 \geq 0$, stems from the equivalence of two geometrical models of the matrix for *K* and *K'* bases the phasors of which are oppositely oriented. Assuming

$$2\delta_{K'}^K = -\pi \quad \text{and} \quad 2\gamma_{K'}^K = \varepsilon_{K'}^K = 0 \tag{K.6}$$

one obtains the rotation (the change-of-PP-vector) basis matrix

$$C_{K',K'}^K = C_{K'}^K C_{K'}^{K'} = C_{K'}^K = \begin{bmatrix} u_{K'}^K & u_{K'}^{K\times} \end{bmatrix} = \begin{bmatrix} j & 0 \\ 0 & -j \end{bmatrix} \tag{K.7}$$

and the following mutual dependence of the Sinclair matrices in the two bases

$$\begin{aligned} A_K &= \widetilde{C}_{K'}^K A_{K'} C_{K'}^K \\ &= \begin{bmatrix} j & 0 \\ 0 & -j \end{bmatrix} \begin{bmatrix} A_2 & -B_1 - jB_2 \\ B_1 + jB_2 & A_1 \end{bmatrix}_{K'} \begin{bmatrix} j & 0 \\ 0 & -j \end{bmatrix} \\ &= -\begin{bmatrix} A_2 & B_1 + jB_2 \\ -B_1 - jB_2 & A_1 \end{bmatrix}_K . \end{aligned} \tag{K.8}$$

The obtained transposition of the matrix with the reversal of its phase corresponds to the rotation of the model by 180^0 about the $Q_K = Q_{K'}$ axis joined with the change-of-sense of the rotation after inversion angle. So, the two models are entirely adequate and there is no reason for the use of two such characteristic models which will correspond to opposite signs of elements of the second diagonal of Sinclair matrices in the two characteristic bases. Therefore, the second condition can be found justified.

Exactly the same argumentation applies to the third condition. Therefore it is sufficient, for $B_2 \geq 0$, to consider $U_K^I \leq 0$ only, what corresponds to $B_1 \geq 0$ in the CCS, and has been additionally explained on an example for $|U_K^I|_{max} = r/2$ in Fig. K.1.

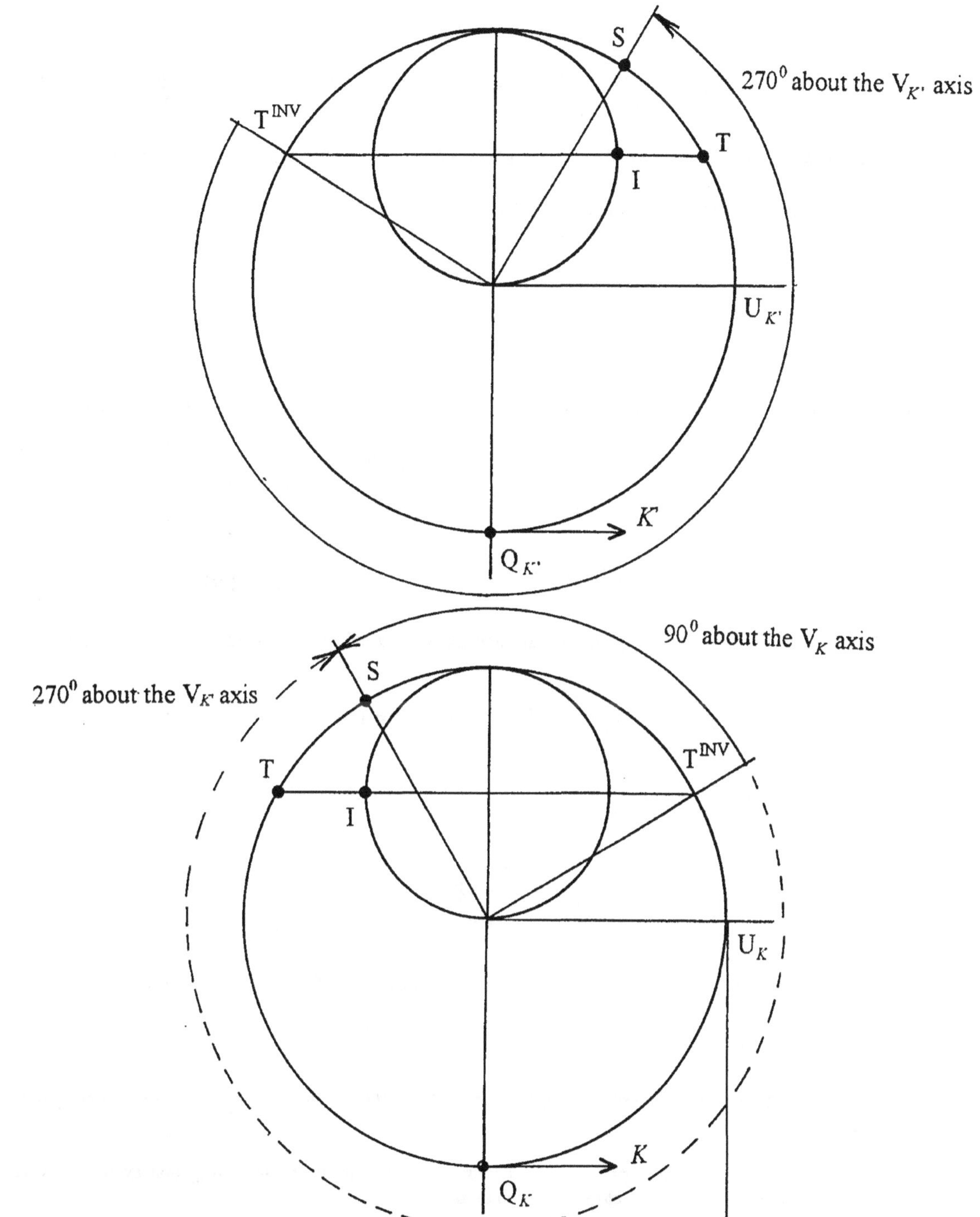

Fig. K. 1. The equivalence of models for oppositely oriented basis phasors, K and K', corresponding to $B_1>0$ and $B_1<0$ (for $B_2=0$). Examples with the I point in the $V_{K'} = 0$ plane

$$(-U^I_K = U^I_{K'} = -Q^I_K = -Q^I_{K'} = r/2).$$

$$\tilde{C}^K_{K'} A_{K'} C^K_{K'} = \begin{bmatrix} j & 0 \\ 0 & -j \end{bmatrix} \begin{bmatrix} A_2 & -B_1 \\ B_1 & 0 \end{bmatrix}_{K'} \begin{bmatrix} j & 0 \\ 0 & -j \end{bmatrix} = -\begin{bmatrix} A_2 & B_1 \\ -B_1 & 0 \end{bmatrix}_K$$

APPENDIX L

THE WANIELIK'S/CARREA REPRESENTATION OF THE POINCARE SPHERE TANGENTIAL POLARIZATION (TP) VECTORS USING THE TP PHASOR NOTATION

L1. Background: the change of basis transformation in the 2-dim. space of the PP vectors

For the Cayley-Klein parameters (see (4.31) and (5.40))

$$a = \cos\gamma e^{-j(\delta+\varepsilon)} \qquad 0^0 \le \gamma \le 90^0$$
$$\text{with} \qquad -90^0 < \delta \le 90^0 \qquad \text{(L1)}$$
$$b = \sin\gamma e^{+j(\delta-\varepsilon)} \qquad -180^0 \le \varepsilon \le 180^0$$

the matrix and vectorial forms of the PP vectors have been determined as follows (see (4.30) and (4.29), or (5.1) to (5.4)):

$$u_H^P = \begin{bmatrix} a \\ b \end{bmatrix}_H^P, \quad u_H^{P\times} = \begin{bmatrix} -b^* \\ a^* \end{bmatrix}_H^P; \qquad aa^* + bb^* = 1, \qquad \text{(L2)}$$

$$u^P = [u^H u^{H\times}]\, u_H^P, \qquad u^{P\times} = [u^H u^{H\times}]\, u_H^{P\times}, \qquad \text{(L3)}$$

fulfilling conditions of unitarity and the 'null-phase orthogonality' (see (4.32) and (6.1)) in the C^2 space:

$$|u^P| = |u^{P\times}| = 1, \quad u^P \bullet u^{P\times *} = 0; \quad u_H^{P\times *} = \begin{bmatrix} 0 & -1 \\ 1 & 0 \end{bmatrix} u_H^P. \qquad \text{(L4)}$$

With those vectors the rotation matrix (see (5.8) and (5.30)) for the C^2 space has been determined,

$$\begin{bmatrix} u_H^P & u_H^{P\times} \end{bmatrix} = \begin{bmatrix} \cos\gamma e^{-j(\delta+\varepsilon)} & -\sin\gamma e^{-j(\delta-\varepsilon)} \\ \sin\gamma e^{j(\delta-\varepsilon)} & \cos\gamma e^{j(\delta+\varepsilon)} \end{bmatrix}_H^P = C_H^P \qquad \text{(L5)}$$

which can be used to present the change-of-basis transformation (5.9) for the PP vectors,

$$C_H^B u_B^P = u_H^P. \qquad \text{(L6)}$$

L2. Extension of the PP vector complex form to its representation in the 3-dim. real space of Stokes parameters

By analogy to the previously presented procedure in the C^2 space, the following matrix and vectorial forms of three mutually perpendicular real vectors can be determined:

$$\mathbf{p}_H^P = \begin{bmatrix} \mathrm{q} \\ \mathrm{u} \\ \mathrm{v} \end{bmatrix}_H^P, \quad \mathbf{q}_{RH}^P = \begin{bmatrix} \mathrm{q}_{R1} \\ \mathrm{q}_{R2} \\ \mathrm{q}_{R3} \end{bmatrix}_H^P, \quad \mathbf{q}_{IH}^P = \begin{bmatrix} \mathrm{q}_{I1} \\ \mathrm{q}_{I2} \\ \mathrm{q}_{I3} \end{bmatrix}_H^P, \qquad \text{(L7)}$$

$$\mathbf{p}^P = \begin{bmatrix} \mathbf{1}_q & \mathbf{1}_u & \mathbf{1}_v \end{bmatrix}_H \mathbf{p}_H^P, \quad \mathbf{q}_R^P = \begin{bmatrix} \mathbf{1}_q & \mathbf{1}_u & \mathbf{1}_v \end{bmatrix}_H \mathbf{q}_{RH}^P, \quad \mathbf{q}_I^P = \begin{bmatrix} \mathbf{1}_q & \mathbf{1}_u & \mathbf{1}_v \end{bmatrix}_H \mathbf{q}_{IH}^P. \qquad \text{(L8)}$$

The above used lower indices 'R' and 'I' have been assumed to correspond with the real and imaginary parts of the Cartan's null vector [138],

$$\mathbf{q} = \mathbf{q}_R + j\mathbf{q}_I = \begin{bmatrix} -2ab \\ a^2 - b^2 \\ j(a^2 + b^2) \end{bmatrix} \equiv \begin{bmatrix} \mathrm{q}_1 \\ \mathrm{q}_2 \\ \mathrm{q}_3 \end{bmatrix}; \quad \mathrm{q}_1^2 + \mathrm{q}_2^2 + \mathrm{q}_3^2 = 0. \tag{L.9}$$

The three just defined real vectors are assumed to satisfy conditions of unitarity and mutual perpendicularity:

$$\begin{gathered} |\mathbf{p}| = |\mathbf{q}_R| = |\mathbf{q}_I| = 1, \\ \mathbf{p} \bullet \mathbf{q}_R = \mathbf{q}_R \bullet \mathbf{q}_I = \mathbf{q}_I \bullet \mathbf{p} = 0, \\ \mathbf{p} \times \mathbf{q}_R = \mathbf{q}_I, \quad \mathbf{q}_R \times \mathbf{q}_I = \mathbf{p}, \quad \mathbf{q}_I \times \mathbf{p} = \mathbf{q}_R. \end{gathered} \tag{L.10}$$

It can be verified that these vectors can form the known rotation matrix (5.45) as follows,

$$\begin{bmatrix} 1 & 0 & 0 & 0 \\ \mathbf{0} & \mathbf{p} & \mathbf{q}_R & \mathbf{q}_I \end{bmatrix}_H^P = D_H^P$$

$$= \begin{bmatrix} 1 & 0 & 0 & 0 \\ 0 & \cos 2\gamma & -\sin 2\gamma \cos 2\varepsilon & \sin 2\gamma \sin 2\varepsilon \\ 0 & \sin 2\gamma \cos 2\delta & \cos 2\gamma \cos 2\delta \cos 2\varepsilon - \sin 2\delta \sin 2\varepsilon & -\cos 2\gamma \cos 2\delta \sin 2\varepsilon - \sin 2\delta \cos 2\varepsilon \\ 0 & \sin 2\gamma \sin 2\delta & \cos 2\gamma \sin 2\delta \cos 2\varepsilon + \cos 2\delta \sin 2\varepsilon & -\cos 2\gamma \sin 2\delta \sin 2\varepsilon + \cos 2\delta \cos 2\varepsilon \end{bmatrix}_H^P \tag{L.11}$$

with the following explicit forms of their components:

$$\mathrm{q}_H^P = \cos 2\gamma_H^P, \quad \mathrm{u}_H^P = (\sin 2\gamma \cos 2\delta)_H^P, \quad \mathrm{v}_H^P = (\sin 2\gamma \sin 2\delta)_H^P, \tag{l.12}$$

$$\begin{aligned} \mathrm{q}_{R1H}^P &= (-\sin 2\gamma \cos 2\varepsilon)_H^P \\ \mathrm{q}_{R2H}^P &= (\cos 2\gamma \cos 2\delta \cos 2\varepsilon - \sin 2\delta \sin 2\varepsilon)_H^P \\ \mathrm{q}_{R3H}^P &= (\cos 2\gamma \sin 2\delta \cos 2\varepsilon + \cos 2\delta \sin 2\varepsilon)_H^P \end{aligned} \tag{L.13}$$

$$\begin{aligned} \mathrm{q}_{I1H}^P &= (\sin 2\gamma \sin 2\varepsilon)_H^P \\ \mathrm{q}_{I2H}^P &= (-\cos 2\gamma \cos 2\delta \sin 2\varepsilon - \sin 2\delta \cos 2\varepsilon)_H^P \\ \mathrm{q}_{I3H}^P &= (-\cos 2\gamma \sin 2\delta \sin 2\varepsilon + \cos 2\delta \cos 2\varepsilon)_H^P \end{aligned} \tag{L.14}$$

determined in the ranges (5.19):

$$0 \le 2\gamma \le \pi$$

$$-\pi < 2\delta \le \pi$$

and

$$-2\pi \le 2\varepsilon \le 2\pi \tag{L.15}$$

The range for the last Euler angle, 2ε, is especially important for ensuring the one to one correspondence between the PP vectors, $\boldsymbol{u}^P$, or the corresponding TP phasors, P, and their real counterparts, $\mathbf{q}_R^P$. These real vectors do not change their orientation after addition of 2π to their 2ε angle. Therefore, to omit the ambiguity, it is always necessary to present on the Poincare sphere, for the $\mathbf{q}_{RH}^P$ vectors, the $2\varepsilon_H^P$ angle which may be contained in the $\pm 2\pi$ range. Only the 4π change of that argument relates that real vector to the same TP phasor (compare [29]).

L3. An exemplary derivation of the $\mathbf{q}_{\mathrm{R}}$ and $\mathbf{q}_{\mathrm{I}}$ vector expressions for the $\mathbf{p}$ vector given

Very simple derivation may start with the evident result for vectorial product of two tangential vectors, $\mathbf{q}_{\mathrm{R}H}^{P_0}$ and $\mathbf{q}_{\mathrm{R}H}^{P}$, the first of which corresponds to the angle $2\varepsilon_H^{P_0}=0$. That product can be presented by the following determinantial equation with the three unknown components, $q_{\mathrm{R}1H}^{P}$, $q_{\mathrm{R}2H}^{P}$, and $q_{\mathrm{R}3H}^{P}$:

$$\begin{aligned}\mathbf{q}_{\mathrm{R}H}^{P_0}\times\mathbf{q}_{\mathrm{R}H}^{P} &= \sin 2\varepsilon_H^P\,\mathbf{p}_H^P \\ &= \sin 2\varepsilon_H^P(\mathbf{1}_\mathbf{q}\cos 2\gamma+\mathbf{1}_\mathbf{u}\sin 2\gamma\cos 2\delta+\mathbf{1}_\mathbf{v}\sin 2\gamma\sin 2\delta)_H^P \\ &= \begin{vmatrix}\mathbf{1}_\mathbf{q} & \mathbf{1}_\mathbf{u} & \mathbf{1}_\mathbf{v}\\ -\sin 2\gamma & \cos 2\gamma\cos 2\delta & \cos 2\gamma\sin 2\delta\\ q_{\mathrm{R}1} & q_{\mathrm{R}2} & q_{\mathrm{R}3}\end{vmatrix}_H^P\end{aligned} \tag{L.16}$$

Comparison of the third and second components of the product yields the expressions for $q_{\mathrm{R}2}$ and $q_{\mathrm{R}3}$ obtained in terms of $q_{\mathrm{R}1}$:

$$\begin{aligned}q_{\mathrm{R}2} &= \frac{-q_{\mathrm{R}1}\cos 2\gamma\cos 2\delta-\sin 2\gamma\sin 2\delta\sin 2\varepsilon}{\sin 2\gamma},\\ q_{\mathrm{R}3} &= \frac{-q_{\mathrm{R}1}\cos 2\gamma\sin 2\delta+\sin 2\gamma\cos 2\delta\sin 2\varepsilon}{\sin 2\gamma}.\end{aligned} \tag{L.17}$$

After simple manipulations one obtains

$$q_{\mathrm{R}2}^2+q_{\mathrm{R}3}^2=1-q_{\mathrm{R}1}^2=\frac{q_{\mathrm{R}1}^2\cos^2 2\gamma+\sin^2 2\gamma\sin^2 2\delta}{\sin^2 2\gamma}$$

and

$$q_{\mathrm{R}1}=-\sin 2\gamma\cos 2\varepsilon. \tag{L.18}$$

The minus sign can be found, for example, when considering $2\gamma=\pi/2$ and $2\varepsilon=0$ in which case $q_{\mathrm{R}1}=-1$. The remaining values (L.13), $q_{\mathrm{R}2}$ and $q_{\mathrm{R}3}$, can be immediately obtained from (L.18) and (L.17). The (L.14) components will then be determined by the vector product of (L.10),

$$\mathbf{p}\times\mathbf{q}_{\mathrm{R}}=\mathbf{q}_{\mathrm{I}}. \tag{L.19}$$

L4. Change-of-basis rules in the 3-dim. Stokes parameter space

Denoting three basis vectors of an ONP PP basis B:

$$\begin{bmatrix}1\\0\\0\end{bmatrix}_B^B\equiv\mathbf{p}_B^B,\quad \begin{bmatrix}0\\1\\0\end{bmatrix}_B^B\equiv\mathbf{q}_{\mathrm{R}B}^B,\quad \begin{bmatrix}0\\0\\1\end{bmatrix}_B^B\equiv\mathbf{q}_{\mathrm{I}B}^B \tag{L.20}$$

the following change-of basis equations can be written

$$\begin{bmatrix}q & q_{\mathrm{R}1} & q_{\mathrm{I}1}\\ u & q_{\mathrm{R}2} & q_{\mathrm{I}2}\\ v & q_{\mathrm{R}3} & q_{\mathrm{I}3}\end{bmatrix}_H^B\begin{bmatrix}1\\0\\0\end{bmatrix}_B^B=\begin{bmatrix}q\\u\\v\end{bmatrix}_H^B=\mathbf{p}_H^B \tag{L.21}$$

$$\begin{bmatrix} q & q_{R1} & q_{I1} \\ u & q_{R2} & q_{I2} \\ v & q_{R3} & q_{I3} \end{bmatrix}_H^B \begin{bmatrix} 0 \\ 1 \\ 0 \end{bmatrix}_B^B = \begin{bmatrix} q_{R1} \\ q_{R2} \\ q_{R3} \end{bmatrix}_H^B = \mathbf{q}_{RH}^B \tag{L.22}$$

$$\begin{bmatrix} q & q_{R1} & q_{I1} \\ u & q_{R2} & q_{I2} \\ v & q_{R3} & q_{I3} \end{bmatrix}_H^B \begin{bmatrix} 0 \\ 0 \\ 1 \end{bmatrix}_B^B = \begin{bmatrix} q_{I1} \\ q_{I2} \\ q_{I3} \end{bmatrix}_H^B = \mathbf{q}_{IH}^B \tag{L.23}$$

The linear combination of the above equations leads to the final change-of-basis expressions:

$$[\mathbf{p} \;\; \mathbf{q}_R \;\; \mathbf{q}_I]_H^B \mathbf{p}_B^P = \mathbf{p}_H^P, \quad [\mathbf{p} \;\; \mathbf{q}_R \;\; \mathbf{q}_I]_H^B \mathbf{q}_{RB}^P = \mathbf{q}_{RH}^P, \quad [\mathbf{p} \;\; \mathbf{q}_R \;\; \mathbf{q}_I]_H^B \mathbf{q}_{IB}^P = \mathbf{q}_{IH}^P \tag{L.24}$$

In the 4-dim. Stokes parameter space the equivalent change-of-basis equations take the form:

$$\begin{bmatrix} 1 & 0 & 0 & 0 \\ \mathbf{0} & \mathbf{p} & \mathbf{q}_R & \mathbf{q}_I \end{bmatrix}_H^B \frac{1}{\sqrt{2}} \begin{bmatrix} 1 \\ \mathbf{p} \end{bmatrix}_B^P = \frac{1}{\sqrt{2}} \begin{bmatrix} 1 \\ \mathbf{p} \end{bmatrix}_H^P = \mathbf{P}_H^P = \mathrm{D}_H^B \mathbf{P}_B^P \tag{L.25}$$

$$\begin{bmatrix} 1 & 0 & 0 & 0 \\ \mathbf{0} & \mathbf{p} & \mathbf{q}_R & \mathbf{q}_I \end{bmatrix}_H^B \begin{bmatrix} 0 \\ \mathbf{q}_R \end{bmatrix}_B^P = \begin{bmatrix} 0 \\ \mathbf{q}_R \end{bmatrix}_H^P = \mathbf{Q}_{RH}^P = \mathrm{D}_H^B \mathbf{Q}_{RB}^P \tag{L.26}$$

$$\begin{bmatrix} 1 & 0 & 0 & 0 \\ \mathbf{0} & \mathbf{p} & \mathbf{q}_R & \mathbf{q}_I \end{bmatrix}_H^B \begin{bmatrix} 0 \\ \mathbf{q}_I \end{bmatrix}_B^P = \begin{bmatrix} 0 \\ \mathbf{q}_I \end{bmatrix}_H^P = \mathbf{Q}_{IH}^P = \mathrm{D}_H^B \mathbf{Q}_{IB}^P \tag{L.27}$$

It is worth noticing that such a compact change-of-basis expression for the Cartan's null vector,

$$\mathbf{Q}_H^P = \mathbf{Q}_{RH}^P + j\mathbf{Q}_{IH}^P = \mathrm{D}_H^B \mathbf{Q}_B^P, \tag{L.28}$$

has been obtained owing to application of the TP phasor notation.

L5. Tangential vectors in terms of the q, u, v parameters and the spatial phase double angle, 2ν. An alternative form of the rotation matrix in the Stokes parameters space.

In the formulae (L.13) and (L.14) trigonometric functions of the Euler angles can be exchanged with the expressions resulting from the equalities (L.11):

$$\begin{aligned} q &= \cos 2\gamma & &\Rightarrow & \sin 2\gamma &= \sqrt{1-q^2} = \frac{u^2+v^2}{\sqrt{1-q^2}} \\ u &= \sin 2\gamma \cos 2\delta & &\Rightarrow & \cos 2\delta &= \frac{u}{\sqrt{1-q^2}} \\ v &= \sin 2\gamma \sin 2\delta & &\Rightarrow & \sin 2\delta &= \frac{v}{\sqrt{1-q^2}} \end{aligned} \tag{L.29}$$

and additionally, from the definition of the double spatial phase delay (see (3.23)),

$$2\nu = 2\varepsilon + 2\delta, \tag{L.30}$$

corresponding to the equivalent alternative form of the PP vector:

$$u=\begin{bmatrix}a\\ b\end{bmatrix}=\frac{1}{\sqrt{1+|\rho|^2}}\begin{bmatrix}1\\ \rho\end{bmatrix}e^{-j2\nu};\quad \rho=\tan\gamma e^{j2\delta}. \tag{L.31}$$

Using (L.30) and (L.29), simple transformations yield:

$$\left.\begin{aligned}\cos 2\nu&=\frac{\mathrm{u}\cos 2\varepsilon-\mathrm{v}\sin 2\varepsilon}{\sqrt{1-\mathrm{q}^2}}\\ \sin 2\nu&=\frac{\mathrm{u}\sin 2\varepsilon+\mathrm{v}\cos 2\varepsilon}{\sqrt{1-\mathrm{q}^2}}\end{aligned}\right\}\Rightarrow\left\{\begin{aligned}\cos 2\varepsilon&=\frac{\mathrm{u}\cos 2\nu+\mathrm{v}\sin 2\nu}{\sqrt{1-\mathrm{q}^2}}\\ \sin 2\varepsilon&=\frac{\mathrm{u}\sin 2\nu-\mathrm{v}\cos 2\nu}{\sqrt{1-\mathrm{q}^2}}\end{aligned}\right. \tag{L.32}$$

With these trigonometric functions of the three Euler angles, 2γ, 2δ, and 2ε, from (L.13) and (L.14) one obtains the tangential vectors components in the form presented by Carrea and Wanielik [72]:

$$\begin{aligned}\mathrm{q}_{\mathrm{R1}H}^{P}&=(-\mathrm{u}\cos 2\nu-\mathrm{v}\sin 2\nu)_H^P\\ \mathrm{q}_{\mathrm{R2}H}^{P}&=\left(\frac{[\mathrm{q}(1+\mathrm{q})+\mathrm{v}^2]\cos 2\nu-\mathrm{uv}\sin 2\nu}{1+\mathrm{q}}\right)_H^P\\ \mathrm{q}_{\mathrm{R3}H}^{P}&=\left(\frac{-\mathrm{uv}\cos 2\nu+[\mathrm{q}(1+\mathrm{q})+\mathrm{u}^2]\sin 2\nu}{1+\mathrm{q}}\right)_H^P\end{aligned} \tag{L.33}$$

and

$$\begin{aligned}\mathrm{q}_{\mathrm{I1}H}^{P}&=(\mathrm{u}\sin 2\nu-\mathrm{v}\cos 2\nu)_H^P\\ \mathrm{q}_{\mathrm{I2}H}^{P}&=\left(\frac{-[\mathrm{q}(1+\mathrm{q})+\mathrm{v}^2]\sin 2\nu-\mathrm{uv}\cos 2\nu}{1+\mathrm{q}}\right)_H^P\\ \mathrm{q}_{\mathrm{I3}}&=\left(\frac{\mathrm{uv}\sin 2\nu+[\mathrm{q}(1+\mathrm{q})+\mathrm{u}^2]\cos 2\nu}{1+\mathrm{q}}\right)_H^P\end{aligned} \tag{L.34}$$

Substitution of those expressions to (L.11) yields an alternative form of the rotation matrix

$$D_H^P=\begin{bmatrix}1&0&0&0\\ \mathbf{0}&\mathbf{p}&\mathbf{q}_\mathbf{R}&\mathbf{q}_\mathbf{I}\end{bmatrix}_H^P=$$

$$=\begin{bmatrix}1&0&0&0\\ 0&\mathrm{q}&-\mathrm{u}\cos 2\nu-\mathrm{v}\sin 2\nu&\mathrm{u}\sin 2\nu-\mathrm{v}\cos 2\nu\\ 0&\mathrm{u}&\dfrac{[\mathrm{q}(1+\mathrm{q})+\mathrm{v}^2]\cos 2\nu-\mathrm{uv}\sin 2\nu}{1+\mathrm{q}}&\dfrac{-[\mathrm{q}(1+\mathrm{q})+\mathrm{v}^2]\sin 2\nu-\mathrm{uv}\cos 2\nu}{1+\mathrm{q}}\\ 0&\mathrm{v}&\dfrac{-\mathrm{uv}\cos 2\nu+[\mathrm{q}(1+\mathrm{q})+\mathrm{u}^2]\sin 2\nu}{1+\mathrm{q}}&\dfrac{\mathrm{uv}\sin 2\nu+[\mathrm{q}(1+\mathrm{q})+\mathrm{u}^2]\cos 2\nu}{1+\mathrm{q}}\end{bmatrix}_H^P \tag{L.35}$$

Also that simple form of the rotation matrix has been obtained using the method of the PP vectors and the TP phasor notation.

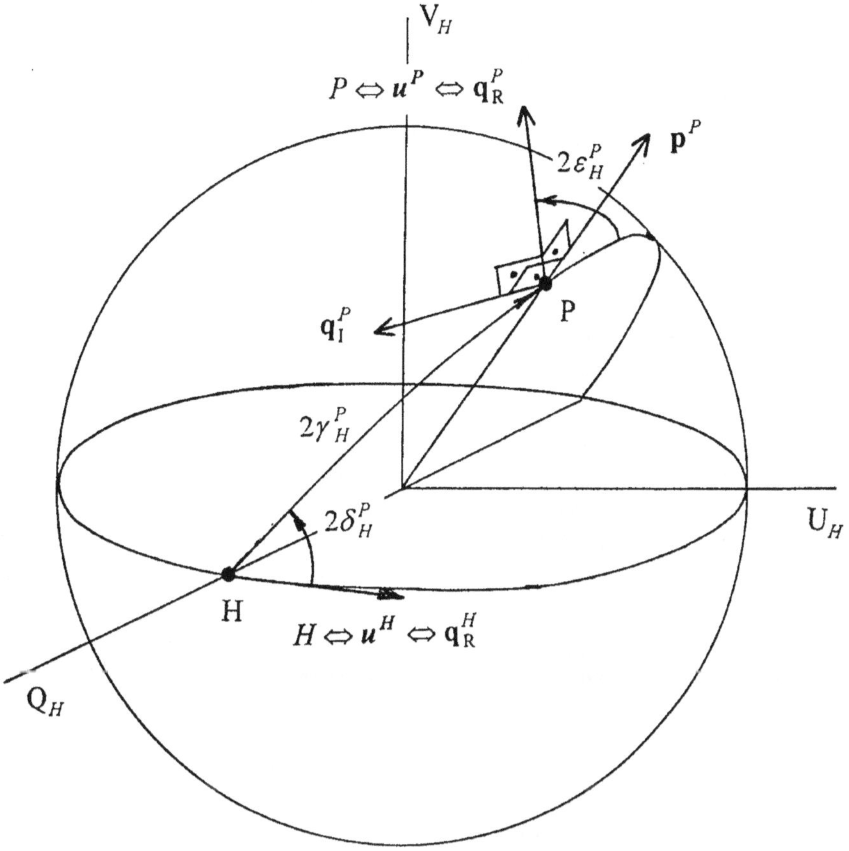

Fig. L.1. The polarization and phase orthonormal vectors triplet (one vector, $\mathbf{p}$, perpendicular to the polarization Poincare sphere and two, $\mathbf{q}_R$ and $\mathbf{q}_I$, tangential)

APPENDIX M

ANGULAR PARAMETERS OF THE TANGENTIAL POLARIZATION (TP) PHASORS

Three kinds of angular parameters of the TP phasors can be distinguished:

- 'analytical' parameters, 2γ, 2δ, 2ε and $2\nu = 2\delta + 2\varepsilon$,
- 'geometrical' parameters, 2α, 2β, and 2χ, and (M.1)
- 'mixed' parameters, $2\eta = 2\chi - 2\varepsilon$, and $2\varphi = 2\nu - 2\chi$.

Among analytical angular parameters two of them, 2γ and 2δ, are polarization dependent, and two, 2ε and $2\nu = 2\delta + 2\varepsilon$, are phase dependent. The argument $2\varepsilon > 0$ means the 'analytical' spatial phase delay of the wave, and $2\nu > 0$ means the 'analytical' spatial phase delay of the first wave's component, both for the wave's PP vector expressed in any ONP PP basis. However, the polarization dependent $2\delta > 0$ angle means also the spatial phase delay but of the first versus the second orthogonal PP vector *component* presented in any ONP PP basis being assumed.

Polarization dependent geometrical angular parameters are 2α and 2β, and 2χ is a phase dependent geometrical parameter. The argument $2\chi > 0$ means the 'geometrical' spatial phase delay of the wave.

The two mixed parameters are both polarization dependent only but can be expressed by differences between wave's analytical and geometrical spatial phase delay parameters, as has been shown above.

Analytical and geometrical parameters can take values in the following ranges:

$$
\begin{array}{ll}
0 \le 2\gamma \le \pi & -\pi/2 \le 2\alpha \le \pi/2 \\
-\pi < 2\delta \le \pi & -\pi < 2\beta \le \pi \\
-2\pi \le 2\varepsilon \le 2\pi & -2\pi \le 2\chi \le 2\pi
\end{array}
\tag{M.2}
$$

They determine the polarization and (spatial) phase delay of a completely polarized plane wave, identically for both opposite directions of propagation (along the propagation axis in a $\pm z$ direction of a right-handed local spatial coordinate system xyz), and for any assumed order of its spatial components, 'natural' (*xy*) or 'reversed' (*yx*), satisfying condition that the second component leads the first one by the spatial phase angle 2δ.

Mutual dependences between some different angular parameters (see Fig. M1) are as follows:

$$
\begin{array}{l}
2\nu = 2\delta + 2\varepsilon = 2\varphi + 2\chi, \\
2\chi = 2\eta + 2\varepsilon, \quad \text{or} \quad 2\eta = 2\chi - 2\varepsilon, \\
2\delta = 2\varphi + 2\eta.
\end{array}
\tag{M.3}
$$

In any ONP PP basis *H*, for example corresponding to linear bases, (*xy*) or (*yx*), these angular parameters are presented in Fig. M1. They determine the PP vectors represented by the TP phasors *P*, situated on the upper or lower part of the polarization sphere in the Stokes parameter coordinate system $Q_H U_H V_H$ corresponding to the ONP PP basis determined by its first TP phasor *H*. In the linear polarization bases these two TP phasors *P* represent elliptical polarizations of opposite handedness. Their positive $2\delta_H^P$ and $2\alpha_H^P$ angles denote the left-handed polarizations for the natural order, *xy*, of the basis vectors, or right-handed - for their reversed order, *yx*.

Angular parameters, as in Fig. M1, are related versus the right spherical triangles HLP. Any pair of parts of such a triangle, its sides or angles, plus one phase angle, completely determine the TP phasor *P*. Also, they determine all other parts of the triangle and other phase angle parameters.

Simple mnemonic Neper's rule for a spherical right triangle (see [133]) enables one to immediately present trigonometric relations between, expressed in degrees, sides and angles of such a triangle. That rule can be formulated as follows.

If three parts of a right triangle are situated side by side, then the cosine of the middle part is equal to the product of cotangents of the extreme parts; if however the parts are situated not side by side, then the cosine of the separately situated part is equal to the product of sinuses of parts situated side by side, with legs of the right triangle being exchanged for their complements with respect to 90°, and with the right angle not treated as a separate part, what means that legs should be treated as situated side by side.

Ten equations presented beneath follow that mnemonic rule (see the right spherical triangles in Fig. M1):

$$\begin{aligned}
\cos(90^0 - 2\alpha) &= \sin 2\alpha = \sin 2\gamma \sin 2\delta = \cot(90^0 - 2\beta)\cot(90^0 - 2\eta) \\
\cos(90^0 - 2\beta) &= \sin 2\beta = \sin 2\gamma \sin(90^0 - 2\eta) = \cot(90^0 - 2\alpha)\cot 2\delta \\
\cos 2\gamma &= \sin(90^0 - 2\alpha)\sin(90^0 - 2\beta) = \cot 2\delta \cot(90^0 - 2\eta) \\
\cos 2\delta &= \sin(90^0 - 2\alpha)\sin(90^0 - 2\eta) = \cot(90^0 - 2\beta)\cot 2\gamma \\
\cos(90^0 - 2\eta) &= \sin 2\eta = \sin 2\delta \sin(90^0 - 2\beta) = \cot 2\gamma \cot(90^0 - 2\alpha)
\end{aligned} \tag{M.4}$$

where, after (M.1),

$$2\eta = 2\chi - 2\varepsilon,$$

$$\sin(90^0 - 2\eta) = \cos(2\chi - 2\varepsilon),$$

$$\cot(90^0 - 2\eta) = \tan(2\chi - 2\varepsilon).$$

For comparison, see also the equalities (4.1), and the last equality (3.32).

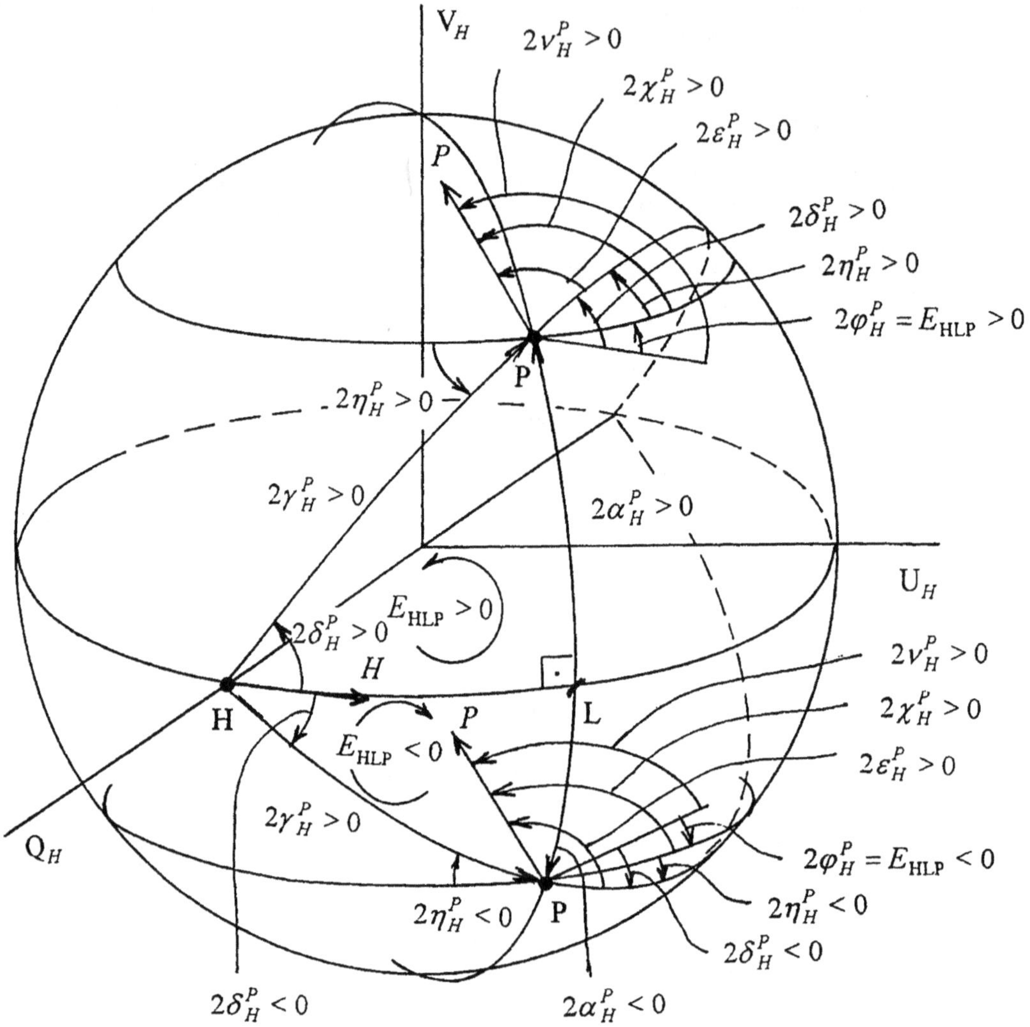

Fig.M1. Angular parameters of the TP phasors P in the ONP PP H basis

APPENDIX N

COMPARISON OF THE PP AND VARIOUS CA VECTOR APPROACHES BEING MET IN THE LITERATURE

A has been proposed by Mott in [118], three local spatial *xyz* coordinate systems, orthogonal and right-handed, will be introduced: *x1y1z1* for the transmitting antenna, *x2y2z2* for the scattering object at its 'output', and *x3y3z3* for the receiving antenna, with all three *z* axes considered as the propagation axes and directed: *z1* from the transmitting antenna to the scattering object, *z2* from the scattering object to the receiving antenna, and *z3* from the receiving antenna to the scattering object. Moreover, the *y2* and *y3* axes will be parallel, what means that *x1* and *x2* axes will be antiparallel.

Beneath, a comparison will be presented of notations used in cases of the complex amplitude (CA) and polarization and phase (PP) vector approaches.

N1. CA and PP vectors, E and E_0, and complex polarization ratios, P and ρ, in linear polarization bases of natural order

Incident (transmitted) electric vectors in the two approaches usually are being presented as follows,

CA: $$E^T(t,z) = E^T e^{j(\omega t - kz)}, \quad E^T = E^T_{x1}(1_{x1} + P^T 1_{y1}), \quad P^T = \frac{E^T_{y1}}{E^T_{x1}}, \tag{N.1}$$

PP: $$E^T(t,z) = E^T_0 e^{j(\omega t - kz)}, \quad E^T_0 = E^T_{0x}(1_x + \rho^T_{(x,y)} 1_y), \quad \rho^T_{(x,y)} = \frac{E^T_{0y}}{E^T_{0x}}, \tag{N.2}$$

PP vs CA: $$E^T_0 = E^T, \quad \rho^T_{(x,y)} = P^T. \tag{N.3}$$

Scattered electric vectors,

CA: $$E^S(t,z) = E^S e^{j(\omega t + kz)}, \quad E^S = E^S_{x2}(1_{x2} + P^S 1_{y2}) = E^S_{x3}(1_{x3} - P^S 1_{y3}), \tag{N.4}$$

with

$$P^S = \frac{E^S_{y2}}{E^S_{x2}} = -\frac{E^S_{y3}}{E^S_{x3}}, \tag{N.5}$$

PP: $$E^S(t,z) = E^S_0 * e^{j(\omega t + kz)}, \quad E^S * = E^S_{0x} * (1_x + \rho^S_{(x,y)} * 1_{y2}), \quad \rho^S_{(x,y)} = \frac{E^S_{0y}}{E^S_{0x}}, \tag{N.6}$$

PP vs CA: $$E^S_0 * = E^S, \quad \rho^S_{(x,y)} = -P^S *. \tag{N.7}$$

N2. Scattering (S1 alignment, or BSA, and linear polarization bases of natural order)

CA: $$E^S_{(x3,y3)} = \begin{bmatrix} E^S_{x3} \\ E^S_{y3} \end{bmatrix} = \begin{bmatrix} S_{x3x1} & S_{x3y1} \\ S_{y3x1} & S_{y3y1} \end{bmatrix} \begin{bmatrix} E^T_{x1} \\ E^T_{y1} \end{bmatrix} = S_{(x3,y3)} E^T_{(x1,y1)}, \tag{N.8}$$

PP: $$E^S_{0(x,y)} * = \begin{bmatrix} E^S_x * \\ E^S_y * \end{bmatrix} = \begin{bmatrix} A_2 & A_3 \\ A_4 & A_1 \end{bmatrix}_{(x,y)} \begin{bmatrix} E^T_{0x} \\ E^T_{0y} \end{bmatrix} = A_{(x,y)} E^T_{0(x,y)}, \tag{N.9}$$

PP vs CA: $$E^T_{0(x,y)} = E^T_{(x1,y1)}, \quad E^S_{0(x,y)} * = E^S_{(x3,y3)}, \quad A_{(x,y)} = S_{(x3,y3)}. \tag{N.10}$$

N3. Transmission in the S1 and P1 alignments (in the BSA and FSA)

CA:

$$V_r = \widetilde{h}^R_{(x3,y3)} E^S_{(x3,y3)} = \begin{bmatrix} h^R_{x3} & h^R_{y3} \end{bmatrix} \begin{bmatrix} E^S_{x3} \\ E^S_{y3} \end{bmatrix} = \begin{bmatrix} h^R_{x3} & h^R_{y3} \end{bmatrix} \begin{bmatrix} -1 & 0 \\ 0 & 1 \end{bmatrix} \begin{bmatrix} E^S_{x2} \\ E^S_{y2} \end{bmatrix}$$
$$= \widetilde{h}^R_{(x3,y3)} \begin{bmatrix} -1 & 0 \\ 0 & 1 \end{bmatrix} E^S_{(x2,y2)} = \widetilde{h}^R_{(x2,y2)} E^S_{(x2,y2)}, \tag{N.11}$$

PP:

$$V_r = \widetilde{h}^R_{0(x,y)} E^S_{0(x,y)}{}^* = \begin{bmatrix} h^R_{0x} & h^R_{0y} \end{bmatrix} \begin{bmatrix} E^S_{0x}{}^* \\ E^S_{0y}{}^* \end{bmatrix} = \begin{bmatrix} h^R_{0x} & h^R_{0y} \end{bmatrix} \begin{bmatrix} -1 & 0 \\ 0 & 1 \end{bmatrix} \begin{bmatrix} E^{So}_{0x} \\ E^{So}_{0y} \end{bmatrix}$$
$$= \widetilde{h}^R_{0(x,y)} C^0_{(x,y)} E^{So}_{0(x,y)} = \widetilde{h}^{Ro}_{0(x,y)}{}^* E^{So}_{0(x,y)}, \tag{N.12}$$

PP vs CA:

$$h^R_{0(x,y)} = h^R_{(x3,y3)}, \quad E^S_{0(x,y)}{}^* = E^S_{(x3,y3)},$$
$$E^{So}_{0(x,y)} = E^S_{(x2,y2)}, \quad h^{Ro}_{0(x,y)}{}^* = h^R_{(x2,y2)}. \tag{N.13}$$

Note regular product of the CA vectors and Hermitian product of the PP vectors.

N4. Propagation (P1 alignment, or FSA, and linear polarization bases of natural order)

CA: $$E^S_{(x2,y2)} = T_{(x2,y2)} E^T_{(x1,y1)} = \begin{bmatrix} -1 & 0 \\ 0 & 1 \end{bmatrix} E^S_{(x3,y3)} = \begin{bmatrix} -1 & 0 \\ 0 & 1 \end{bmatrix} S_{(x3,y3)} E^T_{(x1,y1)} \tag{N.14}$$

PP: $$E^{So}_{0(x,y)} = A^0_{(x,y)} E^T_{0(x,y)} = C^0_{(x,y)}{}^* E^S_{0(x,y)}{}^* = C^0_{(x,y)}{}^* A_{(x,y)} E^T_{0(x,y)} \tag{N.15}$$

PP vs CA:

$$A^0_{(x,y)} = T_{(x2,y2)}, \quad C^0_{(x,y)} = \begin{bmatrix} -1 & 0 \\ 0 & 1 \end{bmatrix},$$
$$A^0_{(x,y)} = C^0_{(x,y)}{}^* A_{(x,y)} \Leftrightarrow T_{(x2,y2)} = \begin{bmatrix} -1 & 0 \\ 0 & 1 \end{bmatrix} S_{(x3,y3)} \tag{N.16}$$

Conjugate value of the $C^0_{(x,y)}$ matrix in the above expression is justified because after change of basis that matrix becomes complex, though always remains symmetric. In any other ONP PP basis B one obtains (see (6.14)):

$$C^0_B = \widetilde{C}^B_{(x,y)} C^0_{(x,y)} C^B_{(x,y)}$$
$$C^0_B{}^* = (\widetilde{C}^B_{(x,y)} C^0_{(x,y)} C^B_{(x,y)})^* = C^{(x,y)}_B C^0_{(x,y)}{}^* C^B_{(x,y)}{}^* \tag{N.17}$$

what results in

$$C^0_{(B)} C^0_{(B)}{}^* = \begin{bmatrix} 1 & 0 \\ 0 & 1 \end{bmatrix} = C^0_{(x,y)} C^0_{(x,y)}{}^* \tag{N.18}$$

and confirms correctness of the transmission equation,

$$V_r = \widetilde{h}^{Ro}_{0(x,y)}{}^* E^{So}_{0(x,y)} = \widetilde{h}^R_{0(x,y)} C^0_{(x,y)} E^{So}_{0(x,y)} = \widetilde{h}^R_{0(x,y)} C^0_{(x,y)} A^0_{(x,y)} E^T_{0(x,y)}$$
$$= \widetilde{h}^R_{0(x,y)} C^0_{(x,y)} C^0_{(x,y)}{}^* A_{(x,y)} E^T_{0(x,y)} = \widetilde{h}^R_{0(x,y)} A_{(x,y)} E^T_{0(x,y)}. \tag{N.19}$$

N5. Stokes four-vectors and Kennaugh matrices

Utilizing expressions for the amplitude transmission equation, and the unitary matrix U,

$$\mathsf{U} = \frac{1}{\sqrt{2}} \begin{bmatrix} 1 & 1 & 0 & 0 \\ 0 & 0 & 1 & -j \\ 0 & 0 & 1 & j \\ 1 & -1 & 0 & 0 \end{bmatrix}, \tag{N.20}$$

one obtains the power (polarimetric) transmission equation for the S1 alignment in the (CA) approach,

CA:

$$\begin{aligned} V_r V_r{}^* &= V_r \otimes V_r{}^* = (\tilde{h}^R_{(x3,y3)} S_{(x3,y3)} E^T_{(x1,y1)}) \otimes (\tilde{h}^R_{(x3,y3)} S_{(x3,y3)} E^T_{(x1,y1)})^* \\ &= (\tilde{h}^R_{(x3,y3)} \otimes \tilde{h}^R_{(x3,y3)}{}^*) \mathsf{U}^* \tilde{\mathsf{U}} (S_{(x3,y3)} \otimes S_{(x3,y3)}{}^*) \mathsf{U} \tilde{\mathsf{U}}^* (E^T_{(x1,y1)} \otimes E^T_{(x1,y1)}{}^*) \\ &= \tilde{\mathsf{G}}^R_{(x3,y3)} \mathsf{K}_{(x3,y3)} \mathsf{G}^T_{(x1,y1)}. \end{aligned} \tag{N.21}$$

When using similar procedure for the (PP) approach, simpler form of the power transmission equation results,

PP:

$$\begin{aligned} V_r V_r{}^* &= V_r \otimes V_r{}^* = (\tilde{h}^R_{0(x,y)} A_{(x,y)} E^T_{0(x,y)}) \otimes (\tilde{h}^R_{0(x,y)} A_{(x,y)} E^T_{0(x,y)})^* \\ &= (\tilde{h}^R_{0(x,y)} \otimes \tilde{h}^R_{0(x,y)}{}^*) \mathsf{U}^* \tilde{\mathsf{U}} (A_{(x,y)} \otimes A_{(x,y)}{}^*) \mathsf{U} \tilde{\mathsf{U}}^* (E^T_{0(x,y)} \otimes E^T_{0(x,y)}{}^*) \\ &= \tilde{\mathsf{G}}^R_{(x,y)} \mathsf{K}_{(x,y)} \mathsf{G}^T_{(x,y)} = \tilde{\mathbf{I}}^R_{\text{eff}(x,y)} \mathsf{K}_{(x,y)} \mathbf{I}^T_{\text{eff}(x,y)}. \end{aligned} \tag{N.22}$$

As is seen from the above expressions, the Stokes four-vectors, G, and Kennaugh matrices, K, are identical for the two approaches. One can prove that these Stokes four-vectors can be expressed also in terms of complex polarization ratios,

$$\rho^T_{(x,y)} = P^T, \quad \text{and} \quad \rho^R_{(x,y)} = P^R \tag{N.23}$$

e.g. (for incident wave) as follows,

$$\begin{aligned} \mathsf{G}^T_{(x,y)} &= \tilde{\mathsf{U}}^* (E^T_{(x1,y1)} \otimes E^T_{(x1,y1)}{}^*) = \tilde{\mathsf{U}}^* (E^T_{0(x,y)} \otimes E^T_{0(x,y)}{}^*) \\ &= \frac{1}{\sqrt{2}} \cdot \frac{|E^T_x|^2 + |E^T_y|^2}{1 + (\rho\rho^*)^T_{(x,y)}} \begin{bmatrix} 1 + \rho\rho^* \\ 1 - \rho\rho^* \\ \rho^* + \rho \\ j(\rho^* - \rho) \end{bmatrix}^T_{(x,y)}, \end{aligned} \tag{N.24}$$

with

$$\begin{bmatrix} E^T_x \\ E^T_y \end{bmatrix} = \begin{bmatrix} E^T_{x1} \\ E^T_{y1} \end{bmatrix} = \begin{bmatrix} E^T_{0x} \\ E^T_{0y} \end{bmatrix}; \quad \rho^T_{(x,y)} = \frac{E^T_y}{E^T_x} = \tan\gamma^T_{(x,y)} \exp(j2\delta^T_{(x,y)}). \tag{N.25}$$

The problem however arises if somebody wants to apply that formula to calculate the Stokes four-vector for the scattered wave taking $\rho^S_{(x,y)} = P^S$ instead of the correct dependence (N.7),

$$\rho^S_{(x,y)} = -P^S{}^*. \tag{N.26}$$

That leads to erroneous results what happens to less experienced researchers.

N6. Comparison of different forms of Stokes four-vectors and Kennaugh matrices being met in the literature

When using the PP vector presentation of the polarimetric power transmission equation, it is a good occasion for discussing its different possible forms, all correct, but the one only being accepted by most of researchers for practical applications. The four possibilities of such (normalized) equations will be considered which in the PP vector approaches take the following forms:

$$P_r = V_r \otimes V_r{}^*$$

$$= (\tilde{u}^R_{(x,y)} \otimes \tilde{u}^R_{(x,y)}{}^*)\mathsf{U}^*\tilde{\mathsf{U}}(A_{(x,y)} \otimes A_{(x,y)}{}^*)\mathsf{U}\tilde{\mathsf{U}}^*(u^T_{(x,y)} \otimes u^T_{(x,y)}{}^*) = \tilde{\mathbf{P}}^R_{(x,y)}\mathsf{K}_{(x,y)}\mathbf{P}^T_{(x,y)} \tag{N.27a}$$

$$= (\tilde{u}^R_{(x,y)} \otimes \tilde{u}^R_{(x,y)}{}^*)\mathsf{U}\tilde{\mathsf{U}}^*(A_{(x,y)} \otimes A_{(x,y)}{}^*)\mathsf{U}^*\tilde{\mathsf{U}}(u^T_{(x,y)} \otimes u^T_{(x,y)}{}^*) \tag{N.27b}$$

$$= (\tilde{u}^R_{(x,y)} \otimes \tilde{u}^R_{(x,y)}{}^*)\mathsf{U}\tilde{\mathsf{U}}^*(A_{(x,y)} \otimes A_{(x,y)}{}^*)\mathsf{U}\tilde{\mathsf{U}}^*(u^T_{(x,y)} \otimes u^T_{(x,y)}{}^*) \tag{N.27c}$$

$$= (\tilde{u}^R_{(x,y)} \otimes \tilde{u}^R_{(x,y)}{}^*)\mathsf{U}^*(\tilde{\mathsf{U}}\mathsf{U})\tilde{\mathsf{U}}^*(A_{(x,y)} \otimes A_{(x,y)}{}^*)\mathsf{U}\tilde{\mathsf{U}}^*(u^T_{(x,y)} \otimes u^T_{(x,y)}{}^*) \tag{N.27d}$$

The first version, (N.27a), is being the most commonly used and recommended for applications. It has been used by this author since 1967 [39], following the Kennaugh's concept [95], but with introduction of the reversed, (y,x), order of the ONP PP basis what results in the right-circular polarization point situated at the upper pole of the Poincare sphere.

The next version, (N.27b), originally proposed by Kennaugh [95], places the right-circular polarization at the upper pole of the Poincare sphere when using the ONP PP basis of natural order, (x,y), with $2\delta^{RC}_{(x,y)} = -90^0$. The corresponding Stokes scattering matrix differs from the now being applied radar Stokes scattering matrix (and now called the 'Kennaugh matrix', K) by opposite signs of elements of its fourth row and column, except of their common last element.

The third version, (N.27c), applies different 'receiving' Stokes four-vectors, represented on the Poincare sphere by points which are of opposite handedness in comparison with four-vectors of incident waves of the same polarization (compare, e.g., [14], neglecting the 'modified form' of the Stokes reflection matrix). Their Kennaugh matrices have opposite signs of elements of the fourth row and analytically are formed from the Sinclair matrices the same way like the Mueller matrices from the Jones matrices.

The last version, (N.27d), applies the Stokes four-vectors of the first version but the Kennaugh matrix of the third version. Therefore, it requires insertion of an additional transformation matrix, $\tilde{\mathsf{U}}\mathsf{U}$, to the transmission equation (see, e.g., [100]).

N7. Comparison of two simple forms of bistatic scattering matrices known as real diagonal and 'canonical' (in the characteristic polarization basis)

By applying special orthogonal polarization bases there appear possibilities of obtaining amplitude bistatic scattering matrices in particularly simple forms.

One of such possibilities leads to real diagonal form, Σ, of the Sinclair matrix. Any Sinclair matrix S can be expressed by the Σ matrix (see Lueneburg and Cloude [111], Section 5),

$$S = V^*\Sigma\tilde{W}^*; \qquad \Sigma = \begin{bmatrix} \lambda_1 & 0 \\ 0 & \lambda_2 \end{bmatrix}, \quad \lambda_1 \geq \lambda_2 \geq 0,, \tag{N.28}$$

when using two 2x2 matrices, V and W, both unitary though not necessarily unimodular. Each of them depends on three real parameters. With two real parameters of the Σ matrix, altogether eight real parameters are in use determining the nonsymmetrical S matrix. That Σ matrix denotes the S matrix transformed to two different orthogonal bases on its input and output, not necessarily the ONP bases.

Another simple 'canonical' form, A_K , takes on the Sinclair matrix expressed in the so-called characteristic ONP PP basis K corresponding to the so-called characteristic coordinate system (CCS) in the Stokes parameter space. In any other ONP PP basis B such a matrix, A_B , can be written in terms of A_K as follows,

$$A_B = \widetilde{C}_K^B A_K C_K^B; \quad A_K = \begin{bmatrix} A_2 & A_3 \\ -A_3 & A_1 \end{bmatrix}_K = \begin{bmatrix} A_2 & B_1 + jB_2 \\ -B_1 - jB_2 & A_1 \end{bmatrix}_{CCS} e^{j\mu}, \tag{N.29}$$

with unitary unimodular 2x2 matrices, C_K^B, dependent on three real parameters, and with five real parameters (altogether eight) in the A_K matrix: $A_2 \geq A_1 \geq 0,\ B_2 \geq 0\ (\text{if}\ B_1 \neq 0),\ B_1$, and μ.

It may be interesting to compare scattering equations with matrices in these two notations, for $S = A_K$. Denoting: $W = [w_1 \quad w_2],\ V = [v_1 \quad v_2]$, with column vectors $w_{1,2} = u_K^{M,N},\quad v_{1,2} = u_K^{M',N'}$, and eigenvalues of power matrices $\sigma_{1,2}^2 = (\lambda^{M,N})^2$, the sets of power and amplitude scattering equations corresponding to each other in the two notations are:

$$\begin{aligned}
\widetilde{S} * S w_{1,2} &= \sigma_{1,2}^2 w_{1,2} &&\Leftrightarrow&& \widetilde{A}_K * A_K u_K^{M,N} = (\lambda^{M,N})^2 u_K^{M,N} \\
S w_{1,2} &= \sigma_{1,2} v_{1,2} * &&\Leftrightarrow&& A_K u_K^{M,N} = \lambda^{M,N} u_K^{M',N'} * \\
S * \widetilde{S} v_{1,2} &= \sigma_{1,2}^2 v_{1,2} &&\Leftrightarrow&& A_K * \widetilde{A}_K u_K^{M',N'} = (\lambda^{M,N})^2 u_K^{M',N'} \\
\widetilde{S} v_{1,2} &= \sigma_{1,2} w_{1,2} * &&\Leftrightarrow&& \widetilde{A}_K u_K^{M',N'} = \lambda^{M,N} u_K^{M,N} *
\end{aligned} \tag{N.30}$$

Of course, the form of equations on the right side will not change if the K basis will be exchanged for any other ONP B basis. However, in case of the K basis we may show a simple geometrical explanation of those equations using the Poincare sphere models of those scattering matrices. For example, consider the transmission equation in the K basis when the receiving antenna is polarimetrically matched to the scattered wave,

$$V_r = \widetilde{u}_K^{M'} A_K u_K^M = \lambda^M \widetilde{u}_K^{M'} u_K^{M'} * = \lambda^M. \tag{N.31}$$

It should be stressed that local characteristic ONP bases, for transmitter and receiver, denoted by the same symbol K, have been determined for elements of scattering matrix corresponding to local spatial bases of arbitrarily chosen rotations about z axes (directed to the scatterer for the BSA or antenna alignment). Keeping that in mind we see that the same received voltage for transmission in opposite direction can be presented by the transposed equation,

$$V_r = \widetilde{V}_r = \widetilde{u}_K^M \widetilde{A}_K u_K^{M'} = \lambda^M \widetilde{u}_K^M u_K^M * = \lambda^M \tag{N.32}$$

That equation, without changing its form, can be rewritten in any ONP basis. Let us take the first vector of a new ONP basis, K'', tangent to the polarization sphere at the same K point on the Q_K axis but rotated about that axis by $+180^0$. The PP vectors of the new basis can be written as:

$$u^{K''} = -ju^K = \begin{bmatrix} u^K & u^{Kx} \end{bmatrix} \begin{bmatrix} -j \\ 0 \end{bmatrix} \tag{N.33a}$$

$$u^{K''x} = ju^{Kx} = \begin{bmatrix} u^K & u^{Kx} \end{bmatrix} \begin{bmatrix} 0 \\ j \end{bmatrix} \tag{N.33b}$$

what determines the change of basis matrix,

$$C_K^{K''} = \begin{bmatrix} -j & 0 \\ 0 & j \end{bmatrix},$$

and the Sinclair matrix in the new basis,

$$\begin{aligned}
A_{K''} &= \widetilde{C}_K^{K''} A_K C_K^{K''} = -\widetilde{A}_K \\
&= -\begin{bmatrix} A_2 & -A_3 \\ A_3 & A_1 \end{bmatrix}_K = \begin{bmatrix} A_2 & -B_1 - jB_2 \\ B_1 + jB_2 & A_1 \end{bmatrix}_{CCS} e^{j(\mu \pm \pi)}.
\end{aligned} \tag{N.34}$$

That means rotation of the matrix polarization sphere model by -180^0 in the opposite direction versus the basis rotation (with the change of the phase μ by π, see also (8.24) and (8.25)).

For $2\delta_K^M \leq 0$ and the PP vector u^M parallel to the eigencircle plane (see Fig.J.1) we have

$$2\varepsilon_K^M = 2\delta_K^P - (2\delta_K^M + 180^0) \quad \Rightarrow \quad \begin{cases} -(\delta_K^M + \varepsilon_K^M) = 90^0 - \delta_K^P, \\ \delta_K^M - \varepsilon_K^M = +90^0 - \delta_K^P + 2\delta_K^M \end{cases} \tag{N.35}$$

where from

$$u_K^M = \begin{bmatrix} \cos\gamma e^{-j(\delta+\varepsilon)} \\ \sin\gamma e^{j(\delta-\varepsilon)} \end{bmatrix}_K^M = j\exp(-j\delta_K^P)\begin{bmatrix} \cos\gamma \\ \sin\gamma e^{j2\delta} \end{bmatrix}_K^M . \tag{N.36}$$

The scattered wave PP vector in the same basis can be found when considering inversion, rotation and change of phase. Angular parameters of its phasor are,

$$\left.\begin{aligned} 2\gamma_K^{M^*} &= 2\gamma_K^M \\ 2\delta_K^{M^*} &= 2\delta_K^M + 180^0 \\ 2\varepsilon_K^{M^*} &= 2\varepsilon_K^M + 180^0 = 2\delta_K^P - 2\delta_K^M \end{aligned}\right\} \Rightarrow \begin{cases} -(\delta_K^{M^*} + \varepsilon_K^{M^*}) = -90^0 - \delta_K^P \\ \delta_K^{M^*} - \varepsilon_K^{M^*} = 90^0 + 2\delta_K^M - \delta_K^P \end{cases} \tag{N.37}$$

The corresponding column vector, when taking into account the possible change of phase, ξ (see (8.24)), is

$$\begin{aligned} u_K^{M^*} &= j\exp(-j\xi)\begin{bmatrix} -1 & 0 \\ 0 & 1 \end{bmatrix} \\ &= C_{M,K}^{M^*} u_K^M, \quad \text{with} \quad C_{M,K}^{M^*} = \exp(-j\xi)\begin{bmatrix} -1 & 0 \\ 0 & 1 \end{bmatrix} \end{aligned} \tag{N.38}$$

After change of basis

$$\begin{aligned} u_{K^*}^{M^*} &= C_{K^*}^K u_K^{M^*} \\ &= \begin{bmatrix} j & 0 \\ 0 & -j \end{bmatrix} \exp(-j\xi)\begin{bmatrix} -1 & 0 \\ 0 & 1 \end{bmatrix} u_K^M \\ &= -j\exp(-j\xi)u_K^M \end{aligned} \tag{N.39}$$

On the other hand,

$$\begin{aligned} u_{K^*}^M &= C_{K^*}^K u_K^M \\ &= C_{K^*}^K \exp(+j\xi)\begin{bmatrix} 1 & 0 \\ 0 & -1 \end{bmatrix} u_K^{M^*} \\ &= j\exp(+j\xi)u_K^{M^*} \end{aligned} \tag{N.40}$$

Taking inverse formulae for the two last results:

$$u_K^M = j\exp(+j\xi)u_{K^*}^{M^*} \tag{N.41}$$

$$u_K^{M^*} = -j\exp(-j\xi)u_{K^*}^M \tag{N.42}$$

and substituting them to the last equation of the set (N.30):

$$\widetilde{A}_K u_K^{M^*} = \lambda^M u_K^M * \tag{N.43}$$

one obtains another equation of the same set (its form does not depend on the ONP PP basis)

$$A_{K^*} u_{K^*}^M = \lambda^M u_{K^*}^{M^*} *. \tag{N.44}$$

The results obtained confirm correctness of the applied procedure of the polarization phasor's transformation when bistatic scattering by its inversion, rotation, and change of phase. They also show full agreement of that procedure with predictions of the existing literature.

APPENDIX O

COMMENTS ABOUT RELATIONS TO THE EXISTING WORKS

O.1. About representation of the complete polarization

To define the complete polarization by the 'polarization ellipse' is a right way but rather in optics only. For radar purposes it is completely inadequate because then we have to deal with waves propagating in two opposite directions, and in such circumstances two different ellipses define the same polarization. Looking from one side at those ellipses (corresponding to outgoing and incoming waves, identically polarized) we observe their same tilt angle but opposite handedness.

It would be an ill-advised suggestion to convince researchers or students that two different ellipses should define, or represent, the exactly the same polarization (for the oppositely propagating waves). That is especially true when there exists another way to define polarization which omits the above mentioned difficulty.

For years it is well known that, in radar applications, the only reasonable way to represent the complete polarization is through the polarization helix, being shifted along the propagation axis in two opposite directions, instead of through the polarization ellipses. The handedness of the polarization helix does not depend on the direction of propagation. Unfortunately, the right-handed polarization helix corresponds to the left-handed polarization. From that point of view the 'optical' definition of the polarization handedness seems to be more reasonable than the IEEE definition, but that is a separate problem. The 'spatial phase' of the polarization helix (and of the wave), for $t = 0$, can be uniquely determined by its shift along the propagation axis: we may speak about the spatial phase advance (lead) or retardation (delay) of the wave if its helix is shifted in positive or negative direction of the propagation axis. That shift denotes also the temporal phase advance for waves propagating in positive direction of the propagation axis, but temporal phase delay for oppositely propagating waves. An essential conclusion is that, in order to properly define the spatial phase of the wave, we note its shift along positive direction of the propagation axis independently of the direction of wave propagation (!).That way we arrive at the necessity to define logically, and of course most simply, the local spatial coordinate systems to which the polarization and spatial phase (PP) vectors of completely polarized waves will be uniquely related.

O.2. About the local spatial coordinate systems (SCS) and their scattering alignments

Evidently, each local SCS can and should be defined most simply as an orthogonal and right-handed coordinate system, the *xyz* for example, with *z* as the propagation axis (instead of the wave number vector ***k***, indicating the direction of propagation when defining the propagation axis).

Some authors are using the names of an 'antenna' or 'wave' local SCS. These terms seem to be rather unfortunate ones, and not only because they bound the propagation axes with directions of propagation. For instance, you may want to choose the *z* propagation axis directed from the target to the transmit antenna. Such an *xyz* local SCS will be neither 'antenna' nor 'wave' coordinate system. Of course, you may call them a 'reversed antenna' or 'reversed wave' SCS, but it would sound strangely. Much simpler will be to call them the coordinate system 'with the *z* axis from the target' (at its side under consideration).

The same can be said about the commonly used terms for the two scattering alignments: 'backward' ('BSA') or 'forward' ('FSA'). In both cases we assume the same *z* axis, for the illuminating wave, directed 'to the target' (though the opposite direction could be also applied). Therefore, the following terms have been proposed here for two 'scattering' and two 'propagation' alignments:

- S1 or S2 scattering alignments for the *z* axes directed to or from the target, and
- P1 or P2 propagation alignments: for the *z* axis toward the target for illumination and from the target for scattering, or - both directions opposite ones.

So, according to the above proposal, all four combinations are taken into account, with the S1 alignment corresponding to the BSA, and P1 - to the FSA. The spherical coordinate system centered at the target would be an example for a reasonable application of the S2 alignment. Any change of the alignment can be performed easily when using the here presented passive transformation of the *z* axis reversal by adequate rotation of the local SCS.

O.3. About polarization and spatial phase (PP) vectors

Other authors use the Jones vectors (in optics) or the directive Jones vectors (in radar) to present polarizations and temporal phases of waves (their complex amplitudes, CA's) or complex heights of antennas radiating and optimally receiving those waves. All scattering, propagation, and passive transformation matrices operate on those CA vectors, forming also their power (Stokes', e.g.) counterparts. The CA vectors represent polarization ellipses and their temporal phases, and as such differently describe same polarizations and phases for oppositely propagating waves. The directive Jones vectors undergo different change of basis transformations and other passive and active transformations (also the Jones to Stokes' vector transformations) because their 'directivity' reflects the dependence of their transformation rules on the direction of propagation along the established propagation axis. Components of those Jones vectors correspond to the local SCS's. The polarimetric transmission equations are based on simple products of those complex Jones vectors satisfying however the condition of opposite directivity of the two vectors being multiplied, and another condition of the same local SCS in which their components must be expressed. A special care is needed when using the elliptical polarization bases which should be of the same directivity as the Jones vectors themselves.

No doubt that the most advisable simplification of such an approach has been done here by introduction of the PP vectors in place of the CA's in the form of the directive Jones vectors. The CA vectors have been expressed by the PP vectors which are independent of direction of propagation and directly correspond to the polarization helices with their spatial phases. In the local SCS *xyz*, the CA's just equal the PP vectors for waves traveling in the positive *z* direction or are equal to their conjugate values for waves traveling in the opposite direction. Such a concept, based on the time symmetry of Maxwell equations, has been proposed as an extension of the Kennaugh's pseudo-eigenvector scattering equation in the S1 alignment: from monostatic scattering and eigenpolarizations to bistatic scattering in each one of the four possible alignments and all possible incident PP vectors. Components of the PP vectors are expressed also in the PP bases, also independent on the direction of propagation, and the polarimetric transmission equations are based on the Hermitian products of the PP vectors expressed in the same basis of both vectors and related to the same local SCS.

O.4. About the Poincare sphere representation of the PP vectors by the tangential polarization (TP) phasors

The PP vectors have been represented on the Poincare polarization sphere by the TP phasors. In the earlier existing works the tangential phasors have been also proposed but they differ by the orientation angle, proportional to the temporal phase angle, while in this approach their orientation angles depend on the double spatial phase angles (compare, e.g., first [138], then [29]). Owing to that fact the addition of phasors representing waves propagating in the same direction but of different polarizations and spatial phases is possible. Phasors are uniquely determined for the time $t = 0$. In time, phasors corresponding to the oppositely propagating waves rotate in opposite directions.

O.5. About the polarization bases

In the existing literature the orthogonal polarization bases are being commonly applied but usually they are limited to the linear and circular bases of null-phase vectors. In this text the so-called characteristic bases are of special interest in which scattering matrices obtain a very simple, canonical form. Also labeling of bases is different. Instead of indicating both orthogonal basis vectors, only the first basis vector is being presented owing to the fact of existing the rule for uniquely defining the second one (in case of the so-called orthogonal null-phase polarization basis - the 'ONP PP basis'). Usually the lower index of the PP vectors and scattering matrices presents a symbol of the TP phasor corresponding to the first basis vector.

In the existing literature the polarization bases are very often determined with an insufficient precision. For example, the (HV) linear basis of horizontal and vertical polarizations may correspond to two different orders of vector components: natural order, (*xy*), or reversed order, (*yx*) , in the right-handed *xyz* coordinate system. Establishing of that order is essential because it determines angular coordinates of the polarization sphere points. Also the phase differences between vector components are commonly defined in such a way that the first component is delayed versus the second one for positive arguments of the polarization ratios (ratios of the second versus the first component). These positive arguments correspond to the upper part of the Poincare sphere. Therefore that upper part, above the equator of linear polarizations, presents left-handed polarizations if natural order of components is being applied. In this text the order of basis vectors is always precisely determined by relation to the local right-handed *xyz* SCS. Usually it is the reversed order corresponding to the right-circular polarizations at the 'north' pole of the Poincare sphere. Then the first basis phasor H, of horizontal linear polarization, is being identified with the *y*-component of the PP vector. Elliptical bases, as rotated versions of the original linear bases, preserve their order.

The ONP PP bases are also called the collinear phasor bases and are most often used as linear or characteristic bases. The circular bases are usually of another type. They are called the parallel phasor bases. Their order can be deduced by inspection of their transformation from the original ONP bases.

O.6. About new problems here considered and their relations to the existing concepts

Consideration of some new problems was possible mainly owing to

- the original concept of the PP vectors (independent of direction of propagation) and their especielly convenient labeling with two indices, upper and lower , indicating the TP phasors representing the PP vector itself and its ONP PP basis, respectively; that concept was based on known property of the time-symmetry of Maxwell equations,
- simple forms, with such a labeling, of scattering and transmission equations, and of the polarization ONP PP basis transformation by its rotation, leading to the original canonical expressions for scattering matrices in the characteristic coordinate systems (CCS),
- introduction of the passive transformation of the propagation z-axis reversal by appropriate rotation of the local SCS, especially useful for changing the scattering matrices alignment.
- using the inversion point concept introduced by Kennaugh for monostatic scattering and here extended to the case of bistatic scattering,

All that resulted in such important issues as:

- formation of the Poincare sphere geometrical models for bistatic scattering matrices,
- determination of mutual locations of special polarization points on the polarization sphere for bistatic scattering,
- developing geometrical constructions leading to designation of those points,
- extension of the Copeland's [36] classification of monostatic scattering targets (linear, isotropic, general) to the case of bistatic scattering by location of the inversion point on its boundary surfaces in the CCS,
- developing the theory of five-parameter lossless polarimetric two-ports,
- most simple decomposition of the partially depolarizing 16-parameter Kennaugh bistatic scattering matrix into four non-depolarizing matrices depending on 7, 5, 3, and 1 parameters,
- the polarization four-sphere concept and its use for cancellation of the partially depolarized clutter.

The approach here presented can also be applied to compare seemingly contradictory results of different authors by analyzing the admitted assumptions about local spatial coordinate systems, polarization bases (also their phases and order), the way of defining Stokes' four-vectors and matrices, and scattering or propagation alignments used.

Apart from the new concepts applied and results obtained this text differs from other existing works by precise definition of bases for all vectors and matrices involved and by simplified notation which drops all coefficients and indices not essential for correct polarimetric presentation of scattering and transmission equations.

APPENDIX P

COMPARISON OF THE HERE APPLIED NOTATION WITH THAT OF INTERNATIONALLY ESTABLISHED NOMENCLATURE ON MATHEMATICAL FORMULATIONS

P1. TP phasors and their use as necessary extension of the internationally accepted nomenclature

In order to understand this Appendix an elementary knowledge about the Poincare sphere is necessary.

Here admitted notation differs from the established international nomenclature on mathematical formulations in one point only. It uses the tangential polarization (TP) phasors to determine not only polarization but also phase of the elliptically polarized waves. Introduction of that new notion leads, and is necessary, to assure precise description of the Poincare sphere transformations. As a side effect, it causes an essential simplification of the form of many mathematical formulae presented in this monograph.

P2. The way of introduction of the TP phasors to the established international nomenclature by the upper and lower indices for Jones vectors and their parameters

In a standard manner, polarization is being determined on the Poincare sphere by a point, say, P at which a TP phasor, denoted as *P*, is tangent to the sphere. Many such phasors can be tangent to the sphere at the same point, precisely speaking - their 'continuum' number. Each one will differ in orientation versus some other phasor, *H*, considered as a 'basis' phasor, tangent to the Poincare sphere in another point, H, by an angle $2\nu_H^P$ being the 'double phase angle', as shown in Fig. M1 (Appendix M, p. 157).

Altogether, three real angular parameters determine the TP phasor (see Fig M1):

- $2\gamma_H^P$, expressing angular distance between points H and P; $0 \le 2\gamma_H^P \le 180^0$,
- $2\delta_H^P$, denoting direction to the point P versus orientation of the basis phasor *H*; $-180^0 < 2\delta_H^P \le +180^0$,
- $2\nu_H^P$, meaning orientation of the *P* phasor versus the phasor *H* shifted along the HP arc parallel, thus preserving its null-phase (or 'basis') orientation; $-360^0 \le 2\nu_H^P \le +360^0$.

The first two angles determine the wave's polarization (a point on the Poincare sphere).

It should be observed that the third, double phase angle, may change in the range 4π. It means that two phasors tangent to the Poincare at the same point, and of the same geometrical orientation, may differ in phase by 180^0. Therefore, it is strictly required always to indicate the way (direction) by which the phasor has been rotated from its null-phase orientation.

Column vector of complex amplitude of the elliptically polarized wave, when using the $e^{+j(\omega t-kz)}$ time/space convention for waves propagating in the +z direction of a local right-handed coordinate system xyz, can be expressed in the established international nomenclature enriched with the 'TP phasor notation' as

$$E_{(x,y)} = \begin{bmatrix} E_x \\ E_y \end{bmatrix} \equiv E_0 u_X^P = \left\{ \frac{1}{\sqrt{1+\rho\rho^*}} \begin{bmatrix} 1 \\ \rho \end{bmatrix} e^{-j\nu} \right\}_X^P \qquad \text{(P.1)}$$

with magnitude of the complex amplitude,

$$E_0 = \sqrt{|E_x|^2 + |E_y|^2} \tag{P.2}$$

with the complex polarization ratio,

$$\rho_X^P = \frac{E_y}{E_x} = \left\{\tan\gamma e^{j2\delta}\right\}_X^P \equiv \tan\gamma_X^P \ \exp(j2\delta_X^P) \tag{P.3}$$

and with the phase delay of the first component,

$$\nu_x = \nu_X^P. \tag{P.4}$$

Observe, please, that introduction of the polarization phasors is necessary. Only enriched with the TP phasor notation the established standard nomenclature allows one to precisely indicate both the wave's polarization and phase (its γ, δ, and ν parameters, determined by the P phasor), and the polarization and phase base of the wave's electric column vector (in this example the (x,y) base, determined by the X phasor).

P3. The use of the indexed vectors and column vectors. Two orders of the polarization bases

As has been shown, the natural formal consequence of introduction of the new notion, the TP phasor, is the use of two indices, upper and lower, determining wave's polarization and phase by one phasor (in the upper index), and amplitude column vector's basis by another phasor (in the lower index).

Standard representation of the complex amplitude of the electric vector, when using such an enriched standard nomenclature, takes the form

$$\begin{aligned} \boldsymbol{E} &= E_x(\boldsymbol{1}_x + \rho_X^P \boldsymbol{1}_y) \\ &= E_0\begin{bmatrix}\boldsymbol{1}_x & \boldsymbol{1}_y\end{bmatrix} u_X^P \\ &= E_0 \boldsymbol{u}^P. \end{aligned} \tag{P.5}$$

Here a difference should be shown between the notions of *coordinates*, x or y, and TP *phasors*, X or Xx, the last orthogonal versus X, by presentation of the first and second basis vector in a similar manner like the $\boldsymbol{E}$ vector has been presented above,

$$\begin{aligned} \boldsymbol{1}_x &= \begin{bmatrix}\boldsymbol{1}_x & \boldsymbol{1}_y\end{bmatrix}\begin{bmatrix}1\\0\end{bmatrix} \\ &= \begin{bmatrix}\boldsymbol{1}_x & \boldsymbol{1}_y\end{bmatrix} u_X^X \\ &= \boldsymbol{u}^X, \end{aligned} \qquad \text{and} \qquad \begin{aligned} \boldsymbol{1}_y &= \begin{bmatrix}\boldsymbol{1}_x & \boldsymbol{1}_y\end{bmatrix}\begin{bmatrix}0\\1\end{bmatrix} \\ &= \begin{bmatrix}\boldsymbol{1}_x & \boldsymbol{1}_y\end{bmatrix} u_X^{Xx} \\ &= \boldsymbol{u}^{Xx}. \end{aligned} \tag{P.6}$$

Immediately it should be explained, why not to use 'Y' instead of 'Xx'.

Observe please, that because the orientation angle of any phasor changes in the range 4π, the phase of the Y phasor may be undetermined without additional instruction indicating the way by which it has been shifted parallel from phasor's X position to its antipodal point Y. It will be assumed that the Xx phasor is being obtained by shifting the X phasor in direction indicated by its arrow. By shifting in opposite direction it will become of opposite phase, and will correspond to $-\boldsymbol{u}^{Xx}$ vector. So, with such assumption, the Xx phasor becomes uniquely determined, contrary to the 'Y' phasor with the 180^0 phase ambiguity.

It doesn't mean that we cannot use the Y phasor at all. Of course, we may apply the (y,x) linear basis, called the basis of the 'reversed order', *always* using the *right-handed* xyz coordinate system. In such a case we can write,

$$E_{(y,x)} = \begin{bmatrix}E_y\\E_x\end{bmatrix} \equiv E_0 u_Y^P = \left\{\frac{1}{\sqrt{1+\rho\rho^*}}\begin{bmatrix}1\\\rho\end{bmatrix} e^{-j\nu}\right\}_Y^P \tag{P.7}$$

with

$$\rho_Y^P = \frac{E_x}{E_y} = \left\{\tan\gamma e^{j2\delta}\right\}_Y^P \equiv \tan\gamma_Y^P \ \exp(j2\delta_Y^P), \qquad \nu_y = \nu_Y^P \tag{P.8}$$

and

$$\begin{aligned} \boldsymbol{E} &= E_y(\boldsymbol{1}_y + \rho_Y^P \boldsymbol{1}_x) \\ &= E_0\begin{bmatrix}\boldsymbol{1}_y & \boldsymbol{1}_x\end{bmatrix} u_Y^P \\ &= E_0 \boldsymbol{u}^P \end{aligned} \tag{P.9}$$

with

$$\begin{aligned} \boldsymbol{1}_y &= \begin{bmatrix}\boldsymbol{1}_y & \boldsymbol{1}_x\end{bmatrix}\begin{bmatrix}1\\0\end{bmatrix} \\ &= \begin{bmatrix}\boldsymbol{1}_y & \boldsymbol{1}_x\end{bmatrix} u_Y^Y \\ &= \boldsymbol{u}^Y \end{aligned} \qquad \text{and} \qquad \begin{aligned} \boldsymbol{1}_x &= \begin{bmatrix}\boldsymbol{1}_y & \boldsymbol{1}_x\end{bmatrix}\begin{bmatrix}0\\1\end{bmatrix} \\ &= \begin{bmatrix}\boldsymbol{1}_y & \boldsymbol{1}_x\end{bmatrix} u_Y^{Yx} \\ &= \boldsymbol{u}^{Yx} \end{aligned} \tag{P.10}$$

again, with the Yx second basis phasor, exactly determined by its first phasor Y.

P4. Unit Jones vectors $\boldsymbol{u}^{Xx}$ and $-\boldsymbol{u}^{Xx}$ expressed in terms of internationally established parameters

It will be instructive to present both vectors, $\boldsymbol{u}^{Xx}$ and $-\boldsymbol{u}^{Xx}$, by their column matrices employing ρ (or γ and δ) and ν parameters. We can write, applying the X = (x,y) polarization basis, for $\gamma_X^{Xx} = 90^0$, and $\delta_X^{Xx} = \nu_X^{Xx} = 0^0$,

$$\begin{aligned} \boldsymbol{u}^{Xx} &= \begin{bmatrix}\boldsymbol{1}_x & \boldsymbol{1}_y\end{bmatrix} u_X^{Xx} = \begin{bmatrix}\boldsymbol{1}_x & \boldsymbol{1}_y\end{bmatrix}\begin{bmatrix}0\\1\end{bmatrix} \\ &= \begin{bmatrix}\boldsymbol{1}_x & \boldsymbol{1}_y\end{bmatrix}\left\{\frac{1}{\sqrt{1+\rho\rho^*}}\begin{bmatrix}1\\ \rho\end{bmatrix} e^{-j\nu}\right\}_X^{Xx} \\ &= \begin{bmatrix}\boldsymbol{1}_x & \boldsymbol{1}_y\end{bmatrix}\begin{bmatrix}\cos 90^0 \\ \sin 90^0 e^{j(2x0^0)}\end{bmatrix} e^{-j0^0} \end{aligned} \tag{P.11}$$

and, defining $-Xx$ as a new phasor with the new phase parameter, ε, by putting $\nu = \delta + \varepsilon$ we obtain for $\gamma_X^{-Xx} = 90^0$, and $\delta_X^{-Xx} = -\varepsilon_X^{-Xx} = 90^0$,

$$\begin{aligned} -\boldsymbol{u}^{Xx} \equiv \boldsymbol{u}^{-Xx} &= \begin{bmatrix}\boldsymbol{1}_x & \boldsymbol{1}_y\end{bmatrix} u_X^{-Xx} = \begin{bmatrix}\boldsymbol{1}_x & \boldsymbol{1}_y\end{bmatrix}\begin{bmatrix}0\\-1\end{bmatrix} \\ &= \begin{bmatrix}\boldsymbol{1}_x & \boldsymbol{1}_y\end{bmatrix}\left\{\frac{1}{\sqrt{1+\rho\rho^*}}\begin{bmatrix}1\\ \rho\end{bmatrix} e^{-j\nu}\right\}_X^{-Xx} \\ &= \begin{bmatrix}\boldsymbol{1}_x & \boldsymbol{1}_y\end{bmatrix}\begin{bmatrix}\cos 90^0 \\ \sin 90^0 e^{j(2x90^0)}\end{bmatrix} e^{-j[90^0+(-90^0)]} \end{aligned} \tag{P.12}$$

P5. The use of indexed transformation matrices

Combining the two obtained results for presentation of vectors in bases of the reversed order we can write alternatively,

$$
\begin{aligned}
\boldsymbol{E} &= E_0\left[\boldsymbol{u}^X \quad \boldsymbol{u}^{Xx}\right] u_X^P \\
&= E_0\left[\boldsymbol{u}^Y \quad \boldsymbol{u}^{Yx}\right] u_Y^P \\
&= E_0\left[\boldsymbol{u}^B \quad \boldsymbol{u}^{Bx}\right] u_B^P
\end{aligned}
\tag{P.13}
$$

The last equality presents the electric column vector in the new basis determined by its first phasor B which corresponds to the first basis unit vector

$$
\begin{aligned}
\boldsymbol{u}^B &= \left[\boldsymbol{u}^X \quad \boldsymbol{u}^{Xx}\right] u_X^B \\
&= \left[\boldsymbol{u}^Y \quad \boldsymbol{u}^{Yx}\right] u_Y^B \\
&= \left[\boldsymbol{u}^B \quad \boldsymbol{u}^{Bx}\right] u_B^B, \quad \text{with } u_B^B = \begin{bmatrix} 1 \\ 0 \end{bmatrix}.
\end{aligned}
\tag{P.14}
$$

Fundamental internationally established transformations with the use of an additional phasor notation take the following forms:

<u>The orthogonality transformation,</u>

$$
\begin{aligned}
u_X^{Bx} &\equiv \left\{ \frac{1}{\sqrt{1+\rho\rho^*}} \begin{bmatrix} -\rho^* \\ 1 \end{bmatrix} e^{jv} \right\}_X^B, \\
&= \begin{bmatrix} 0 & -1 \\ 1 & 0 \end{bmatrix} (u_X^B)^*
\end{aligned}
\tag{P.15}
$$

with

$$
u_X^B = \left\{ \frac{1}{\sqrt{1+\rho\rho^*}} \begin{bmatrix} 1 \\ \rho \end{bmatrix} e^{-jv} \right\}_X^B
\tag{P.16}
$$

Observe, please, that here the orthogonal vector has been defined uniquely, what is not the common case in the international nomenclature but very convenient because it allows for exact determination of the orthogonal polarization (and phase!) basis by its first vector only.

Without use of the phasor notation we had,

$$
\begin{aligned}
u_{(x,y)}^{x} &\equiv \left\{ \frac{1}{\sqrt{1+\rho\rho^*}} \begin{bmatrix} -\rho^* \\ 1 \end{bmatrix} e^{jv} \right\}_{(x,y)} \\
&= \begin{bmatrix} 0 & -1 \\ 1 & 0 \end{bmatrix} (u_{(x,y)})^*
\end{aligned}
\tag{P.17}
$$

with the unit vector being transformed,

$$
u_{(x,y)} = \left\{ \frac{1}{\sqrt{1+\rho\rho^*}} \begin{bmatrix} 1 \\ \rho \end{bmatrix} e^{-jv} \right\}_{(x,y)}.
\tag{P.18}
$$

The disadvantage of such an incomplete notation is that polarization and phase of both vectors should be specified if they have to be used in scalar product with other PP vectors.

Change of basis (passive) transformation takes the form,

$$u_X^P = \left\{ \frac{1}{\sqrt{1+\rho\rho^*}} \begin{bmatrix} 1 & -\rho^* \\ \rho & 1 \end{bmatrix} e^{-j\nu} \right\}_X^A u_A^P \equiv \begin{bmatrix} u_X^A & u_X^{A\times} \end{bmatrix} u_A^P$$
$$\equiv C_X^A u_A^P \tag{P.19}$$

or, without phasor notation,

$$u_{(x,y)} = U_{2,\,(A,B)\to(x,y)} u_{(A,B)} \tag{P.20}$$

where problem appears with an unambiguous description of the unitary transformation matrix U_2 because here the A and B symbols of basis vectors do not represent phasors and in each case ought to be precisely determined.

Active transformation of the Jones unit vector corresponding to the tangential phasor P into similar vector corresponding to the tangential phasor A, in the circular (R,L) basis for example, takes the form (compare formula (5.23), p. 24):

$$u_R^A = \left\{ \frac{1}{\sqrt{1+\rho\rho^*}} \begin{bmatrix} 1 & -\rho^* \\ \rho & 1 \end{bmatrix} e^{-j\nu} \right\}_R^A \left\{ \frac{1}{\sqrt{1+\rho\rho^*}} \begin{bmatrix} 1 & -\rho^* \\ \rho & 1 \end{bmatrix} e^{-j\nu} \right\}_P^R u_R^P$$
$$\equiv C_R^A C_P^R u_R^P \tag{P.21}$$
$$\equiv C_{P,R}^A u_R^P$$

The same transformation when using the 'international nomenclature' without phasor notation would be

$$u_{(R,L)}^A = U_{2,\,(A,B)\to(R,L)} U_{2,\,(R,L)\to(P,Q)} u_{(R,L)}^P\,. \tag{P.22}$$

That form is not only longer but involves more basis vectors (six: R,L,P,Q,A,B, instead of three: R,P,A) with the same problem of precise description of the unitary matrices U_2.

P6. Few convenient new symbols, C, D, and U

Immediately the question arises why to use symbols C instead of U_2? There are two reasons justifying such a modification. The first one is dictated by the possibility of simplification of formulae by using simpler symbols, $C=U_2$ and then $D=U_4$, the last for transformations in the Stokes parameter space. Another reason has appeared after applying the symbol U for transformation the Jones to Stokes vector, also in the 4-dimensional space of Stokes parameters. Such an unitary matrix used to be expressed by some authors through $A = \sqrt{2}\,\widetilde{U}^*$ (see Boerner in [13] and [23]), or $Q = \sqrt{2}\,\widetilde{U}^*$ (see Mott in [38]).

Those symbols have been used in this monograph to present other entities. Following van de Hulst in [83], the symbol A has been chosen to present amplitude matrices (2x2 complex matrices). However, not the Jones 'propagation' matrices are here denoted by A, what exactly has been proposed by van de Hulst, but the Sinclair 'scattering' matrices. The Jones matrices have been denoted by A^0 in order to strongly indicate the exact mutual dependence between the Sinclair and Jones matrices which can be expressed by the simple transformation equation in any orthogonal polarization B basis,

$$A_B^0 = C_B^0 * A_B\,. \tag{P.23}$$

Such a transformation, presented in the (precisely named) 'orthogonal polarization and phase basis of collinear phasors' - see Section 7.5 , has been called the 'spatial coordinate system reversal by its rotation' and presents one of essential results of this work. Owing to it, the up to date existing problem of ambiguity in such a dependence (in any polarization basis) has been solved definitely.

Similarly, the Q symbol has here been reserved rather to describe one of the Stokes parameters.

Moreover, the unitary U matrix here being applied has been chosen as strictly unitary, satisfying the equality

$$U^{-1} = \widetilde{U}^* \quad \text{and} \quad U\widetilde{U}^* = diag(1,1,1,1) \tag{P.24}$$

what appears very convenient when introducing, e.g., definitions of Stokes vectors or Kennaugh matrices by proper modifications of transmission equations (for instance, see Section N6, formulae (N.27)).

P7. Other indexed transformation matrices

For the Stokes parameter space one example will be given explaining the problem.

Rotation matrices in the Stokes parameter space. Transformation matrices just mentioned, U and A, are:

$$U = \frac{1}{\sqrt{2}}\begin{bmatrix} 1 & 1 & 0 & 0 \\ 0 & 0 & 1 & -j \\ 0 & 0 & 1 & j \\ 1 & -1 & 0 & 0 \end{bmatrix}, \qquad A = \begin{bmatrix} 1 & 0 & 0 & 1 \\ 1 & 0 & 0 & -1 \\ 0 & 1 & 1 & 0 \\ 0 & j & -j & 0 \end{bmatrix} \tag{P.25}$$

Rotation matrices in the Stokes parameter space, when using international nomenclature enriched with the phasor notation, expressed in any orthogonal R basis of collinear phasors, take the following unambiguous form,

$$D_R^P = \widetilde{U}^* (C_R^P \otimes C_R^P{}^*)\, U \,. \tag{P.26}$$

Without application of phasor notation, the corresponding expression is of the form,

$$U_{4,\,(P,Q)\to(R,L)} = A\,(U_{2,\,(P,Q)\to(R,L)} \otimes U_{4,}{}^*{}_{(P,Q)\to(R,L)})\, A^{-1} \,. \tag{P.27}$$

Bases vectors P, Q, R, and L here applied, not being phasors, need precise description which should take into account their phases (vectors always have phases, though sometimes assumed tacitly, whereas phasors are completely determined by three real parameters each - very simply!).

Transformation by reversal of the spatial coordinate system. Such transformation can be used, for example, to convert the Sinclair into Jones matrix. Using the international nomenclature with the TP phasors the following expression for the Jones versus Sinclair matrix has been presented above,

$$A_B^0 = C_B^0{}^* A_B \,. \tag{P.28}$$

Such expression was never used without the phasor notation except of its presentation in orthogonal linear or circular bases of collinear phasors. In linear basis it reads (see, e.g., Mott [38], p.316, formula (6.88))

$$\boldsymbol{T} = \begin{bmatrix} -1 & 0 \\ 0 & 1 \end{bmatrix} \boldsymbol{S} \tag{P.29}$$

with $\boldsymbol{T} = A_X^0$, $\boldsymbol{S} = A_X$ and $\begin{bmatrix} -1 & 0 \\ 0 & 1 \end{bmatrix} = C_X^0{}^*$; $X \Leftrightarrow (x,y)$. In the B basis the following transformation should be applied in which the C_X^0 matrix undergoes similar transformation like the Sinclair matrix, and:

$$C_B^0{}^* = (\widetilde{C}_X^B C_X^0 C_X^B)^* \,. \tag{P.30}$$

In the circular L basis of collinear phasors and of natural order, with $\delta_X^L = \gamma_X^L = \nu_X^L = 90^0$ (and $\varepsilon_X^L = 0^0$) one obtains

$$C_L^0* = (\tilde{C}_X^L C_X^0 C_X^L)* = (\begin{bmatrix} 0 & j \\ j & 0 \end{bmatrix}\begin{bmatrix} -1 & 0 \\ 0 & 1 \end{bmatrix}\begin{bmatrix} 0 & j \\ j & 0 \end{bmatrix})* = \begin{bmatrix} -1 & 0 \\ 0 & 1 \end{bmatrix} = C_X^0 * \quad \text{(P.31)}$$

In another circular L basis of collinear phasors and of natural order, with $\delta_X^L = \gamma_X^L = -\varepsilon_X^L = 90^0$ (and $\nu_X^L = 0^0$) one obtains different matrix

$$C_L^0* = (\tilde{C}_X^L C_X^0 C_X^L)* = (\begin{bmatrix} 0 & -1 \\ 1 & 0 \end{bmatrix}\begin{bmatrix} -1 & 0 \\ 0 & 1 \end{bmatrix}\begin{bmatrix} 0 & 1 \\ -1 & 0 \end{bmatrix})* = \begin{bmatrix} 1 & 0 \\ 0 & -1 \end{bmatrix} = -C_X^0 *. \quad \text{(P.32)}$$

However, in the Stokes parameter space, these two amplitude transformation matrices produce the same 'power, polarimetric' transformation matrix of the form

$$D_L^0 = \tilde{U}(C_X^0 \otimes C_X^0 *)U = \begin{bmatrix} 1 & 0 & 0 & 0 \\ 0 & 1 & 0 & 0 \\ 0 & 0 & -1 & 0 \\ 0 & 0 & 0 & 1 \end{bmatrix} = D_X^0, \quad \text{(P.33)}$$

well known from the international literature and easily obtainable with the use of formulae employing the phasor notation.

APPENDIX Q
POLARIMETRIC INTERPRETATION OF KNOWN MATHEMATICAL THEORIES

This monograph has been written also for those readers who may be not familiar with 'higher mathematics'. Only the knowledge of complex numbers and of fundamental rules of matrix calculus is sufficient to read this text. For other readers however, aquainted with the Riemann geometry and spinors ([130], [116], [117], [106], [125]), the following polarimetric interpretation of known mathematical theories may be helpful in immediate understanding the way in which here the Poincare sphere transformations have been presented.

Q1. Application of tangential planes on manifolds, flags, geodetics, parallel transport of vectors on manifolds, to definition of the TP phasors and PP vectors and their use in polarimetric transmission equations.

Topologically, tangential phasors (or spinors) are considered as elements of the two-folded (complex) Riemann surface of constant curvature called here the polarization (Poincare) sphere of tangential phasors (TP phasors). Great circles of that TP phasors sphere, or their segments, are called geodetics on such a Riemann surface. Phasors 'shifted parallel', without rotation, along those geodetics by 2π (along one closed loop interpreted as a great circle of the Poincare sphere) take their initial value multiplied by -1, and only shifted by the 4π distance return to their initial value. Similarly phasors, only rotated in their tangential plane, take their initial value after rotation by the 4π angle (or its multiple). Generally, one shift of the phasor along any closed loop on that Riemann surface, when remaining tangent to that loop, can be interpreted as a sum of 'parallel shifts' along elementary segments of great circles plus rotations at their nodes (on that loop) what results in the total change of phasor's orientation by 2π what consists of one half of the solid angle subtended by that loop plus the sum of angles of rotation at all nodes. This is also one half of the phasor's phase change equal to π. Orientations of the TP phasors can be presented on that two-folded Riemann surface of constant curvature by the so-called 'flags' (compare 'Gravitation' [116] by Misner, Thorne and Wheeler), of orientations in the range of 4π.

Complex amplitudes of waves propagating in the +z direction of the right-handed xyz local spatial coordinate system are equal to the polarization and phase (PP) vectors. They can be presented by the TP phasors on the Poincare sphere and represented, e. g., by nondotted contravariant spinors (or by dotted covariant spinors). Complex amplitudes of waves propagating in the -z direction can be represented by dotted contravariant spinors (or nondotted covariant spinors), accordingly, because they are equal to conjugate values of the PP vectors. However, their TP phasors always correspond to the PP vectors, not to their conjugate versions. The only difference between the TP phasors representing oppositely propagating waves (or oppositely oriented antennas) is that the received voltage, being expressed by the Hermitian product of the two PP vectors, has its phase argument equal to one half of the difference: of the orientation angle of the TP (TP+) phasor, corresponding to the '+z oriented' wave (antenna), minus the orientation angle of the TP (TP-) phasor of the '-z oriented' antenna (wave). That reflects the fact that, in the Hermitian product, the PP vector corresponding to the TP- phasor appears always in its complex conjugate form.

So, the Sinclair scattering matrices transform the PP vectors of incident waves to complex conjugate PP vectors of scattered waves. This is because the incident waves are propagating in the +z direction of the local z-axis, oriented to the scatterer, and the scattered waves are propagating in the -z direction of their local z-axis, also oriented to the scatterer. Applying conventional nomenclature, Sinclair matrices transform complex amplitudes (CA's) of incident waves (alias: positively directed Jones vectors) into CA's of scattered waves (alias: negatively directed Jones vectors).

The two-way transmission equation in the BSA, with the Sinclair matrix of the scatterer, is being presented by the Hermitian product of two PP vectors, of an antenna and scattered wave, in which that of the scattered wave, oriented along -z direction of the propagation axis, takes the complex conjugate form.

Direct transmission between two antennas is being presented as the two-way transmission with the so-called 'Sinclair scattering matrix of free space', otherwise called the matrix of transformation by reversal of the spatial coordinate system by 180^0 rotation about its axis perpendicular to the propagation axis. Such a matrix transforms the PP vector of a wave radiated by the transmit antenna, in the +z direction of its local coordinate system, to the conjugate PP vector of the 'scattered' wave, propagating in the -z direction of the receiving antenna. That matrix is of course symmetrical, to fulfill the requirements of reciprocity [53].

Q2. Comparison with spinor notation

Spinor notation may start with determination of the real numbers (see, e.g., 'Gravitation' [116], eqns. (41.64))

$$\begin{aligned} \varepsilon_{12} &= \varepsilon^{12} = -\varepsilon_{21} = -\varepsilon^{21} = 1 \\ \varepsilon_{11} &= \varepsilon^{11} = \varepsilon_{22} = \varepsilon^{22} = 0 \end{aligned} \tag{Q.1}$$

defining the alternating symbols

$$\varepsilon^{AB} = -\varepsilon^{BA}, \qquad \varepsilon_{AB} = -\varepsilon_{BA} \tag{Q.2}$$

being used to define the rule of rising and lowering the spinor's index (in [116], eqns. (41.65) and (41.66)):

$$\{ (\xi_A = \xi^B \varepsilon_{BA} = -\xi^B \varepsilon_{AB}) \leftrightarrow \xi^A = \xi_B \varepsilon^{AB} \} \leftrightarrow \{\xi_1 = -\xi^2,\ \xi_2 = \xi^1\}, \tag{Q.3}$$

where the spinor itself can be considered as corresponding to the two-dimensional PP vector, elements of which are complex numbers. Two contravariat spinors can form a normalized basis (compare in [116] eqns. (41.96) and (41.97) for $2r = 1$) consisting of the first basis spinor, corresponding to the unit PP vector,

$$\xi^A \Leftrightarrow \begin{bmatrix} a \\ b \end{bmatrix} = u \leftrightarrow \xi_A \Leftrightarrow \begin{bmatrix} -b \\ a \end{bmatrix} = u^{\times} * = C^{\times} u; \quad C^{\times} = \begin{bmatrix} 0 & -1 \\ 1 & 0 \end{bmatrix}, \tag{Q.4}$$

and its 'mate', or second basis spinor, which corresponds to the *uniquely determined* orthogonal unit PP vector

$$\eta^B \Leftrightarrow \begin{bmatrix} -b * \\ a * \end{bmatrix} = u^{\times} \leftrightarrow \eta_B \Leftrightarrow \begin{bmatrix} -a * \\ -b * \end{bmatrix} = u^{\times\times} * = -u *. \tag{Q.5}$$

They are linked by the equation (see [116], eqn. (41.84))

$$\xi^A \eta^B - \eta^A \xi^B = \varepsilon^{AB} \Leftrightarrow \det \begin{bmatrix} a & -b * \\ b & a * \end{bmatrix} = \det \begin{bmatrix} u & u^{\times} \end{bmatrix} = +1. \tag{Q.6}$$

A scalar product of spinors should be defined enabling one to express the received votage in a spinor language. That requires a special care because scalar product of any spinor with itself desappears:

$$\begin{aligned} 0 = \xi^A \xi_A &= \begin{bmatrix} a & b \end{bmatrix} \begin{bmatrix} -b \\ a \end{bmatrix} = \tilde{u} u^{\times} * \\ &= \tilde{u} C^{\times} u = \xi^A \varepsilon_{BA} \xi^B = -\xi_B \xi^B \end{aligned} \tag{Q.7}$$

and only scalar product of the two basis vectors equals one (see [116], eqns. (41.81)):

$$\begin{aligned} 1 = \xi_A \eta^A &= \begin{bmatrix} -b & a \end{bmatrix} \begin{bmatrix} -b * \\ a * \end{bmatrix} = \tilde{u}^{\times} * u^{\times} \\ &= -\xi^A \eta_A = \begin{bmatrix} -a & -b \end{bmatrix} \begin{bmatrix} -a * \\ -b * \end{bmatrix} = \tilde{u} u^*. \end{aligned} \tag{Q.8}$$

Therefore, taking (see also (Q.5))

$$\begin{aligned} &\xi^A \Leftrightarrow u^R \\ &\zeta^B \Leftrightarrow u^{To\times} \leftrightarrow \zeta_B \Leftrightarrow u^{To\times\times} * = -u^{To} * \end{aligned} \tag{Q.9}$$

we arrive at the desired equation for the normalized received voltage (compare (6.13), (6.5) and, in [116], the definition (41.67) of scalar product for spinors):

$$\begin{aligned} V_r &= \tilde{u}^R u^{To} * = -\tilde{u}^R u^{To\times\times} * = -\xi^B \zeta_B \\ &= \tilde{u}^R \tilde{C}^{\times} C^{\times} u^{To} * = \tilde{u}^{R\times} * u^{To\times} = \xi_A \zeta^A \\ &= \tilde{u}^R C^{\circ} u^T = \xi^A S_{AB} \tau^B; \quad -\zeta_A = S_{AB} \tau^B; \quad S_{AB} = S_{BA} \Leftrightarrow C^{\circ} = \tilde{C}^{\circ}. \end{aligned} \tag{Q.10}$$

The 'coordinate system reversal by rotation matrix' transformed from the linear, H, to any ONP PP basis B, according to (6.14) is:

$$C^{\circ} \equiv C_B^{\circ} = \tilde{C}_H^B C_H^{\circ} C_H^B; \qquad C_H^{\circ} = \begin{bmatrix} -1 & 0 \\ 0 & 1 \end{bmatrix}. \tag{Q.10a}$$

Q 3. Complex antenna height and complex antenna receiving height. The received voltage reciprocal equation

Following Booker/Kales ([26], 1951), or Stutzman ([135], 1993, p.138, (6.52)), and using concept of the PP vectors, the received voltage

$$V = V_0 e^{j\omega t}; \quad V_0 \equiv V_r \in C^1 \tag{Q.11}$$

can be expressed through its complex value V_0 by means of the equations

$$\begin{aligned} V_0 &= E_0 h\,(\boldsymbol{u}^R \bullet \boldsymbol{C}^\circ \bullet \boldsymbol{u}^T) \\ &= E_0 h\,(\boldsymbol{u}^R \bullet \boldsymbol{u}^{T\circ} *) = E_0 h\,(\boldsymbol{u}^{R\circ} * \bullet \boldsymbol{u}^T) \\ &= \boldsymbol{E}_0 \bullet \boldsymbol{h}^\circ * \end{aligned} \tag{Q.12}$$

in which:

- $\boldsymbol{u}^R$ and $\boldsymbol{u}^T$ are the unit PP vectors, and also the directional Jones vectors, of an antenna and wave, respectively, oriented in +z directions in their own local right-handed xyz coordinate systems; by means of those vectors the antenna height complex vector, the antenna *receiving height complex vector* (compare with Hollis et al. in [80], Chapter 3, or/and with the IEEE Standard [90]), and the electric vectors of the incoming wave, propagating in two directions along the z-axes of the local coordinate systems, can be expressed in succession as follows:

$$\begin{aligned} &\boldsymbol{h}^+ = \boldsymbol{h} \equiv \boldsymbol{h}^R = h\,\boldsymbol{u}^R, \\ &\boldsymbol{h}^{\circ-} = \boldsymbol{h}^\circ * \equiv \boldsymbol{h}^{R\circ} * = h\,\boldsymbol{u}^{R\circ} *; \qquad \boldsymbol{u}^{R\circ} * = \boldsymbol{C}^\circ \bullet \boldsymbol{u}^R = \boldsymbol{u}^R \bullet \boldsymbol{C}^\circ, \\ &\boldsymbol{E}^+ = \boldsymbol{E}_0^+ e^{j\omega t}; \qquad \boldsymbol{E}_0^+ = \boldsymbol{E}_0 \equiv \boldsymbol{E}_0^T = E_0 \boldsymbol{u}^T, \\ &\boldsymbol{E}^{\circ-} = \boldsymbol{E}_0^{\circ-} e^{j\omega t}; \qquad \boldsymbol{E}_0^{\circ-} = \boldsymbol{E}_0^\circ * \equiv \boldsymbol{E}_0^{T\circ} * = E_0 \boldsymbol{u}^{T\circ} *; \qquad \boldsymbol{u}^{T\circ} * = \boldsymbol{C}^\circ \bullet \boldsymbol{u}^T. \end{aligned} \tag{Q.13}$$

These vectors are the *directive* Jones vectors [13], $\boldsymbol{h}^+$, $\boldsymbol{h}^{\circ-}$, $\boldsymbol{E}_0^+$, $\boldsymbol{E}_0^{\circ-}$, expressed in terms of the corresponding unit PP vectors, $\boldsymbol{u}^R$, $\boldsymbol{u}^{R\circ}$, $\boldsymbol{u}^T$, $\boldsymbol{u}^{T\circ}$, which are *independent of the direction of wave propagation, or antenna orientation, versus the z-axes of their local coordinate systems.* Such an independence is essential for presentation of the corresponding TP phasors on the same Poincare sphere, what enables one to see the angle between phasors of the incoming wave and the receiving antenna; the cosine of one half of that angle is the magnitude of the normalized received voltage. The sign of the voltage's phase depends on which factor of the scalar product is conjugated.

- $\boldsymbol{C}^\circ$ is an operator (dyadic) reversing: (1) direction of z-axis of the local coordinate system (by its 180^0 rotation) and (2) direction of time, or in other words: direction of propagation/orientation versus actual direction of the reversed z-axis, what has been expressed by conjugate value of the transformed vector. In space, direction of propagation/orientation remains unchanged.

(General remark: polarization and phase (PP) vectors are always related to their local spatial coordinate system in such a sense that *they change* under reversal of the propagation z-axis of that system, the reversal performed by 180^0 rotation of that system about an axis perpendicular to the z-axis. **Such a change of the PP vector can be compared with transformation of that vector by a scattering matrix.** In this case the matrix corresponds to the $\boldsymbol{C}^\circ$ dyadic and can be considered as a Sinclair scattering matrix of 'free space' between antennas).

$\boldsymbol{C}^\circ$ is a symmetric dyadic. This satisfies a demand which ensures full reciprocity (!) of the first equation of (Q.12). The received voltage cannot depend on direction of propagation between antennas. Their complex heights can be presented by the positively directed Jones vectors $\boldsymbol{h}^{R+} = h\,\boldsymbol{u}^R$ and $\boldsymbol{h}^{T+} = h\,\boldsymbol{u}^T$. Antennas are looking at each other, but both vectors of their complex heights are expressed in their own local coordinate systems with the z-axes directed out of them. Therefore the free space between the antennas should be considered as a hypothetical 'target' in local coordinate systems on its both sides with z-axes directed to the target. Its Sinclair scattering matrix is given by (Q.10a).

A comparison of the two approaches, employing the directive Jones vectors and polarization and spatial phase (PP) vectors, on the example of expression for the received voltage, can be presented by the two following equations, respectively, when assuming *the propagation z-axis being oriented in the direction of wave's propagation or in the opposite direction*:

$$\begin{aligned} V_0 &= \boldsymbol{E}_0^{+} \bullet \boldsymbol{h}^{\mathrm{o}-} = \boldsymbol{E}_0 \bullet \boldsymbol{h}^{\mathrm{o}}\,* \\ &= \boldsymbol{E}_0^{\mathrm{o}-} \bullet \boldsymbol{h}^{+} = \boldsymbol{E}_0^{\mathrm{o}}\,* \bullet \boldsymbol{h}. \end{aligned} \tag{Q.14}$$

Appendix R
Maxwell Equations in Radar Polarimetry

In *radar* polarimetry, solutions of Maxwell equations should lead to determination of complex amplitude vectors, electric and magnetic, of two plane electromagnetic waves *propagating in opposite directions.* Those vectors are functions of wave's polarization (two real parameters) and its *spatial* phase (one real parameter). ***Polarization and spatial phase parameters do not depend on wave's direction of propagation.*** However, complex amplitude vectors of waves propagating in opposite directions are different functions of those parameters.

Consider the first two, real, time dependent Maxwell equations, for plane waves propagating in the isotropic, homogeneous, source-free, linear medium along the z-axis of an xyz coordinate system in both directions, $+z$ or $-z$:

$$\nabla \times \mathcal{H}(t,z) = \varepsilon \frac{\partial \mathcal{E}(t,z)}{\partial t}, \qquad \varepsilon = const \tag{R.1}$$

$$\nabla \times \mathcal{E}(t,z) = -\mu \frac{\partial \mathcal{H}(t,z)}{\partial t}, \qquad \mu = const \tag{R.2}$$

where

$$\nabla = \mathbf{1}_x \frac{\partial}{\partial x} + \mathbf{1}_y \frac{\partial}{\partial y} + \mathbf{1}_z \frac{\partial}{\partial z}$$

For the harmonic electric and magnetic vectors of those waves introduce their space-dependent complex amplitudes, $\boldsymbol{E}_0^{\pm}(z)$ and $\boldsymbol{H}_0^{\pm}(z)$, defined as follows for waves propagating in the two directions:

$$\begin{aligned} \mathcal{E}(t,z) &= \mathcal{E}^{\pm}(t,z) = \mathrm{Re}\{\boldsymbol{E}_0^{\pm}(z)e^{j\omega t}\} \\ \mathcal{H}(t,z) &= \mathcal{H}^{\pm}(t,z) = \mathrm{Re}\{\boldsymbol{H}_0^{\pm}(z)e^{j\omega t}\}. \end{aligned} \tag{R.3}$$

At first, consider electric vectors of waves propagating in the $+z$ direction. Define also their complex amplitudes, $\boldsymbol{E}_0$ and $\boldsymbol{H}_0$, *independent* of the space coordinate z according to the equalities:

$$\boldsymbol{E}_0^{+}(z) = \boldsymbol{E}_0(z) = \boldsymbol{E}_0 e^{-jkz}, \tag{R.4}$$

$$\boldsymbol{H}_0^{+}(z) = \boldsymbol{H}_0(z) = \boldsymbol{H}_0 e^{-jkz}. \tag{R.5}$$

The first of those complex amplitudes, $\boldsymbol{E}_0$, called the 'Jones vector' (in its column matrix form), can be expressed in terms of a unit complex vector $\boldsymbol{u}$ and vector's magnitude E_0 as follows,

$$\boldsymbol{E}_0 = E_0 \boldsymbol{u} = E_0(\mathbf{1}_x u_x + \mathbf{1}_y u_y); \qquad u_x u_x{}^* + u_y u_y{}^* = 1. \tag{R.6}$$

In turn, the unit $\boldsymbol{u}$ vector in its Jones vector form can be expressed in terms of 'analytical parameters' of polarization, γ and δ, and phase, ε,

$$u_{(x,y)} \equiv \begin{bmatrix} u_x \\ u_y \end{bmatrix} = \begin{bmatrix} \cos\gamma e^{-j(\delta+\varepsilon)} \\ \sin\gamma e^{j(\delta-\varepsilon)} \end{bmatrix}_{(x,y)}. \tag{R.7}$$

Complex Maxwell equations for complex amplitudes dependent on spatial coordinate z are of the form

$$\nabla \times \boldsymbol{H}_0(z) = j\omega\varepsilon \boldsymbol{E}_0(z), \tag{R.8}$$

$$\nabla \times \boldsymbol{E}_0(z) = -j\omega\mu \boldsymbol{H}_0(z). \tag{R.9}$$

In order to present the vector $\boldsymbol{H}_0$ in the form similar to that in (R.6) we should first observe that

$$k = \omega\sqrt{\varepsilon\mu}\,, \qquad E_0 = H_0\sqrt{\frac{\mu}{\varepsilon}}\,, \tag{R.10}$$

what results in

$$\frac{kE_0}{\omega\mu} = H_0\,, \qquad \frac{kH_0}{\omega\varepsilon} = E_0\,. \tag{R.11a,b}$$

Now, from (R.9) and making use of (R.6), we obtain

$$\begin{aligned} \nabla \times \boldsymbol{E}_0(z) &= E_0 \begin{vmatrix} \boldsymbol{1}_x & \boldsymbol{1}_y & \boldsymbol{1}_z \\ \partial/\partial x & \partial/\partial y & \partial/\partial z \\ u_x e^{-jkz} & u_y e^{-jkz} & 0 \end{vmatrix} = jkE_0 e^{-jkz}\left(\boldsymbol{1}_x u_y - \boldsymbol{1}_y u_x\right) \\ &= -j\omega\mu\, \boldsymbol{H}_0(z) \end{aligned} \tag{R.12}$$

and with (R.11a) we have

$$\begin{aligned} \boldsymbol{H}_0(z) &= -\frac{kE_0}{\omega\mu} e^{-jkz}\left(\boldsymbol{1}_x u_y - \boldsymbol{1}_y u_x\right) \\ &= H_0 e^{-jkz}\left(\boldsymbol{1}_x(-u_y) + \boldsymbol{1}_y u_x\right) \\ &= H_0(\boldsymbol{u}^{\times}*) e^{-jkz} = \boldsymbol{H}_0 e^{-jkz} = \boldsymbol{H}_0^{+}(z), \end{aligned} \tag{R.13}$$

where

$$\boldsymbol{u}^{\times}* = \begin{bmatrix}\boldsymbol{1}_x & \boldsymbol{1}_y\end{bmatrix}\begin{bmatrix} -u_y \\ u_x \end{bmatrix} = \begin{bmatrix}\boldsymbol{1}_x & \boldsymbol{1}_y\end{bmatrix}\begin{bmatrix} 0 & -1 \\ 1 & 0 \end{bmatrix}\begin{bmatrix} u_x \\ u_y \end{bmatrix} = \begin{bmatrix}\boldsymbol{1}_x & \boldsymbol{1}_y\end{bmatrix}(u^{\times}_{(x,y)}*), \tag{R.14}$$

and where the magnetic complex amplitude vector of the forward propagating electromagnetic plane wave is

$$\boldsymbol{H}_0 = H_0(\boldsymbol{u}^{\times}*) = H_0\left(\boldsymbol{1}_x(-u_y) + \boldsymbol{1}_y u_x\right)\,. \tag{R.15}$$

Also the Maxwell equation (R.8) is satisfied for vectors (R.4,6) and (R.13):

$$\begin{aligned} \nabla \times \boldsymbol{H}_0(z) &= H_0 \begin{vmatrix} \boldsymbol{1}_x & \boldsymbol{1}_y & \boldsymbol{1}_z \\ \partial/\partial x & \partial/\partial y & \partial/\partial z \\ -u_y e^{-jkz} & u_x e^{-jkz} & 0 \end{vmatrix} = jkH_0 e^{-jkz}\left(\boldsymbol{1}_x u_x + \boldsymbol{1}_y u_y\right) \\ &= j\omega\varepsilon\, \boldsymbol{E}_0(z), \end{aligned} \tag{R.16}$$

where from, when using (R.11b),

$$\boldsymbol{E}_0(z) = \frac{kH_0}{\omega\varepsilon} e^{-jkz}\boldsymbol{u} = E_0\boldsymbol{u}\, e^{-jkz} = \boldsymbol{E}_0^{+}(z)\,, \tag{R.17}$$

according to (R.4) and (R.6).

For complex amplitudes (R.4), (R.6) and (R.5), (R.15), the 'polarimetric' expression for the Poynting vector of a wave propagating in the $+z$ direction can be found as follows:

$$S^{+} = \tfrac{1}{2}\mathrm{Re}\left(E_0^{+}(z) \times H_0^{+}{}^{*}(z)\right) = \tfrac{1}{2}\mathrm{Re}\left(E_0 \times H_0{}^{*}\right)$$

$$= \tfrac{1}{2}E_0 H_0 \,\mathrm{Re}\left(u \times u^{\times}\right) = \tfrac{1}{2}E_0 H_0 \begin{vmatrix} 1_x & 1_y & 1_z \\ u_x & u_y & 0 \\ -u_y{}^{*} & u_x{}^{*} & 0 \end{vmatrix} \qquad \text{(R.18)}$$

$$= \tfrac{1}{2}E_0 H_0 1_z \;.$$

The corresponding real electric and magnetic vectors for $t = z = 0$ (see Fig. R.1) are:

$$\mathcal{E}^{+}(0,0) = E_0 \,\mathrm{Re}\, u, \qquad \text{(R.18a)}$$
$$\mathcal{H}^{+}(0,0) = H_0 \,\mathrm{Re}(u^{\times}{}^{*}).$$

Summarizing, complex expressions for electric and magnetic vectors of the 'forward' propagating electromagnetic plane wave take forms

$$E^{+}(t,z) = E_0 u \, e^{j(\omega t - kz)} = E_0 \, e^{j(\omega t - kz)} \qquad \text{(R.19)}$$

$$H^{+}(t,z) = H_0 (u^{\times}{}^{*}) e^{j(\omega t - kz)} = H_0 e^{j(\omega t - kz)} . \qquad \text{(R.20)}$$

The corresponding expressions for waves propagating in the $-z$ (backward) direction can be found when inspecting Maxwell equations which are conjugate versus (R.8)-(R.9),

$$\nabla \times H_0{}^{*}(z) = -j\omega\varepsilon E_0{}^{*}(z) \qquad \text{(R.21)}$$

$$\nabla \times E_0{}^{*}(z) = j\omega\mu H_0{}^{*}(z) \;. \qquad \text{(R.22)}$$

They can be rewritten in the form

$$\nabla \times (-H_0{}^{*}(z)) = j\omega\varepsilon E_0{}^{*}(z) \qquad \text{(R.21')}$$

$$\nabla \times E_0{}^{*}(z) = -j\omega\mu(-H_0{}^{*}(z)) \;. \qquad \text{(R.22')}$$

Defining the following complex amplitudes for backward propagating waves

$$E_0^{-}(z) = E_0{}^{*}(z) = E_0{}^{*} e^{jkz} \qquad \text{(R.4')}$$

$$H_0^{-}(z) = -H_0{}^{*}(z) = -H_0{}^{*} e^{jkz} \qquad \text{(R.5')}$$

we conclude that $E_0{}^{*}(z)$ and $-H_0{}^{*}(z)$ vectors fulfill equations (R.8)-(R.9) when being used in place of vectors $E_0(z)$ and $H_0(z)$. Therefore, the new couple of vectors represents an electromagnetic wave too, though propagating in opposite direction because of the exponential term with the positive imaginary exponent, $+jkz$. The corresponding expressions for electric and magnetic complex vectors of that wave take the following forms, corresponding to those as in formulae (R.19) and (R.20) for forward propagating waves:

$$E^{-}(t,z) = E_0 u^{*} \, e^{j(\omega t + kz)} = E_0{}^{*} e^{j(\omega t + kz)} \qquad \text{(R.23)}$$

$$H^{-}(t,z) = H_0(-u^{\times}) \, e^{j(\omega t + kz)} = -H_0{}^{*} e^{j(\omega t + kz)} . \qquad \text{(R.24)}$$

All that can be verified by direct inspection. Indeed,

$$\nabla \times \boldsymbol{E}_0 * (z) = E_0 \begin{vmatrix} \boldsymbol{1}_x & \boldsymbol{1}_y & \boldsymbol{1}_z \\ \partial/\partial x & \partial/\partial y & \partial/\partial z \\ u_x * e^{jkz} & u_y * e^{jkz} & 0 \end{vmatrix} = -jkE_0 e^{jkz}\left(\boldsymbol{1}_x u_y * - \boldsymbol{1}_y u_x *\right) \tag{R.25a}$$

$$= -j\omega\mu\,(-\boldsymbol{H}_0 * (z)),$$

what results in

$$-\boldsymbol{H}_0 * (z) = \frac{kE_0}{\omega\mu} e^{jkz} (-\boldsymbol{u}^\times) = H_0 (-\boldsymbol{u}^\times)\, e^{jkz}$$

$$= \boldsymbol{H}_0^- e^{jkz} = \boldsymbol{H}_0^- (z) \tag{R.26}$$

Similarly,

$$\nabla \times (-\boldsymbol{H}_0 * (z)) = H_0 \begin{vmatrix} \boldsymbol{1}_x & \boldsymbol{1}_y & \boldsymbol{1}_z \\ \partial/\partial x & \partial/\partial y & \partial/\partial z \\ u_y * e^{jkz} & -u_x * e^{jkz} & 0 \end{vmatrix} = jkH_0 e^{jkz}\left(\boldsymbol{1}_x u_x * + \boldsymbol{1}_y u_y *\right) \tag{R.25b}$$

$$= j\omega\varepsilon\, \boldsymbol{E}_0 * (z),$$

and

$$\boldsymbol{E}_0 * (z) = \frac{kH_0}{\omega\varepsilon} e^{jkz} \boldsymbol{u}^* = E_0 \boldsymbol{u} *\ e^{jkz}. \tag{R.27}$$

For complex amplitudes (R.4')-(R.5') and (R.27)-(R.26), the 'polarimetric' expression for the Poynting vector of a wave propagating in the –z direction can be found as follows:

$$\boldsymbol{S}^- = \tfrac{1}{2}\mathrm{Re}\left(\boldsymbol{E}_0^- (z) \times \boldsymbol{H}_0^- (z) *\right) = -\tfrac{1}{2}\mathrm{Re}\left(\boldsymbol{E}_0 * \times \boldsymbol{H}_0\right)$$

$$= -\tfrac{1}{2} E_0 H_0\, \mathrm{Re}\left(\boldsymbol{u} * \times \boldsymbol{u}^\times *\right) = -\tfrac{1}{2} E_0 H_0 \begin{vmatrix} \boldsymbol{1}_x & \boldsymbol{1}_y & \boldsymbol{1}_z \\ u_x * & u_y * & 0 \\ -u_y & u_x & 0 \end{vmatrix} \tag{R.28}$$

$$= -\tfrac{1}{2} E_0 H_0 \boldsymbol{1}_z\ .$$

The corresponding real electric and magnetic vectors for $t = z = 0$ (see Fig. R.2) are:

$$\mathcal{E}^- (0,0) = E_0\, \mathrm{Re}\, \boldsymbol{u}^*,$$

$$\mathcal{H}^- (0,0) = H_0\, \mathrm{Re}(-\boldsymbol{u}^\times). \tag{R.28a}$$

The most important results of those considerations are being expressed by the equalities:

$$\boldsymbol{E}^+ (t,z) = E_0 \boldsymbol{u}\, e^{j(\omega t - kz)} = \boldsymbol{E}_0\, e^{j(\omega t - kz)} \tag{R.19}$$

and

$$\boldsymbol{E}^- (t,z) = E_0 \boldsymbol{u} *\ e^{j(\omega t + kz)} = \boldsymbol{E}_0 * e^{j(\omega t + kz)}\ . \tag{R.23}$$

When neglecting the exponential wave terms, they present complex amplitudes of electric vectors and indicate that those amplitudes become conjugate for waves propagating in the –z direction. Complex amplitudes can be also expressed in terms of their directive Jones (column) vectors as in the following equalities (where the orthogonal $\boldsymbol{u}^X$ vectors can be uniquely determined with the help of the (R.14) transformation formula):

$$u^P = \begin{bmatrix} 1_x & 1_y \end{bmatrix} u^P_{(x,y)} = \begin{bmatrix} u^B & u^{B\times} \end{bmatrix} u^P_B ,$$
$$u^P * = \begin{bmatrix} 1_x & 1_y \end{bmatrix} u^P_{(x,y)} * = \begin{bmatrix} u^B & u^{B\times} \end{bmatrix} * (u^P_B) * \tag{R.29}$$

The directive Jones vectors were for the first time introduced by Graves [79]. He indicated different (mutually conjugate) rules of their change under the basis transformation. Indeed:

$$(u^P_H)^+ \equiv u^P_H = \begin{bmatrix} u^B_H & u^{B\times}_H \end{bmatrix} u^P_B ,$$
$$(u^P_H)^- \equiv (u^P_H)* = \begin{bmatrix} u^B_H & u^{B\times}_H \end{bmatrix} * \; (u^P_B)*. \tag{R.30}$$

By means of those directive Jones vectors the scattering equation which uses the Sinclair scattering matrix (in the back-scattering alignment, BSA) can be presented in the two following equivalent forms:

$$S_B (u^T_B)^+ = \lambda^T (u^S_B)^- , \tag{R.31}$$

or

$$S_B u^T_B = \lambda^T (u^S_B) *. \tag{R.32}$$

Complex amplitude of the received voltage (when neglecting the space attenuation) can be expressed by the two-way transmission equation

$$V_r = \tilde{u}^R_B S_B u^T_B = \lambda^T \tilde{u}^R_B (u^S_B) * \tag{R.33}$$

or by angular functions dependent on mutual locations and orientations of polarization phasors R and S tangent to the polarization (Poincare) sphere (compare with (4.19)).

Jones electric vectors of forward propagating waves (along positive direction of the z-axis), or forward directed antennas, can be called **'polarization and spatial phase vectors'** (the 'PP vectors'). **Jones electric vectors for negative directions (-z) are being expressed by the *conjugate* PP vectors (!).**

A form similar to (R.33) one obtains for the **'reciprocal equation' for direct transmission between two antennas:**

$$V_r = \tilde{u}^R_B C^0_B u^T_B = \tilde{u}^R_B (u^{T0}_B)* = (\tilde{u}^{R0}_B) * u^T_B \tag{R.34}$$

with the **z-reversal matrix by the spatial coordinate system rotation**:

$$C^0_B = \begin{bmatrix} \tilde{u}^B_{(x,y)} \\ \tilde{u}^{B\times}_{(x,y)} \end{bmatrix} \begin{bmatrix} -1 & 0 \\ 0 & 1 \end{bmatrix} \begin{bmatrix} u^B_{(x,y)} & u^{B\times}_{(x,y)} \end{bmatrix} = \tilde{C}^0_B = (C^0_B *)^{-1}, \; \det C^0_B = -1 \tag{R.35}$$

Another modification of equation (R.33)

$$V_r = \tilde{u}^R_B S_B u^T_B = \tilde{u}^R_B C^0_B C^0_B * S_B u^T_B = \tilde{u}^{R0}_B * J_B u^T_B \tag{R.36}$$

leads to the definition of the **'receiving polarization' in the wave's coordinate system** (compare IEEE Std. 145-1983, p. 6 in [90], or p. 3D.7 in [80]) expressed through the conjugate PP vector of an antenna, or its Jones vector, for 'negative orientation' versus the z-axis,

$$u^{R0}_B * = C^0_B u^R_B , \tag{R.37}$$

and to determination of **mutual dependence between the corresponding Sinclair and Jones matrices**

$$J_B = C^0_B * S_B . \tag{R.38}$$

The 'full Stokes four-vector' of the completely polarized wave (or antenna), of the unit 'total power' represented by its first component, should possess the three remaining components being equal to rectangular coordinates of the polarization point on the Poincare sphere of unit radius. That property should be independent of the direction of wave's propagation (or antenna orientation) in order to obtain the received power proportional to cosine square of half an angle between polarization points of the receiving antenna and the incident wave. Therefore, the full **Stokes four-vector** can best be presented with the use of the PP column vector as follows,

$$g_{(x,y)} \equiv \mathrm{P}_{0(x,y)} = \sqrt{2}\tilde{\mathrm{U}}^*(E_{0(x,y)} \otimes E_{0(x,y)}^*) = \sqrt{2}\tilde{\mathrm{U}}^* E_0^2 (u_{(x,y)} \otimes u_{(x,y)}^*)$$

$$= E_0^2 \begin{bmatrix} 1 & 0 & 0 & 1 \\ 1 & 0 & 0 & -1 \\ 0 & 1 & 1 & 0 \\ 0 & j & -j & 0 \end{bmatrix} \begin{bmatrix} u_x u_x^* \\ u_x u_y^* \\ u_y u_x^* \\ u_y u_y^* \end{bmatrix} = E_0^2 \begin{bmatrix} 1 \\ \mathrm{q} \\ \mathrm{u} \\ \mathrm{v} \end{bmatrix}_{(x,y)} ; \quad \mathrm{q}^2 + \mathrm{u}^2 + \mathrm{v}^2 = 1, \tag{R.39}$$

with

$$\mathrm{U} = \frac{1}{\sqrt{2}} \begin{bmatrix} 1 & 1 & 0 & 0 \\ 0 & 0 & 1 & -j \\ 0 & 0 & 1 & j \\ 1 & -1 & 0 & 0 \end{bmatrix}; \qquad \det \mathrm{U} = -j\,. \tag{R.40}$$

That formula is valid for both incident and scattered waves (propagating in the $+z$ or $-z$ direction of each local xyz coordinate system). Here it is expressed in the orthogonal linear, (x,y), polarization basis. For any other orthogonal null-phase (ONP) PP basis, the change of basis transformations (R.30) can be applied. **The development of expression (R.39) immediately follows the equation for the received power**, also based on (R.33):

$$\begin{aligned} P_r &= |V_r|^2 = V_r \otimes V_r^* = (\tilde{u}_B^R S_B u_B^T) \otimes (\tilde{u}_B^R S_B u_B^T)^* \\ &= (\tilde{u}_B^R \otimes \tilde{u}_B^{R*})\mathrm{U}^*\tilde{\mathrm{U}}(S_B \otimes S_B^*)\mathrm{U}\tilde{\mathrm{U}}^*(u_B^T \otimes u_B^{T*}) \\ &= \tfrac{1}{2}\tilde{g}_B^R \mathrm{K}_B g_B^T \\ &= \tfrac{1}{2}\tilde{g}_B^R \mathrm{D}_B^0 \mathrm{D}_B^0 \mathrm{K}_B g_B^T = \tfrac{1}{2}\tilde{g}_B^{R0} \mathrm{M}_B g_B^T . \end{aligned} \tag{R.41}$$

That equation determines both the Kennaugh and Mueller matrices, their mutual dependence:

$$\mathrm{M}_B = \mathrm{D}_B^0 \mathrm{K}_B; \qquad \mathrm{D}_B^0 = \tilde{\mathrm{U}}(C_B^0 \otimes C_B^{0*})\mathrm{U} = \tilde{\mathrm{D}}_B^0 = (\mathrm{D}_B^0)^{-1}, \;\; \det \mathrm{D}_B^0 = -1\,, \tag{R.42}$$

as well as the **'receiving full Stokes four-vector' in the wave's coordinate system,**

$$g_B^{R0} = \mathrm{D}_B^0 g_B^R . \tag{R.43}$$

Rewriting the change of basis equation (R.30) in the new form **with the unitary unimodular (amplitude) change of basis matrix,**

$$u_H^P = \begin{bmatrix} u_H^B & u_H^{B\times} \end{bmatrix} u_B^P = C_H^B u_B^P; \quad \det C_H^B = +1, \quad (C_H^B)^{-1} = (\tilde{C}_H^B)^* = C_B^H\,, \tag{R.44}$$

its Stokes version can be found **with the real (Stokes) change of basis matrix,**

$$g_H^P = \mathrm{D}_H^B g_B^P; \quad \mathrm{D}_H^B = \tilde{\mathrm{U}}^*(C_H^B \otimes C_H^{B*})\mathrm{U}, \quad \det \mathrm{D}_H^B = +1, \quad (\mathrm{D}_H^B)^{-1} = \tilde{\mathrm{D}}_H^B = \mathrm{D}_B^H . \tag{R.45}$$

Observe, please, simple and easy to remember form of the change of basis formulae (R.44) and (R.45).

The above presented formulae have shown how the polarimetric form of Maxwell equations, (R.8-9) and (R.21'-22'), through the definition of the PP vector E_0 as in (R.6), appearing in expressions (R.19) and (R.23), enables one to develop fundamental equations of radar polarimetry.

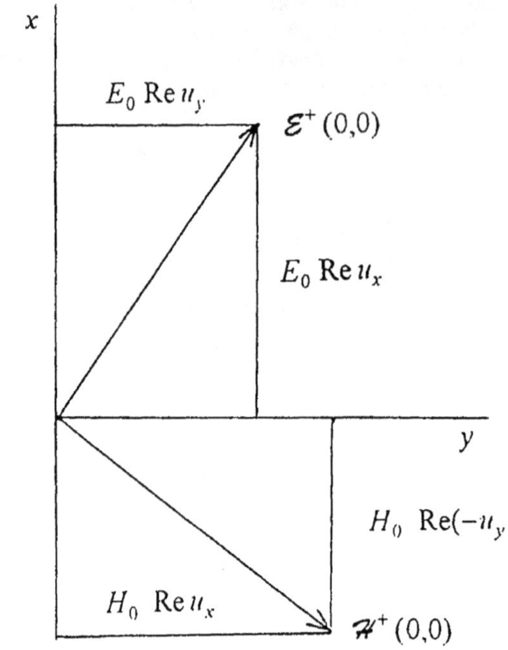

Fig. R1. Real electric and magnetic vectors of a wave propagating in the +z direction for $t = z = 0$.

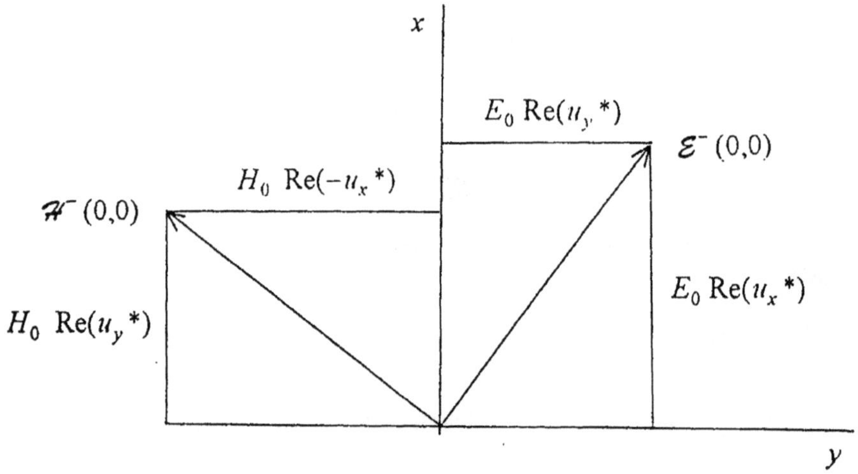

Fig. R2. Real electric and magnetic vectors of a wave propagating in the -z direction for $t = z = 0$.
The same spatial phase of the electric vector as in Fig. R1.

REFERENCES

[1] Agrawal, A. P., 'A Polarimetric Rain Back-scattering Model Developed for Coherent Polarization Diversity Radar Applications', Ph.D. thesis, UIC-GC, University of Illinois, Chicago, IL, December 1986.

[2] Agrawal, A. P., W-M. Boerner, 'Redevelopment of Kennaugh's target characteristic polarization state theory using the polarization transformation ratio formalism for the coherent case', IEEE Trans. Geosci. Remote Sensing, Vol. GRS-27(1) (1989) pp. 2-14.

[3] Azzam, R. M. A. and N. M. Bashara, 'Elipsometry and Polarized Light', North-Holland, Amsterdam, 1977.

[4] Barnes, R.M.,Roll-Invariant Decompositions for the Polarization Covariance Matrix', Polarimetric Technology Workshop, Redstone Arsenal, Alabama, August 1988.

[5] Bebbington, D. H. O., 'Target vectors - spinorial concepts', Second International Workshop on Radar Polarimetry, IRESTE, Nantes, France, September 9-10, 1992, pp. 26-36.

[6] Bebbington, D. H. O., 'Geometrical concepts in optimal polarimetry: Stokes formalism in a Minkowski space', Radar Polarimetry, SPIE Vol. 1479, San Diego, 1992, pp. 126-136.

[7] Bebbington, D. H. O., G. Wanielik, M. C. Chandra, 'Target vector reality criteria', European Microwave Conference, Munich, October 1999.

[8] Bebbington, D. H. O., E. Krogager, M. Hellman, ' Vectorial Generalization of Target Helicity', EUSAR 2000 - 3rd European Conference on Synthetic Aperture Radar, 23-25 May 2000, Munich, Germany, pp.531-534.

[9] Bebbington, D. H. O., 'Analytical foundations of polarimetry', I, to be published.

[10] Beckman, P., 'Depolarization of Electromagnetic Waves', The Golem Press, Boulder, Colorado, 1968.

[11] Beran, M. J. and G. B. Parrent, Jr., 'Theory of partial coherence', Prentice-Hall, Inc., 1964.

[12] Boerner, W-M., 'Use of polarization in electromagnetic inverse problems' Radio Science, vol. 16 (6) (1981) pp. 1037-1045.

[13] Boerner W-M., H. Mott, E. Lueneburg, C. Livingstone, B. Brisco, R. Brown, J.S. Paterson (autors); S.R. Cloude, E.Krogager, J.S. Lee, D.L. Schuler, J.J. van Zyl, D. Randall and P. Budkewitsch (contributing authors) 'Polarimetry in remote sensing - Basic and applied concepts,' Chapter 5 (94 p.) in R.A. Reyerson, ed. *'The manual of remote sensing'*, 3rd Edition, ASPRS Publishing, Bethesda, MD, 1997.

[14] Boerner W-M., M. B. El-Arini, Ch-Y. Chan, and P. M. Mastoris, 'Polarization Dependence in Electromagnetic Inverse Problems," *IEEE Trans. Antennas Propagat.*, vol. 29, pp. 262-271, 1981.

[15] Boerner W-M., B. Y. Foo, H. J. Eom, 'Interpretation of the Polarimetric Co-Polarization Phase Term ($\varphi_{HH} - \varphi_{VV}$) in High Resolution SAR Imaging Using the JPL CV-990 Polarimetric L-Band SAR Data', Special IGARSS Issue of *IEEE Transactions on Geoscience and Remote Sensing,* 25 (1): 77-82.

[16] Boerner W-M., et al. (eds.), 'Inverse Methods in Electromagnetic Imaging', Proceedings of NATO-ARW, (18-24 Sept. 1983, Bad Windsheim, FR Germany), Parts 1&2, NATO-ASI C-143, (1500 pages), D. Reidel Publ. Co., Jan. 1985.

[17] Boerner W-M., et al. (eds.), 'Direct and Inverse Methods in Radar Polarimetry', NATO-ARW, 18-24 Sept. 1988, Proc., Chief Editor, 1987-1991, (1,938 pages) NATO-ASI Series C: Math & Phys. Sciences, Vol. C-350, Parts 1&2, D. Reidel Publ. Co., Kluwer Aca. Publ., Dordrecht, NL, 1992 Feb. 15. Jan. 1985.

[18] Boerner, W-M., W. L. Yan, and A. Q. Xi, 'Basic equations of radar polarimetry and its solutions: the characteristic radar polarization states for the coherent and partially polarized cases', Proc. SPIE Conf. on Polarimetry: Radar, Infrared, Visible, Ultraviolet, and X-Ray, Huntsville, Alabama, vol. 1317, 1990, pp. 16-79.

[19] Boerner, W-M., and A. Q. Xi, 'The characteristic radar target polarization state theory for the coherent monostatic and reciprocal case using the generalized polarization transformation formulation', AEU, vol. 44(4) (1990) pp. 273-281.

[20] Boerner, W-M., Wei-Ling Yan, An- Qing Xi, and Yoshio Yamaguchi, 'Basic concepts of radar polarimetry', [17], pp.155-245.

[21] Boerner, W-M., Chuan-Li Liu, Xin Zhang, 'Comparison of Optimization Procedures for 2x2 Sinclair, 2x2 Graves, 3x3 Covariance, and 4x4 Mueller (Symmetric) Matrices in Coherent Radar Polarimetry and its Application to Target Versus Background Discrimination in Microwave Remote Sensing and Imaging', EARSeL, Int'l Journal in Advances in Remote Sensing, Vol. 2. No. 1 - 1, pp. 55-82, 1993.

[22] Boerner W-M., J. S.Verdi, H.Mott, E. A. Lueneburg, M. Tanaka, Y. Yamaguchi, „Development of polarimetric vector signal and tensor image processing in wideband interferometric POL-RAD/SAR signature analysis'.

[23] Boerner W-M., 'Recent Advances in Polarimetric-Interferometric SAR Remote Sensing - Theory & Technology and its Application in Environmental Stress Assessment', MIKON-2000, XIII International Conference on Microwaves, Radar and Wireless Communications, Wroclaw, Poland, 2000 May 22-24, SPECIAL POLARIMETRY WORKSHOP, Sunday, 2000 May 21, 125p.

[24] W-M. Boerner, E. Lueneburg, H. Mott, Z.H. Czyz, J.J. van Zyl, P. Dubois, S.R. Cloude, M. Tanaka, Y. Yamaguchi, A.I. Kozlov [1995], "Formulation of Unique Sets of Polarimetric Radar Scattering Matrices for the Forward Propagation (Similarity) Versus Backward-Scattering (Con-Similarity) Arrangements and the Development of the Associated Optimal Polarimetric Contrast Enhancement Coefficients," Proceedings of the Third International Workshop on Radar Polarimetry - JIPR '95, Nantes, France, March 21-23, 1995, p. 115.

[25] W-M. Boerner, H. Mott, M. Tanaka, Y. Yamaguchi, 'Determination of optimal polarimetric contrast using the partially coherent Kennaugh, covariance, and Graves matrices in POL-SAR image analysis'.

[26] Booker, H. G., V. H. Rumsey, G. A. Deschamps, M. I. Kales, and J. I. Bohnert, 'Techniques for Handling Elliptically Polarized Waves with Special Reference to Antennas', Proceedings of the IRE, Vol. 39, May 1951, pp. 533-522.

[27] Born, M. and E. Wolf, 'Principles of Optics', 6th ed., Pergamon Press, 1980.

[28] Brosseau, C., Fundamentals of Polarized Light - A Statistical Optics Approach, John Wiley/Sons, New York, 1998.

[29] Carrea, L., and G. Wanielik, 'Geometrical representation of a monochromatic electromagnetic wave using the tangential vector approach', MIKON'2000, wroclaw, Poland, May 22-24, 2000, pp. 87-90.

[30] Chamberlain, N. F., 'Recognition and analysis of aircraft targets by radar, using structural pattern representations derived from polarimetric signatures', Ph. D. dissertation, The Ohio State University, 1989.

[31] Chandrasekhar, S., 'Radiative transfer', New York: Dover, 1960.

[32] Cloude, S. R., 'The Physical Interpretation of Eigenvalue Problems in Optical Scattering Polarimetry', in Proceedings of SPIE '97 International Symposium, San Diego, California, 30 July - 1 August 1997, Vol. 3121: Polarization: Measurements, Analysis, and Remote Sensing, pp. 88-99.

[33] Cloude, S. R., 'Group theory and polarization algebra', Optik - Wissenschaftlige Verlagsgesellschaft

mbH, Stuttgart, 75, No. 1 (1986), pp.26-36.

[34] Cloude, S. R, and E. Pottier, 'A review of target decomposition theorems in radar polarimetry', IEEE Trans. on GRS, vol 34(2), pp. 498-518, Mar. 1996.

[35] Cloude, S. R, and E. Pottier, 'An Entropy-Based Classification Scheme for Land Applications of Polarimetric SAR', IEEE Trans. GRS, vol. 35(1), pp. 68-78, 1997.

[36] Copeland, J. R., 'Radar target classification by polarization properties', Proc. IRE, Vol. 48, 1960, pp. 1290-1296.

[37] Courant, R. and D. Hilbert, 'Methods of Mathematical Physics', Vol. I and II, Interscience Publishers, New York, 1953 and 1963.

[38] Czyż, Z.H., 'Analysis of chain connection of four-terminal networks using the method of multiple reflections' (in Polish), Prace PIT, No.15, 1955, Warsaw, Poland, pp. 1-11.

[39] Czyz, Z. H., 'The utility of applying regulated elliptical polarization in radio detecting and ranging devices from the point of view of the effectivity of those devices', Ph.D. thesis in Polish, Warsaw University of Technology, Warsaw, Poland, 1967 (also translated into English by the ONR, USA, 1997).

[40] Czyz, Z. H., 'Amplitude and Power Representation of Elliptically Polarized Antennas and Waves' (in Polish), Prace PIT, Warsaw 1969, No. 63, pp. 11-22.

[41] Czyz, Z. H., 'Analysis of Polarization Properties of Nondepolarizing Targets' (in Polish), Prace PIT, Warsaw 1970, No.65, pp. 23-36.

[42] Czyz, Z. H., 'Detailed Reconstruction of Kennaugh's Geometrical Interpretation of Bistatic Scattering Dependence on Incident Polarization', ICAP'85, IEE Conf. Publ. No. 248, pp. 375-378.

[43] Czyz, Z. H., 'Geometrical Interpretation of Polarization Transformation when Bistatic Scattering by a Stable Object' (In Polish), 7th National Microwave Conference - MIKON'86, Vol. I, pp. 144-146.

[44] Czyz, Z. H., 'On Some Properties of the Bistatic Power Scattering Matrix for the Stable Object in its Characteristic Orthogonal Polarization Basis', URSI Int. Symp. on Electromagnetic Theory, Budapest, Hungary, August 25-29, 1986, Part B, pp. 628-630.

[45] Czyz, Z. H., 'Polarization of Radar Scatterings' (in Polish), Prace PIT, Supplement No. 5, Warsaw, Poland, 1986, 154 p.

[46] Czyz, Z. H., 'Bistatic Radar Target Classification by Polarization Properties', ICAP'87, IEE Conf. Publ. No. 274, Part 1, pp. 545-548.

[47] Czyz, Z. H., 'Addition of partially coherent waves using Stokes vector representation', ICAP'89, IEE Conf. Publ. No. 301, Part 2, pp.396-399.

[48] Czyz, Z. H., 'Reconstruction of Bistatic Scattering Matrix for Three Special Polarizations', MIKON'91, Rydzyna, Poland, May 20-24, 1991, pp. 314-317.

[49] Czyz Z. H., 'Comparison of Polarimetric Radar Theories," ICEAA'91, Torino, Italy, 17-20 September 1991, pp. 291-294

[50] Czyz, Z. H., 'Polarization Properties of Nonsymmetrical Matrices - A Geometrical Interpretation', Part VII of Polish Radar Technology', IEEE Trans. Aerospace and Electronic Systems, vol. 27, pp. 771-777 and 781- 782, 1991.

[51] Czyz Z. H., 'Characteristic Polarization States for Nonreciprocal Coherent Scattering Case," ICAP'91, IEE Conf. Publ. No. 333, Part 1, pp. 253-256.

[52] Czyz Z. H., 'Comparison of fundamental Approaches to radar polarimetry', [129], pp. 99-116.

[53] Czyz Z. H., 'An Alternative Approach to 'Foundations of Radar Polarimetry', [129], pp.247-266.

[54] Czyz, Z. H., 'The simplest decomposition of the Mueller bistatic scattering Matrix of a distributed target into point target components', Second International Workshop on Radar Polarimetry - JIPR '92, Nantes, France, Sept. 8-10, 1992, pp. 61-68.

[55] Czyz, Z. H., 'Synthesis of a Bistatic Scattering Matrix for a Point Target of Desired Polarimetric Properties', Second International Workshop on Radar Polarimetry - JIPR '92, Nantes, France, Sept. 8-10, 1992, pp. 46-51.

[56] Czyz, Z. H., 'New Concept of Virtual Polarization Adaptation', ICAP'93, IEE Conf. Publ. No.370, P.II, pp. 890-893.

[57] Czyz, Z. H., 'Complete Cancellation of an Unpolarized Clutter in a Mono- or Bistatic Radar', PIERS'93, Pasadena, CA, July 12-16, 1993, p. 201.

[58] Czyz, Z. H., 'Alternative Approaches to Polarimetric Signal and Image Processing', PIERS'94 - Proceedings of the 1994 Progress in Electromagnetics Research Symposium, European Space Agency, Noordwijk, The Netherlands, 11-15 July 1994, Kluwer, CD-ROM edition.

[59] Czyz, Z. H., 'Fundamental Transformations and Decompositions in Radar Polarimetry', PIERS'94 - Proceedings of the 1994 Progress in Electromagnetics Research Symposium', European Space Agency, Noordwijk, The Netherlands, 11-15 July 1994, Kluwer, CD-ROM edition.

[60] Czyz, Z. H., ''Basic Theory of Radar Polarimetry - An Engineering Approach," MIKON '94 - X International Microwave Conference, Książ, Poland, May 30 - June 2, 1994, vol. 3 - Invited Papers, pp. 69-86.

[61] Czyz, Z. H., 'Polarization properties of the coherent lossless TEM transmission channel', IGARSS'95, Firenze, Italy, July 11-14, 1995, vol. III, pp. 2008-2011.

[62] Czyz, Z.H., 'Mutual Polarizations in Bistatic Radar Scattering', Prace PIT, Warsaw, Poland, Supplement 17/19, 1994, pp.45-49. (In Polish).

[63] Czyz, Z. H., 'Polarymetryczny odbiornik optymalny' ('Optimum polarimetric receiver') Polish Patent PL 169175 B1. Approved 28.06.1996.

[64] Czyz, Z.H., 'Advances in the Theory of Radar Polarimetry', Prace PIT, No.117, Vol.XLVI, Warsaw, Poland, 1996, pp.21-28. (In English).

[65] Czyz, Z.H., and W-M. Boerner, 'Scattering and cascading matrices of the lossless reciprocal polarimetric two-port in their general forms', Proc. of ISAP'96, Chiba, Japan, Sept. 24-27, 1996, pp.1037-1040.

[66] Czyz, Z. H., 'Enginering Approach to Bistatic Radar Polarimetry - Summary of Results', MIKON '96, Warsaw, Poland, May 27-30, 1996. Conference Proceedings, Vol. 2, pp 519-522.

[67] Czyz, Z. H., "Basic Theory of Radar Polarimetry - An Engineering Approach' (in English), Prace PIT, No.119, 1997, Warsaw, Poland , pp.15-24.

[68] Czyz, Z. H., 'Fundamental properties of the polarimetric two-ports' (in Polish), Prace PIT, No.119, 1997, Warsaw, Poland, pp. 25-37.

[69] Czyz, Z. H., W-M. Boerner, 'Scattering and cascading matrices of the lossless reciprocal polarimetric two-port in microwave versus millimeterwave and optical polarimetry', in: Proc. of SPIE, vol. 3120, 'Wideband Interferometric Sensing and Imaging Polarimetry', San Diego, California, 28-29 July 1997

pp. 373-384.

[70] Czyz, Z. H., 'Coherent and Noncoherent Polarimetric Radar Receivers Completely Canceling the Partially Polarized Clutter', 12th International Conference on Microwaves and Radar - MIKON'98, Kraków, Poland, May 20-22, 1998, pp.128-132.

[71] Czyz, Z. H., 'Suboptimum noncoherent polarimetric radar receiver completely canceling the partially polarized clutter', Workshop on Advances in Radar Methods, Baveno, Italy, July 20-22, 1998, pp.89-91, and in 'Collection of slides', end of Oral Session - Theme II, Ultra Wide Band and Multi Spectral Radars.

[72] Czyz, Z. H., 'Analysis of the Cascade Connection of the Polarimetric Two-Ports', Prace PIT, No.122, 1998, Warsaw, Poland, pp.6-22. (In Polish, AE).

[73] Czyz, Z. H., 'Polarimetric radar receivers canceling partially polarized clutter', Fourth International Workshop on Radar polarimetry, Nantes, France, July 13-17, 1998, pp.126-135.

[74] Czyz, Z. H., 'Constant co-polarization echo curves on the Poincare sphere', XIII International Conference on Microwaves, Radar, and Wireless Communications - MIKON 2000, Wroclaw Poland, May 22-24, Vol.1, pp. 13-16. See also: Journal of Telecommunications and Information Technology.

[75] Davidovitz, M. and W-M. Boerner, 1986, 'Extension of Kennaugh's Optimal Polarization Concept to the Asymmetric Matrix Case, IEEE Trans. on Antennas and Propagation, Vol. AP-34 (4), pp. 569-574.

[76] Deschamps, G. A. and P. E. Mast, 1973, 'Poincare sphere representation of partially polarized fields', IEEE Trans. on Antennas and Propagation, Vol. AP-21 (4), pp. 474-478.

[77] Germond , A-L., 'Theorie de la Polarimetrie Radar en Bistatique'. PhD Thesis, 28 Jan.1999, IRESTE, Universite de Nantes, France.

[78] Giuli, D., 'Polarization Diversity in Radars', Proc. IEEE, Vol. 74, No.2, Feb. 1986, pp.245-269.

[79] Graves, C., 'Radar polarization power scattering matrix', Proc. IRE, Vol. 44, 1956, pp. 248-252.

[80] Hollis, J. S., T. J. Lyon, and L. Clayton, 'Microwave Antenna Measurements', Scientific Atlanta Inc., Atlanta, 1970.

[81] Holm, W.A. and Barnes, R.M.,On Radar Polarization Mixed Target State Decomposition Techniques', IEEE 1988 National Radar Conference.

[82] Horton, M.C., and R.J. Wenzel, 'General theory and design of optimum quarter-wave TEM filters', IEEE Trans. on Microwave Theory and Techniques, vol. MTT 13, No. 3, pp. 316-327, 1965.

[83] van de Hulst, H. C., 'Light Scattering by Small Particles', Wiley, 1957.

[84] Huynen, J. R., 'Measurements of the target scattering matrix', Proc. IEEE, Vol. 53, 1965, pp. 936-946.

[85] Huynen, J. R., 'Phenomenological theory of radar targets', Ph.D. dissertation, Technical Univ., Delft, The Netherlands, 1970.

[86] Huynen, J. R., 'Physical Reality of Radar Targets, Part I and Part II., P.Q. Research, 10531 Blandor Way, Los Altos Hills, California 94024, Report P.Q.R. No. 106, 20p., May 1992.

[87] Huynen, J. R., 'Phenomenological theory of radar targets', in 'Electromagnetic Scattering', Academic Press, 1978, pp. 653-712.

[88] Huynen, J. R., 'Lexicograthic Radar Target Analysis', P.Q. Research, 10531 Blandor Way, Los Altos Hills, California 94024, Report P.Q.R. No. 108, 11p., March 1994.

[89] Huynen, J. R., 'A New Extended Target Decomposition Scheme', P.Q. Research, 10531 Blandor Way, Los Altos Hills, California 94024, Report P.Q.R. No. 109, 7p., March 1994.

[90] IEEE Standard Number 145-1983, „Definitions of Terms for Antennas', IEEE Transactions on Antennas and Propagation, AP-31 (6), November 1983.

[91] IEEE Standard Number 149-1979, 'Standard Test Procedures'.

[92] Jones, R. C., 'A new calculus for the treatment of optical systems', 'I. Description and discussion', J. Opt. Soc. Am., vol. 31 (July 1941), pp. 488-493; 'II. Proof of the three general equivalence theorems', ibid. pp. 493-499; 'III. The Stokes theory of optical activity', ibid. pp. 500-503.

[93] Jones, R. C., 'A new calculus for the treatment of optical systems', V., 'A more general formulation and description of another calculus', J. Opt. Soc. Am. 37, Feb. 1947, pp. 107-112.

[94] Kanareykin, D. B., N. F. Pawlov and W. A. Potekchin, 'Polarization of radar signals' (in Russian), Sov. Radio, Moscow, 1966.

[95] Kennaugh, E. M., 'Research studies on the Polarization Properties of Radar Targets'. The Ohio State University ElectroScience Laboratory, Commemorative Volumes I - II.

[96] Kennaugh, E.M., 'Polarization Dependence of RCS - A Geometrical Interpretation', IEEE Trans. on Antennas and Propagation, AP-29, March 1981, pp. 412-413.

[97] Ko, H. C., 'On the reception of quasi-monochromatic partially polarized radio waves', Proc. IRE, Vol. 50, Sept. 1962, pp. 1950-1957.

[98] Kong, J. A., 'Polarimetric Remote Sensing', Vol. 3 of PIER, Elsevier, 1990.

[99] Korn, G. A., and T. M. Korn, ' Mathematical Handbook for Scientists and Engineers', McGraw, New York, 1961.

[100] Kostinsky, A. B., and W-M. Boerner, 'On Foundations of Radar Polarimetry', IEEE Trans. on Antennas and Propagation, Vol. AP-34, No. 12, December 1986, pp. 1395-1403, with 'Comments' by H. Mieras, and 'Authors Reply', ibid., pp. 1470-1473.

[101] Kraus, John D., and K. R. Carver, 'Electromagnetics', Second Edition, McGraw-Hill, 1973.

[102] Krogager, E., 'Aspects of Polarimetric Radar Imaging', Ph.D. dissertation, Technical Univ., Lyngby, Denmark; Danish Defence Research Establishment, 1993.

[103] Krogager, E., Z. H. Czyz, 'Properties of the sphere, diplane, helix decomposition', in Proceedings of Third International Workshop on Radar Polarimetry (JIPR'95), Univ. Nantes, France, IRESTE, pp. 106-114, March 1995.

[104] Landau, L. D., and E. M. Lifschitz, 'The Classical Theory of Fields', 4th Edition, Pergamon Press, (Oxford), 1975.

[105] Lindell, I. V., 'Coordinate-free representations of the polarisation of time harmonic vectors', Helsinki University of Technology, Radio Laboratory, Report S66, 1974.

[106] Lopuszanski J., Rachunek spinorow (Spinor Calculus), PWN, Warszawa , Poland 1985.

[107] Lueneburg E. and W.-M. Boerner, 'Optimal polarizations in radar polarimetry', PIERS'94, Noordwijk, The Netherlands, July 11-15, 1994, CD Kluwer Publishers, pp. 1813-1816.

[108] Lueneburg E., 'Optimal polarizations in radar polarimetry', Kleiheubacher Berichte, Vol. 38 (1995) pp. 635-645.

[109] Lueneburg E., 'Canonical bases and Huynen decomposition', Proc. Third International Workshop on

Radar Polarimetry (JIPR), IRESTE, Nantas, France, March 21 - 23, 1995, pp. 75-83.

[110] Lueneburg, E., 'Comments on 'The Specular Null Polarization Theory', IEEE Trans. Geoscience and Remote Sensing, Vol. 35, 1997, pp. 1070-1071.

[111] Lueneburg, E. and R. S. Cloude, 'Bistatic Scattering', SPIE International Symposium on 'Wideband Interferometric Sensing and Imaging', San Diego CA, 28-29 July, 1997, Vol. 3120, pp. 58-68.

[112] Lueneburg, E. and S. R. Cloude, 'Radar versus optical polarimetry', SPIE International Symposium on 'Wideband Interferometric Sensing and Imaging', San Diego CA, 28-29 July, 1997, Vol. 3120, pp. 361-372.

[113] Lueneburg, E. and W.-M. Boerner, Homogeneous and inhomogeneous Sinclair and Jones matrices, SPIE International Symposium on 'Wideband Interferometric Sensing and Imaging', San Diego CA, 28-29 July, 1997, Vol. 3120, pp. 45-54.

[114] Lueneburg, E., 'Polarimetric Target Matrix Decompositions and the Karhunen-Loeve Expansion', IGARSS'99, June 28-July 2, 1999, Hamburg, Germany.

[115] Lueneburg, E. and R. S. Cloude, 'Contractions, Hadamard Products and their Application to Polarimetric Radar Interferometry', IGARSS'99, June 28-July 2, 1999, Hamburg, Germany.

[116] Misner, Ch. W., K. S. Thorne, and J. A. Wheeler, 'Gravitation', Chapter 41, W. H. Freeman and Co., New York, 1973.
For the symplectic geometry see also:
Shaw, R., 'Linear Algebra and Group representations', Vol. I, Chapter 4.4, Academic Press, 1982.

[117] Morse, P. M., and H. Feshbach, 'Methods of theoretical physics', McGraw, New York 1953.

[118] Mott, H., 'Antennas for Radar and Communications', John Wiley & Sons, New York 1992.

[119] Mueller, H., J. Opt. Soc. Am., Vol. 38, 1948, p. 661.

[120] O'Neil, E. L., 'Introduction to Statistical Optics', Addison-Wesley, Reading, Mass., 1963.

[121] Pancharatnam S., „Generalized Theory of Interference and Its Applications", Part 1: 'Coherent Pencils', Proc. Ind. Acad. Sci., 1956, 44A, pp. 247-262..

[122] Pancharatnam S., „Generalized Theory of Interference and Its Applications", Part 2: 'Partially Coherent Pencils', Proc. Ind. Acad. Sci., 1956, 44A, pp. 398-417.

[123] Pancharatnam S., „Partial Polarisation, Partial Coherence and their Spectral Description for Polychromatic Light" - Part I and II, Proc. Ind. Acad. Sci., pp. 218-243.

[124] Papathanassiou, K. P., Polarimetric SAR Interferometry, Ph. D. Thesis, Tech. Univ. Graz, 1999.

[125] Penrose, R., and W. Rindlar, 'Spinors and Space-Time', Vol. I and II, Cambridge University Press, 1984.

[126] Perrin F., „Polarization of Light Scattered by Isotropic Opalescent Media", Journal of Chemica Physics, Vol. 10, July 1942, pp. 415-427.

[127] Poincare, H., 'Theorie Mathematique de la Lumiere, II-12', Paris: Georges Carre Publ. Co., (1892, pp. 282-285.

[128] Pottier, E., 'Contribution a la polarimetrie radar: de l'approche fondamentale aux applications', Habilitation á Diriger des Recherches, Universite de Nantes, France, 1998.

[129] Proceedings of the NATO Advanced Research Workshop on Direct and Inverse Methods in Radar

Polarimetry', W-M., Boerner et al (eds), Bad Windsheim, Germany, September 18-24, 1988; Kluwer Academic Publishers, Dordrecht 1992; NATO ASI Series C: Mathematical and Physical Sciences - vol. 350.

[130] Rashevskyj, P. K., 'Riemann geometry and tensor analysis', in Russian, Moskow, 1953.
Polish translation:
Raszewski, P. K., 'Geometria Riemanna i analiza tensorowa', PWN Warszawa, 1958.
See also:
Weber, H., 'B. Riemann: Gesammelte Matchematische Werke', 2nd ed., paperback reprint, Dover, New York, and
Clifford, W. K., *Nature*, 8, 14 (1873).

[131] Sinclair, G., 'The Transmission and Reception of Elliptically Polarized Waves', Proc. of IRE, Vol. 38, Feb. 1950, pp. 148-151.

[132] Shurcliff, W., 'Polarized Light: Production and Use', Harvard University Press, Cambridge, 1962.

[133] Stepanov, N. N., „Sfericzeskaja trigonometrija", OGIZ, Leningrad-Moskow, 1948.

[134] Stokes, G. G., 'On the composition and resolution of streams of polarized light from different sources', Trans. Cambridge Phyl. Soc., Vol. 9, 1852, pp. 399-416.

[135] Stutzman W.L., Polarization in Electromagnetic Systems. Artech House, Boston-London 1993.

[136] Tragl, K., E. Lueneburg and A. Schroth, 'A polarimetric covariance matrix concept for random radar targets', Int'l Conf. Antennas and Propagation (ICAP'91), Warwick, UK, 15-18 April 1991.

[137] Ulaby F.T., Elachi Ch. (Editors), 'Radar Polarimetry for Geoscience Applications'. Artech House 1990.

[138] Wanielik, G., ' Signaturuntersuchungen an einem polarimetrischen Pulsradar', Fortschr.-Ber. VDI Reihe 10, Nr. 97., Duesseldorf VDI-Verlag 1988.

[139] Wanielik, G., and D. J. R. Stock, 'Radar polarization jamming using the superposition of two fully polarized waves', Int. Conf. RADAR'87, IEE Conf, Publ., No. 281, pp. 330-332.

[140] Wei, P.S., J. R. Huynen and T. C. Bradley, 'Transformation of polarization bases for radar scattering', Electronics Letters, Vol. 22(1), 1985.

[141] Yang, J., Y. Yamaguchi, H. Yamada, M. Sengoku, S. M. Lin, 'Stable Decomposition of Mueller Matrix', IEICE Trans. Comm., Vol.E81-B(6), pp. 1261-1268, June 1998.

[142] Yang, J., 'On theoretical aspects of radar polarimetry', Doctoral Thesis, Niigata University, Japan, September 1999.

[143] Zebker H.A., and J. J. van Zyl, 'Imaging radar polarimetry', Chapter 5. in J.A. Kong (Ed.): Polarimetric Remote Sensing, New York: Elsevier 1990.

[144] van Zyl, J.J. and H. Zebker, 1990, 'Imaging Radar Polarimetry' in 'Polarimetric Remote Sensing', PIER3, Kong J. A., ed. Elsevier, New York: 277-326.

Also available from Wexford Press ...

Introduction to Synthetic Aperture Radar (SAR) Polarimetry

by
Wolfgang-Martin Boerner

A comprehensive overview of the basic principles of radar polarimetry is presented. The relevant fundamental field equations are first provided. The importance of the propagation and scattering behavior in various frequency bands, the electrodynamic foundations such as Maxwell's equations, the Helmholtz vector wave equation and especially the fundamental laws of polarization are presented in the first section. Main poins are the polarization Ellipse, the polarization ratio, the Stokes Parameter, and the Stokes and Jones vector formalisms as well as its presentation on the Poincare sphere and on relavent map projections. The Polarization Fork descriptor and the associated van Zyl polarimetric power density and Agrawal polarimetric phase correlation signatures are introduced also in order to make understandable the polarization state formulations of electromagnetic waves in the frequency domain. The different relevant matrices, the respective terms like Jones Matrix, S-matrix, Muller M-matrix, Kennaugh K-matrix, etc. and its interconnections are defined and described together with change determined for the coherent and partially coherent cases, respectively. Concludes with worked examples and references for further reading.

ISBN: 978-1-934939-06-2

Wexford

www.ingramcontent.com/pod-product-compliance
Lightning Source LLC
Chambersburg PA
CBHW061402210326
41598CB00035B/6073